Report 164 1997

Design of containment systems for the prevention of water pollution from industrial incidents

P A Mason BSc PhD
H J Amies BSc MSc CEng MICE
P R Edwards DPhil MA(Oxon) MInstWM
G Rose CEng FICE FIStructE
G Sangarapillai MSc CEng MICE

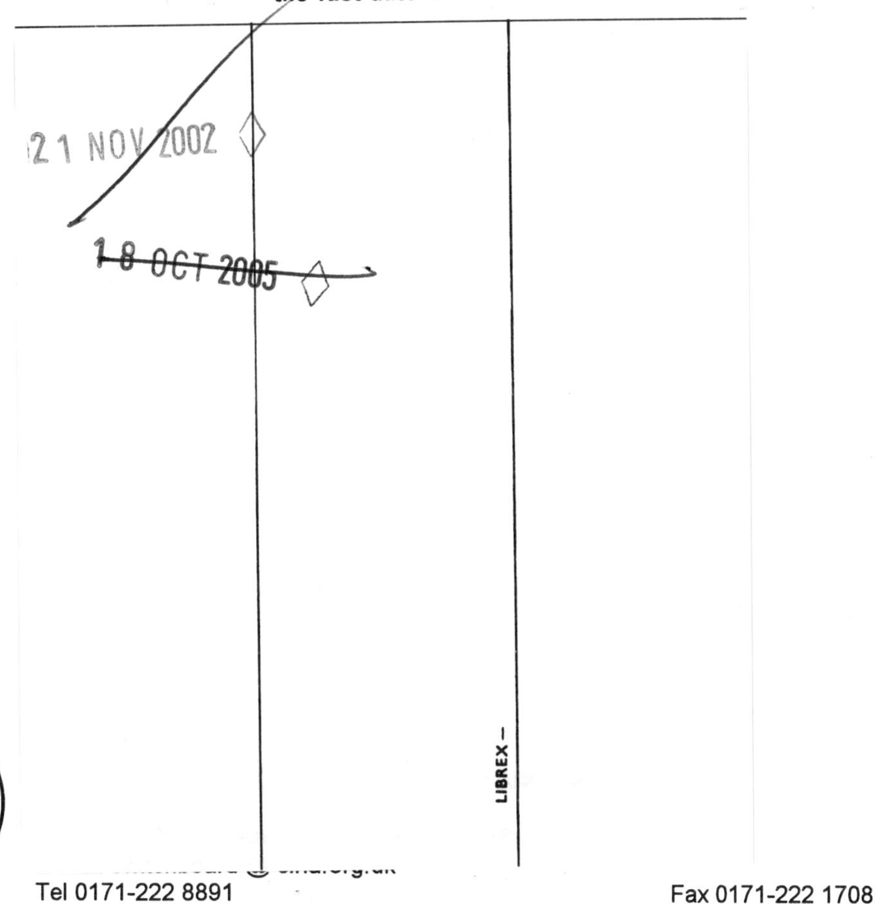

CIRIA

Tel 0171-222 8891 Fax 0171-222 1708

Summary

Industrial sites which manufacture, handle or store hazardous substances have the potential to cause pollution of surface and groundwater if these substances are released into the environment. Incidents which can cause release of hazardous substances include fire and runoff of associated firefighting water, spillage during handling, failure of primary containment and vandalism. Once released, hazardous substances can enter the water environment through surface or foul drainage systems, by direct runoff into a water course or by infiltration into the ground.

The aim of this report is to give guidance on the design of systems to prevent uncontrolled or unforeseen loss of containment of materials which could cause water pollution.

As well as providing detailed technical advice on the planning, design and construction of secondary containment systems the report provides guidance to help designers choose the appropriate type and capacity of system with reference to assessed risk.

P A Mason, H J Amies, P R Edwards, G Rose, G Sangarapillai
Design of containment systems for the prevention of water pollution from industrial incidents
Construction Industry Research and Information Association
CIRIA Report 164, 1997

ISBN 0 86017 476 X
ISSN 0305 408 X

Keywords		
pollution, containment, CIMAH, incident, fire, spillage, risk, bund, lagoon, secondary containment tank		
Reader interest	**Classification**	
site operators, system designers, contractors, equipment suppliers, regulatory bodies, local authorities, fire services, insurers	AVAILABILITY	Unrestricted
	CONTENT	Guidance based on best current practice
	STATUS	Committee-guided
	USER	Site operators

Published by CIRIA, 6 Storey's Gate, Westminster, London, SW1P 3AU.

Foreword

This report results from CIRIA Research Project 493: *Design of containment systems for the prevention of water pollution from industrial incidents* produced as part of CIRIA's water engineering programme. The objective of the project was to develop detailed technical guidance on the planning, design and construction of secondary containment systems for the prevention of water pollution following industrial incidents.

The report was written under contact to CIRIA by Paul Mason of ADAS. Contributions to the text were also made by Jim Amies, Phil Edwards, Gordon Rose and Gopal Sangarapillai.

CIRIA's Project Managers were Nicola Harding and Judy Payne.

Following established CIRIA practice, the research was guided by a Steering Group which comprised:

Mr Brian Fuller	Brian Fuller Associates (Chairman)
Dr David Aston	Hickson Timber Products Ltd
Mr Eric Clark	North Yorkshire Fire Brigade *representing* the Chief and Assistant Chief Fire Officers' Association
Mr Barney Franklin	Home Office
Mr John Holding	Bayer plc
Dr Huw Jones	Environment Agency
Mr Bruce McGlashan	Environment Agency
Mr Chris Newstead	Simon Storage Group *representing* the Independent Tank Storage Association
Dr William Parish	Department of Environment Chemicals and Biotechnology Division
Mr Andrew Sangster	The Institute of Petroleum
Mr David Spencer	Home Office
Mr Derek Wall	White Young Consulting Engineers
Dr Andrea Young	Department of Environment Chemicals and Biotechnology Division

Corresponding members

Dr Emer Bell	Hinton & Higgs (Ireland) Ltd
Mr Ian Buckland	Health and Safety Executive
Mr Martin Clark	Environment Agency
Dr Jeremy Hodge	The Loss Prevention Council
Dr Charles Kirk	The Loss Prevention Council
Ms Dorothy Maxwell	Hinton & Higgs (Ireland) Ltd
Mr Angus McRoberts	Department of Environment (Northern Ireland)
Mr John Seddon	BASIS Registration Ltd
Mr Rod Wallis	Scottish Environment Protection Agency

Financial support for the project was provided by the Department of the Environment and the Environment Agency.

As of April 1996, the functions of the National Rivers Authority (NRA), Her Majesty's Inspectorate of Pollution (HMIP), and the Waste Regulation Authorities (WRAs) were taken over, in England and Wales, by the Environment Agency. The equivalent functions in Scotland were taken over by the Scottish Environment Protection Agency, and in Northern Ireland by the Environment and Heritage Service.

Contents

FIGURES

TABLES

BOXES

Abbreviations

API	American Petroleum Institute
BASIS	British Agrochemicals Inspection Scheme
BATNEEC	Best available technique not entailing excessive cost
BOD	Biochemical oxygen demand
BSCS	British soil classification system
BS(I)	British Standards (Institution)
CEA	European Insurance Committee
CIMAH	Control of Industrial Major Accident Hazards Regulations
DoE	Department of the Environment
DoE(NI)	Department of the Environment - Northern Ireland
EU	European Union
EPCS	Enhanced primary containment system
ETP	Effluent treatment plant
FFW	Fire fighting water
FMEA	Failure, mode and effect analysis
FPA	Fire Protection Association
FTA	Fault tree analysis
GGBFS	Ground granulated blast furnace slag
GCL	Geosynthetic clay liner
HAC	High alumina cement
HAZOP	Hazard and operability study
HRA	Hot rolled asphalt
HSE	Health and Safety Executive
IP	Institute of Petroleum
IPC	Integrated Pollution Control
LHPBFC	Low heat Portland blast furnace cement
LPC	Loss Prevention Council
LPG	Liquefied petroleum gas
MAFF	Ministry of Agriculture Fisheries and Food
NRA	National Rivers Authority - (now part of the Environment Agency)
NWC	National Water Council
OPC	Ordinary Portland cement
PFA	Pulverised fuel ash
PBFC	Portland blast furnace cement
PPGs	Pollution prevention guidance notes (NRA publications)
RHPC	Rapid hardening Portland cement
SEPA	Scottish Environmental Protection Agency
SRPC	Sulphate resisting Portland cement
SSC	Super-sulphated cement
UKEA (SRD)	United Kingdom Atomic Energy Authority (Safety and Reliability Directorate)
VCI	Verband der Chemischen Industrie
WWTP	Waste water treatment plant

Glossary

Black List (also known as List I)	Certain substances and groups of substances selected on the basis of their toxicity, persistence and bioaccumulation in the aquatic environment, pollution by which must be eliminated
Bund (or catch-pit)	A facility (including walls and a base) built around an area where potentially polluting materials are handled, processed or stored, for the purposes of containing any unintended escape of material from that area until such time as remedial action can be taken
Catch-pit	An alternative term for a small bund. Also, a small chamber used to collect runoff from catchment areas, usually incorporating a sediment collection sump, used as an integral part of a transfer system
Catchment	An area surrounding an installation which collects any spills and runoff water. Also used to describe a land area which intercepts rain water
Collar bund	A bund with relatively high walls which are built close to the primary storage container
Combined containment	A secondary containment system comprising both *local* and *remote* elements
Containment (abbreviation of secondary containment)	The interception and storage (usually short-term) of polluting material that may be released or escape from the primary containment.
Containment system (abbreviation of secondary containment system)	A facility designed to prevent the escape of material in the event of failure of the primary containment
Containment system classification	A methodology for categorising secondary containment systems according to structural integrity, in-built redundancy and other features, to match the different requirements of *high*, *moderate* and *low* site risk situations
Double-skinned tank (or double- or twin-walled tank)	A tank constructed with a double rather than single skin to provide additional protection against leakage
Fire fighting water	Water (and other water-based agents) used to extinguish or prevent the spread of fire during an incident

Freeboard	An allowance, in the form of increased height of containment wall, for additional capacity over and above the minimum design requirement
Geomembrane	An impermeable liner, usually a synthetic thermoplastic or thermosetting polymer
Geotextile	A synthetic polymer or natural fibre matting, used to reinforce certain soils and to strengthen geomembranes
Grey List (also known as List II)	Certain substances and groups of substances selected on the basis of their potential to pollute the aquatic environment, pollution by which must be reduced
Groundwater	Water contained in a saturated rock or other permeable strata
Hazard	A situation that in particular circumstances could lead to environmental pollution
Incident	An event that results in a loss of primary containment and consequent release of a hazardous substance to the environment
Infiltration (or percolation)	The downward movement of water under the influence of gravity until it reaches the water table
***In-situ* concrete**	Concrete placed wet on site
Inspection chamber (or manhole)	A chamber used as part of a drainage system usually at pipe junctions and changes in direction or gradient
Integrally-bunded tank	A primary storage tank built within another tank to provide additional protection against leakage. The walls of the inner and outer tanks are usually connected structurally to provide mutual support
Interceptor (or separator)	A device used in drainage systems to separate liquids with different densities (e.g. oil and water) and liquids from solids
Lagoon	Secondary containment structure with earth-banked walls and an earth floor
List I	See Black List
List II	See Grey List
Local containment	A form of secondary containment designed to contain escaped material at source (e.g. a bund)
Mass concrete	Unreinforced concrete placed wet on site

Pathway	The route by which released pollutants may travel from the *source* to the aquatic environment in the absence of secondary containment
Percolation	The downward movement of water under the influence of gravity until it reaches the water table
Permeability	The capacity of soil, rock and man-made materials and structures for transmitting water
Porosity	The ratio of the volume of voids in a rock or soil to the total volume
Pre-cast concrete	Prefabricated concrete sections assembled on site
Primary containment	The facilities used for the storage and handling of raw materials, product, waste and effluent
Pumping chamber	A sump or below-ground tank fitted with a pump for conveying effluent to a remote containment facility
Receiving water	Surface water and groundwater that could be affected in the event of release of pollutant during an incident
Receptor	The humans, animals, fish, plants and biota which would be affected by the escape of pollutants to the receiving water
Red List	A list, published by the DoE in 1988, of those substances considered most dangerous in the aquatic environment and whose discharge must be strictly controlled
Remote containment	A form of secondary containment where escaped material is transferred away from the area of the primary containment to a remote storage facility
Retention period	The defined design period of retention in secondary containment
Risk	The probability that a polluting event occurs during a stated period of time
Rugosity	Surface roughness of a pipe, channel or catchment area
Runoff	That part of precipitation which flows across the ground surface into surface waters
Sacrificial area	An area that has a specific function (e.g. a car park) not related to containment, but which is designed so that it may be used for secondary containment in an emergency
Secondary containment	A system which operates independently from the primary containment, designed to contain spillage in the event of failure of the primary system

Self -bunding tank	A form of proprietary double-skinned primary containment tank which incorporates features to contain overfill spillage within the outer skin
Separator	A device used in drainage systems to separate liquids with different densities (e.g. oil and water) and liquids from solids
Site hazard rating	An assessment of the overall potential of a site to cause environmental harm
Site risk rating	An assessment of overall risk obtained by combining site hazard rating with frequency of loss of primary containment
Sloshing	A mode of failure of a bund where reflected waves of liquid suddenly released from the primary containment overtop the bund wall
Source	The hazardous materials present on a site
System reliability rating	An assessment of the probability that a secondary containment system will perform as intended
Transport potential	An assessment of the ease and rapidity with which a pathway can transport pollutant to receiving waters
Transfer system	The method for collecting and conveying spills and runoff to a remote containment facility

1 Introduction

1.1 OBJECTIVES

The aim of this report is to give guidance on the design of containment systems to prevent uncontrolled or unforeseen loss of materials which could cause water pollution.

Materials are handled, produced and stored on industrial sites within a system of 'primary' containers such as process vessels, storage tanks, pipework and drums. This report is concerned with the design of 'secondary' containment, a term used to describe a second system of physical protection which is provided to contain the materials in the event of a failure of the primary system. Secondary containment most commonly takes the form of a physical structure surrounding the primary system, i.e. a bund, but other facilities such as lagoons, modified drainage systems and emergency basins for fire fighting water often form part of a secondary containment system. At some sites more than two levels of physical protection may be provided. Throughout this report the word containment is used broadly to describe the physical systems used to contain water polluting substances in the event of an escape from the primary containment. The emphasis is on containment following acute incidents rather than on gradual, continuing or chronic pollution caused by seepage or leakage of pollutants.

Other than for the most hazardous and toxic materials, there is a widespread lack of containment of chemicals, products and wastes across many industrial sectors. Where containment is provided, quality, robustness and standards of design and construction are extremely variable, both between and within industrial sectors and from plant to plant within the same company.

In the aftermath of some well documented and widely publicised incidents involving fires at chemical plants, which have led in some cases to severe pollution problems (see case studies at Section 6) there is increasing pressure on industry to provide containment systems that will prevent hazardous materials entering the aquatic environment following accidents. This report is concerned specifically with secondary containment measures to help protect the aquatic environment. It is recognised that there are other ways to protect the environment, e.g. through management controls, but these are outside the scope of this report.

The installation of physical containment structures is likely to be an expensive item in the design of new plants and even more so in improving existing facilities. This report provides information to help with the planning, design and installation of cost-effective protective measures in line with the principles of BATNEEC, i.e. Best Available Technique Not Entailing Excessive Cost.

The report is aimed at regulatory bodies and local authorities, fire services, site operators, system designers, contractors, equipment suppliers and insurers. The term regulatory bodies includes particularly the Environment Agency in England and Wales, the Scottish Environmental Protection Agency (SEPA), and the Environment and Heritage Service in Northern Ireland.

The report provides:

- background technical and legislative information
- an introduction to existing hazard and risk assessment methodologies relevant to pollution control on industrial sites
- guidance to help designers choose the appropriate type and form of containment system, with reference to assessed hazard and risk
- guidance on assessing the required volumetric capacity for containment systems
- detailed technical guidance on the planning, design and construction of containment systems.

1.2 SCOPE

1.2.1 General

The report is applicable to the containment of a wide range of material with the potential to pollute water. It is applicable to sites ranging from very large chemical plants with complex containment requirements through to small sites with perhaps just an individual tank.

The technical guidance concentrates on two issues:

(a) the containment of hazardous substances which are used, stored or produced on site as part of the process, and

(b) the containment of contaminated fire fighting water and rain water.

The scope of the report includes:

- the protection of the aquatic environment - all inland and coastal waters
- all materials with the potential to pollute water, including chemicals, products, reagents, fuels and wastes produced, used or stored on an industrial scale
- technical information on the design and construction of physical structures, including bunds, lagoons, retention basins, catchments, tanks, drainage systems and oil separators
- protection of storage areas, process areas, isolated tanks, drum stores, loading and off-loading points, in-site pipelines and warehouses
- the retention of contaminated waters, particularly fire fighting water
- protection against pollution resulting from on-site traffic incidents
- installations at both new and existing sites.

1.2.2 Exclusions

While the report has a broad scope it does not cover the following:

- the transportation of materials off site by road, rail, sea or air
- gradual or continuing pollution, e.g. leaching from contaminated land
- abandoned sites
- off-shore installations
- spills from pipelines between industrial premises
- pollution of water as a consequence of atmospheric emissions
- radioactive substances, hazardous biological organisms or chemicals used in small quantities such as in research laboratories

- sewage and sewage effluents, farm wastes and related materials
- the recovery, recycling or disposal of contaminated chemicals, wash waters, effluents, contaminated fire waters , etc.
- monitoring, clean-up or treatment of surface or groundwaters after the occurrence of an incident
- details of the management and maintenance of containment systems (although the need for good management practices and the role of protective and warning devices to prevent or detect spillages is included).

1.3 LAYOUT AND CONTENTS OF REPORT

The report is in 14 Sections.

Sections 1 to 6 contain background information applicable to water pollution, industrial installations controlling hazardous substances, and systems designed to contain water endangering substances. These first sections of the report include: a review of current legislation and technical guidance, a general description of types of containment systems and their use in the UK and elsewhere, and a number of case studies which illustrate the types of incident that can happen, the consequences of inadequate containment and the remedial works that have been carried out.

Section 7 develops a methodology for assessing environmental hazard and/or risk associated with a site, and the operations carried out on it, and links the output to the containment design process. In essence, it recommends better standards of containment design and construction for sites with higher hazard or risk ratings and indicates how this may be achieved in practice. It presents necessarily a simplified view of a very complex topic and stresses the need for consultation with the relevant authorities throughout the assessment process. The aim of this Section is to help engineers select and design appropriate containment systems according to the assessed hazard and/or risks, where it has already been decided by others that some form of secondary containment is necessary.

Section 8 provides guidance on systems' design and selection, drawing on the hazard/risk classification system developed in Section 7.

Section 9 provides guidance on the required capacity for containment, giving the option of a much more rigorous assessment approach than is generally used at the moment.

Sections 10 onwards provide detailed design guidance and some model designs and specifications for a range of containment systems.

2 Water pollution from industrial incidents

2.1 BACKGROUND TO THE PROBLEM

2.1.1 Pressures for change

In recent years there has been a number of well publicised incidents of pollution resulting from the escape of materials from industrial premises. Perhaps the most widely known is the Sandoz warehouse fire which had very damaging consequences for the quality of water in the Rhine. Other major accidents, such as the incidents in Seveso and Bhopal, and the various huge losses of oil to the marine environment from tanker and oil rig accidents, whether affecting air, water or people directly, have heightened public awareness of the problems of pollution.

These incidents have translated into pressures on industry to comply with the legislation in force, on the Government to bring in more stringent legislation and on the statutory authorities to increase and improve their policing of the legislation and to act strongly in enforcing it through court action.

Containment of all materials which can pollute when allowed into surface or groundwaters or onto the land is likely to come under increasingly tighter control. Legislation in the UK is available, but it is non-prescriptive, leaving the implementation to the statutory authorities and the final judgement to be made by the courts. In addition to the powers already available under the Water Resources Act 1991, the Control of Industrial Major Accident Hazards Regulations (CIMAH) 1984 and Part I of the Environmental Protection Act 1990, there will be increasing demand for more specific and detailed legislation unless industry can make significant advances under the current statutory regime to reduce uncontrolled discharges.

2.1.2 Pollution potential of industrial premises

The UK chemical, petrochemical and allied industries consist of a very large number of companies operating at sites which vary greatly in size and age. Generally the larger companies produce basic inorganic and organic chemicals which are used by smaller companies to produce a wide range of derivatives. A further group of companies is involved with peripheral activities such as formulation, packaging, transportation and warehousing. It is an old established industry, having grown with the industrial revolution. The industrial premises tend to be located around rivers and estuaries which provide the means for raw materials and products to be transported, cooling water for power generation and what was regarded as a convenient means of disposing of aqueous effluents. Some sites are in environmentally sensitive locations.

The basic commodity chemicals made by the large companies are distributed within the industry itself or sold to other industrial sectors such as power generation, oils and fuels, food and drinks, pharmaceuticals, plastics, textiles, dyes, paints, engineering, metals, minerals, construction, paper, mining and quarrying, water and sewage treatment and waste disposal. The outcome is the production of an enormous range of chemicals, products and wastes, many having the potential to pollute the aquatic environment in the event of loss of containment.

It is important to note that materials with the potential to cause serious damage to water do not stem simply from the major manufacturers and users of chemicals and the related industries. Spillages, leaks and accidents leading to environmental releases arise also from small industrial and commercial premises and from retail sites. These incidents may be smaller in size but they occur more widely. There is a wide range of non-toxic substances such as milk or fruit juice which, if spilled in sufficiently large quantities into a watercourse, would kill fish by virtue of their ability to stimulate microbial growth which depletes river water of oxygen.

Existing operations are often located at sites which have been in use for similar purposes since the last century. The drainage systems are often antiquated and it is not uncommon to find that surface water, domestic effluent and trade effluent drains are all interconnected before discharge to surface waters or to public sewers. This can add to the difficulty of containing, isolating and recovering materials when incidents occur. Solutions are likely to be more difficult on the older and frequently more congested sites, where there may be insufficient space to install containment works.

2.1.3 Water quality control

Experience has shown that the protection provided by containment measures can fail for a variety of reasons (see Section 2.2.5). The provision of effective containment systems is a requirement which the statutory authorities are increasingly likely to require for new or significantly modified installations. Containment is therefore a major issue at all industrial premises where materials with a significant pollution potential are produced, used, handled or stored.

The current rate of incidents resulting in water pollution is too high to be consistent with the achievement of a long-term improvement in the quality of UK surface and groundwaters. Improving surface water and groundwater quality is an objective to which Government, the UK statutory authorities, pressure groups and the public attach great significance. As river quality improves, incidents which adversely affect water quality become more unacceptable. In the period 1990-92 the quality of some 15% of rivers and canals in England and Wales was poor or bad (NRA, 1993). The long-term aim of the Environment Agency is to improve water quality so that no waters fall into these categories.

Under normal circumstances, substances enter surface waters in the form of consented discharges which are controlled and planned. Under abnormal circumstances unplanned, uncontrolled discharges take place for a variety of reasons. The damage to the aquatic environment caused by such discharges can be acute or chronic, short- or long-term and is likely to be far more severe than any environmental harm caused during normal conditions.

In England and Wales, the Environment Agency is empowered to grant consents on discharges of trade and sewage effluents to rivers and estuaries and to groundwater, to impose wide ranging conditions in the consents and to take legal action in the case of infringements. In Scotland the Scottish Environmental Protection Agency (SEPA), and in Northern Ireland the Environment and Heritage Service, have similar powers. The powers of the UK authorities are normally aimed at 'end-of-pipe' controls, which only enables them to advise on containment measures such as bunding. However, where it appears to the Environment Agency that polluting matter is likely to enter waters they may carry out operations to prevent it, for which the site operator will be required to pay.

The Agency also has a duty to monitor and protect the quality of groundwater and in 1991 published a policy document (NRA, 1991) stating its aims and objectives and how they are to be achieved. This contains references to restrictions which may be imposed in sensitive areas on the production, use and storage of chemicals and wastes, on waste disposal operations, on 'prescribed' processes (as defined in the Environmental Protection Act, 1990) and on oil and petroleum storage and transport.

The Agency has a further duty with respect to emissions from those major industrial sectors prescribed in Regulations for Integrated Pollution Control (IPC) (HMSO, 1990). It assesses the need for, and imposes requirements on, the provision of bunds or related containment measures when authorising new or existing industrial processes. It can specify precisely the types of measures that a company must provide in order to be granted an authorisation. In general terms IPC requires existing facilities to be brought up to BATNEEC standards.

Regulations controlling specifically the storage of silage, slurry and agricultural fuel oil were made in 1991 (HMSO, 1991). It is possible that Regulations controlling industrial fuel oil storage in particular, and the storage of dangerous substances in general, will follow, specifying in broad terms the conditions to be applied to the containment of polluting substances (ENDS, 1991, 1993, 1994).

2.1.4 Penalties and liabilities

Loss of containment of pollutants results not only in environmental damage but it can also have a severe effect on the company concerned. In the past, fines imposed for water pollution offences were trivial compared with the cost of installing protective measures. Fines have recently increased greatly, with the £1 million fine of Shell UK for an oil spill to the River Mersey in August 1989 being an example of the penalty likely to be incurred for causing major environmental damage. Environmental legislation also empowers courts to imprison directors and managers of companies if pollution is proved to have resulted from their negligence. Several cases of imprisonment have been imposed for waste management offences (ENDS, 1992, 1993, 1994) and this trend may spread into other areas of environmental management.

In addition to prosecution for criminal acts, there is an increasing trend for companies, individuals and the statutory authorities to use civil proceedings for the recovery of costs incurred in cleaning up after pollution incidents. These costs can be extremely high, particularly in cases where the clean-up of contaminated land or groundwater is involved.

Although companies regard legal compliance as their major priority and are concerned about the direct penalties resulting from court action, many organisations are now striving to improve their public image and are becoming more open in the publication of information relating to their environmental performance.

It is difficult to assess all the liabilities and resulting financial implications arising from the contamination of land caused by loss of containment. The widely reported case of Cambridge Water Company versus Eastern Counties Leather (ENDS, 1993a) in which spillages of a chlorinated solvent migrated into an aquifer, and then to a borehole used for potable water supply, has had a profound effect on the insurance industry. It will become more difficult for companies to sell contaminated sites and more expensive to obtain insurance cover for potentially polluting activities in view of the great expense of cleaning up soil and groundwater.

A programme of risk assessment carried out for Shell UK (ENDS, 1993b) revealed that a third of its 1100 petrol stations are contaminated by leaks from underground tanks to a 'greater or lesser extent'. Clean-up costs at one small site are estimated to be £250k, exceeding the site's market value. It is reported that some petrol retailers are having difficulty obtaining financing because there is insufficient security for the loan in the event of a costly pollution incident. There is specialist insurance available for underground storage tanks.

Given the uncertainties, it may be difficult, or perhaps impossible to obtain insurance cover for all the potential liabilities. Insurance premiums are likely to rise sharply for companies that cause major or highly publicised incidents. One way in which companies can protect themselves against future liabilities is by providing high integrity containment systems for materials known to endanger water. The rate of increase in premiums may be lower for those with good facilities and management.

2.1.5 Spill prevention policies and practices

In the UK there is no legislation prescribing specific conditions to the containment of chemicals and other water polluting materials, other than for highly flammable liquids and liquified petroleum gas (LPG), and for certain agricultural wastes. There is very little practical advice on the design of containment systems. As a result, those containment facilities that are in place have been installed on an *ad hoc* basis, often in response to incidents of chemical spills or fires, or to pressure from the statutory bodies. In contrast, some other countries, notably Germany, have legislation which is much more specific regarding the measures to be taken when storing dangerous substances.

Policy on containment varies from company to company. The non-prescriptive approach adopted in Britain is sometimes not shared by multi-national European or American companies, who use physical structures far more extensively to comply with the more stringent and specific legislation in the countries where they are based. UK firms which are subsidiaries of larger foreign based companies may have installed containment systems because the parent company has a policy which requires it, irrespective of the environmental situation.

The Environment Agency (and equivalent bodies in Scotland and Northern Ireland) has the greatest role to play in developing a consistent policy on bunding and applying it across UK industry. The Agency's policy is that all liquids capable of causing damage in the aquatic environment should be bunded. This approach is sometimes contested by companies, particularly in the chemical industry, who take the view that bunding is an expensive solution which should be provided only where necessary, and that other measures give equivalent protection and are more cost effective.

2.1.6 Risk assessment and management systems

Almost irrespective of the debate between those who believe bunds are always necessary and those who believe that environmental and site factors should be taken into account in assessing the need for containment, companies have only a limited amount of capital for investment. They need to know how to rank their containment problems relative to other issues in order to spend capital in the most effective way.

Risk assessment methods are therefore used by some companies, using a variety of models varying from rudimentary to sophisticated.

The impact of a spill depends not only on the material involved, but on a range of other factors. The risk models may take into account the environmental sensitivity of the location, the nature of the chemicals, products, materials, or wastes present, and the likelihood of loss of containment. The outcome of an assessment may be that a small tank of a particular material in a sensitive area is bunded, whereas a larger tank of the same material in a less sensitive area on the same site is not; a very small amount of a chemical on a small, inland trout stream is bunded, whereas a very large tank of the same chemical on the banks of a major estuary is not. Risk assessment is explained and developed further in Section 7 of this report.

The environmental management standard BS 7750 (BSI, 1994a) specifies requirements for management systems aimed at ensuring compliance with a company's environmental policy and objectives. As part of the procedure, an assessment of the environmental impact of a company's activities must be carried out. BS 7750 establishes a mechanism for examining all aspects of activities at a site in order to compile a register of significant effects. The loss of containment of water endangering substances should be a prime consideration.

Although this report deals with protective measures in the form of physical structures, the reasons for incidents indicate that such facilities must be properly managed and backed up by other ancillary systems to give the highest degree of overall protection. All of these measures, physical, operational and managerial should be supported by contingency or incident response plans to reduce the impact of any incident that does occur.

2.2 NATURE AND EXTENT OF INCIDENTS

2.2.1 Introduction

In considering major incidents the word 'major' is normally taken to relate to the quantity of material involved and consequent loss of human life. If, however, as in this report, the word is taken to refer to the environmental impact of incidents then the problem of classifying incidents becomes much more difficult because some very small-scale releases have caused severe repercussions; e.g. the Seveso incident in which only a few kilograms of dioxins were released.

In a guidance note on the CIMAH Regulations (DoE, 1991) it is suggested that a major incident in the environment is defined as follows:

> 'An incident is considered a major incident if it causes permanent or long-term damage to particular unique, rare or otherwise valued components of the man-made or natural environment, or if there is widespread environmental loss or damage.'

2.2.2 Global view of major incidents

The following overview of major incidents world-wide which have affected the environment is based on a review carried out by the Centre for Exploitation of Science and Technology (Bond, 1991). Although incidents involving spills of materials to the marine environment and those resulting from transport accidents are not included in the scope of this report, the information below has been provided to help put the present work in perspective.

Oil pollution

Losses of oil to the marine environment following tanker accidents, spills from off-shore oil wells and discharges from coastal facilities form the largest single group, both in terms of the number of incidents, the quantity of material spilled and their environmental impact. It is estimated that 230,000 tonnes of oil spilled from the Amoco Cadiz, with estimates of the total financial cost of the environmental damage put at $250 million. In 1979 an oil well in the Gulf of Mexico blew out and was not successfully capped for 10 months, resulting in a loss of oil estimated to be in the range 0.4 to 1.4 million tonnes. At any time some 30 million barrels of oil per day are being moved over considerable distances by sea and a similar amount is being moved in pipelines.

Gas use and distribution

HSE data for the period 1980-1990 showed that of some 600 incidents world-wide there were 128 incidents in the UK involving gas, including both 'fuel' gases such as LPG and 'chemical' gases such as chlorine. The major incident in the UK was the Piper Alpha disaster in 1990 in which 167 people were killed as a result of an explosion of LPG. Incidents have occurred as a result of the transfer of gas in pipelines, the build-up of gas in sewers and natural gas accumulation in buildings (e.g. Abbeystead). About 4,000 million m^3 of LPG (a mixture of ethane, propane and butane) are used annually in the UK. About 50,000 million m^3 of natural gas (methane) are used by domestic and industrial consumers annually. There have been some very serious and high profile incidents involving gas, particularly the transport of LPG.

Chemicals

About 130 million tonnes per year of basic petrochemicals (ethylene, propylene, butadiene, benzene, toluene, xylenes and their derivatives) are produced world-wide, with about 4% of the total being produced in the UK.

The Organisation for Economic Cooperation and Development (OECD) (1989) has compiled a database for the period 1970-1989 listing industrial accidents which have had severe environmental consequences. The database relates mainly to the production and transport of fuels and chemicals. Events of major environmental significance include Bhopal, Seveso and Sandoz. The major UK entry is the Flixborough incident in 1974. The database shows that 104 major incidents involving high volume basic chemicals occurred in the UK during the period 1980 to 1989. Approximately 35% of these were the result of transport accidents. The chemicals involved in the reported incidents are listed in Table 2.1.

Table 2.1 *Industrial incidents which caused severe environmental consequences (UK, 1980-1989)* (Source: OECD)

Chemicals involved	Number of incidents
Ethylene, propylene, butadiene, benzene, toluene, xylenes	14
Sulphuric, hydrochloric, nitric and hydrofluoric acids	40
Sodium and potassium hydroxides	9
Ammonia	19
Chlorine	22

2.2.3 Number and type of incidents in the UK

A substantial proportion of the number of recorded incidents of water pollution caused by industry result from the escape of liquids. Some of these occur because of poor operational procedures or from the inadequacy of operational control devices, indicating a need for improved systems of quality control and improvements to engineering and service control systems. Other failures occur as a result of structural failures or from incidents such as fire.

Information on the number, type and severity of incidents is compiled and published annually by the Environment Agency. The following data have been taken from the Agency's 1995 and earlier reports (Environment Agency, 1995).

In 1995 there were 23,463 reported and substantiated incidents of pollution in England and Wales. A breakdown of incidents by source is shown in Figure 2.1.

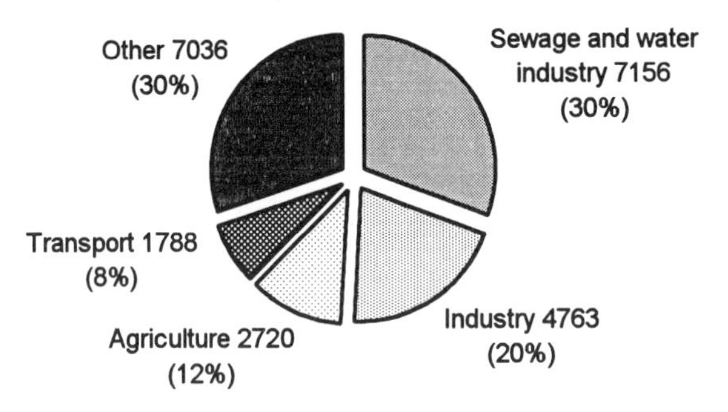

Figure 2.1 *Distribution of substantiated pollution incidents in England and Wales by source, 1995* (Source: Environment Agency)

The total number of reported incidents has been rising steadily since 1981 when about 12,500 incidents in total were reported (some of which may not have been substantiated). Care must be taken in interpreting these data since the apparent increase in incidents may not necessarily mean that the situation is worsening but may simply reflect better policing by the authorities combined with more vigilance in reporting by the general public.

Figure 2.2 shows the distribution of total substantiated incidents in 1995 by type of pollutant.

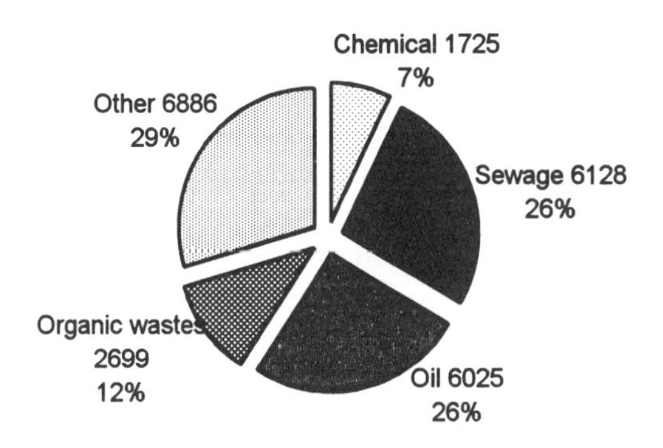

Figure 2.2 *Distribution of substantiated pollution incidents in England and Wales by type of pollution, 1995* (Source: Environment Agency)

There were a total of 199 'category 1' or 'major' incidents of which the highest number, 62 (31%), stemmed from industrial premises. Within category 1, oil pollution accounted for 45 (22%) and chemicals for 48 (24%) of the incidents.

Figure 2.3, based on the Agency's 1993 survey, shows the distribution of total industrial pollution incidents by source for those incidents where the origin of the pollution was sufficiently well defined.

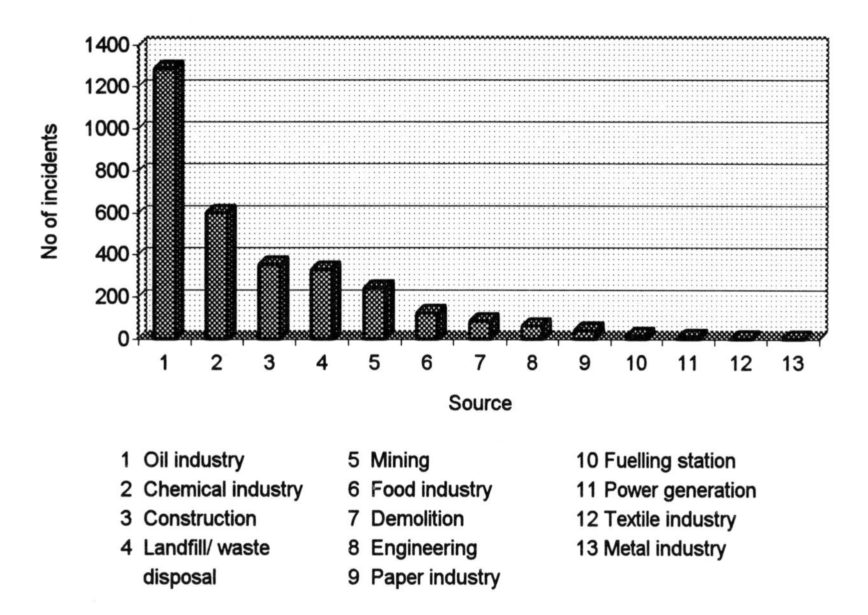

1 Oil industry
2 Chemical industry
3 Construction
4 Landfill/ waste disposal
5 Mining
6 Food industry
7 Demolition
8 Engineering
9 Paper industry
10 Fuelling station
11 Power generation
12 Textile industry
13 Metal industry

Figure 2.3 *Distribution of substantiated industrial pollution incidents in England and Wales by source, 1993* (Source: Environment Agency)

Figure 2.4 shows the distribution of classified chemical incidents by type of chemical.

In 1993 there were 2039 substantiated incidents of chemical pollution (8% of the total for the year). A large group of chemical incidents (73%) remain 'unclassified' since the nature of the incident could not be precisely defined. Pesticides and detergents accounted for the largest proportion of the 75 chemical incidents falling into category 1.

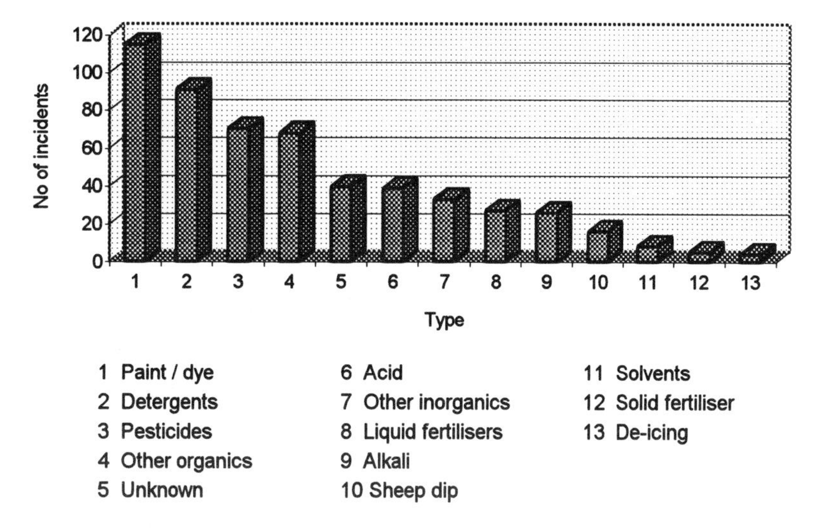

1 Paint / dye
2 Detergents
3 Pesticides
4 Other organics
5 Unknown
6 Acid
7 Other inorganics
8 Liquid fertilisers
9 Alkali
10 Sheep dip
11 Solvents
12 Solid fertiliser
13 De-icing

Figure 2.4 *Distribution of substantiated chemical incidents in England and Wales by type of chemical, 1993* (Source: Environment Agency)

During 1993, other unclassified pollutants, such as inert suspended solids, accounted for 7558 (30%) of the total incidents, of which 65 (1%) were category 1 incidents (20% of the total category 1 incidents).

The Environment Agency's statistics do not reveal how many incidents involved sudden loss of containment as opposed to pollution of a gradual nature.

Over the period 1989-93 an analysis of prosecutions for water pollution offences published in the ENDS Report (1994) revealed that the main offenders in that period were water companies (N.B. sewage and related substances are outside the scope of this report) and the minerals, construction, food and chemical industries. Many materials in bulk liquid form stored by the food industry, e.g. milk, vinegar, apple juice, sugar solution and blood, have a high polluting potential.

A further group of incidents stems from warehouse fires. The Fire Protection Association (Ward, 1984) analysed 19 of the most serious incidents in the period 1977-83 involving fires in warehouses where chemicals were stored. The incident in Woodkirk, Yorkshire, in 1982 was probably the most serious of these in terms of water pollution, resulting in a large quantity of herbicides being washed into the drains and severe pollution of the Rivers Aire, Calder and Ouse (Stansfield, 1983). It is not clear how many of the other incidents led to the discharge of contaminated fire fighting water to surface waters.

Tyre fires are also fairly common. The slow pyrolysis of rubber may result in the release of heavy metals such as zinc and cadmium, and an oily material, possibly containing traces of phenols and polynuclear aromatic hydrocarbons, which is washed out with the fire fighting water or seeps into the ground. The environmental impact of fires of this type and on fire safety guidelines has been described in a joint publication by the Home Office/Scottish Office (1994).

Further information on causes of pollution incidents is included in an HSE Report (1994).

2.2.4 Environmental impact

In general terms, polluting materials affect water quality by changing its physical, chemical or biological properties to a degree that threatens aquatic biota, fisheries, birds, aesthetic properties, human safety, water abstraction, personal property such as boats and structures such as piers, marinas and harbours. The effect of a spillage depends on the size of the spill and the nature of the material. Very small quantities of some materials can have severe effects. A few litres of oil can have a major visual effect on a small stream; a few litres of a chemical can affect drinking water quality; a few hundred grams of a very highly toxic chemical, such as a pesticide, can kill fish for many miles downstream.

Box 2.1 gives examples of spillages to water which have occurred in the UK, chosen to illustrate the types of companies, the types of materials and the nature of the environmental effects. Other examples are included as case studies in Section 6 of this report.

2.2.5 Causes of incidents

Although incidents are regularly reported in the press and technical literature, the reports almost invariably concentrate on the environmental impact of the event and the clean up measures, rather than on the reasons for it.

Table 2.2 summarises the causes of loss of containment of liquids (not necessarily involving pollution of surface waters) which occurred at one large chemical complex in NW England. There were a total 170 such incidents at the complex over the last three years.

Box 2.1 *Examples of environmental damage caused by loss of containment*

The potable water supply of two million people was rendered unfit for consumption for several days by the spill of only a few gallons of phenol to the River Dee from an engineering works in Chirk, Clwyd. The water was abstracted downstream for supply as potable water and in the purification process was chlorinated, producing chlorophenols with a very low taste threshold. The incident cost about £20 million in remedial measures and £400 000 in compensation.

A spill of 6 000 litres of caustic soda from a creamery in May 1994 killed thousands of fish and eels along a 11-kilometre stretch of the River Ellen in Cumbria.

In March 1990 a combination of vandalism and arson at a timber company in Surrey caused 23 000 litres of wood preserving solution containing the highly toxic chemicals tributylin oxide and lindane to enter the River Bourne. About 15 kilometres of the river were affected by the incident which killed about 15 000 fish. Other aquatic life was damaged and water supply companies downstream had to close their intakes until the plume of pollution had passed.

In October 1991 a spill of about 27 000 litres of surfactants occurred from a chemical works in Middleton, Lancashire, of which about 900 litres reached the River Irk. The river was covered with foam up to 10 metres thick.

Some 450 000 litres of sediment-laden water were discharged from the lagoon of a quarrying company into the West Okement river near Okehampton in February 1991. The discharge caused the river to turn chocolate colour, coated the river bed with sediments for several kilometres and killed much of the river's animal life.

A major oil company was fined for causing pollution by oil from its refinery in Essex in 1991. Six hundred gallons of crude oil were discharged into a creek flowing into the Thames, affecting the creek, oiling birds and causing a large fish kill.

A spill of 100 kilograms of dye from a wallpaper manufacturer in Blackburn in July 1993 turned 10 kilometres of the River Darwen red.

A fire at a Shropshire chemical and waste treatment works in February 1993 was extinguished only after several days. The fire fighting water, contaminated with pesticides and heavy metals was prevented from entering surface waters. The NRA believes that chemicals were, however, flushed into groundwater and represent a threat to the sandstone aquifer beneath the site which supports several drinking water abstractions.

Following vandalism in the yard of a transport and distribution company, 1 000 litres of a surfactant, used in a pre-treatment stage in dying in the carpet industry, escaped to a surface drain. All fish and most invertebrates were killed in a severn-kilometre stretch of river.

Following a pipe fracture, 2 000 litres of heating oil escaped into a bund at industrial premises. A rainwater drain device was overtopped and oil escaped to a surface drain, with serious consequences for a nearby fish farm.

Table 2.2 *Incidents at a plant which resulted in loss of containment* (Source: private communication)

Cause of incident	Percentage of total incidents
Transfer of materials through on-site pipework	37
Failure of storage tanks	7
Tankers (i.e. loading and off-loading)	7
Compressors / oil bowsers	6
Valves	5
Pumps	2
Other plant and equipment	7
Human error	10
Miscellaneous	19

The ENDS Reports contain brief descriptions of incidents which polluted surface and groundwaters. During the period May 1990 to September 1994 there are reports of 43 incidents where the prime cause of the incident is briefly described. Of these, seven resulted from defective bunds or valves, six from other plant and equipment, six from fires, five from poor working practices including human error, four from loading/off-loading activities, three from tank failures, three from pipework and nine for 'miscellaneous' reasons.

Analysis of unpublished information obtained from Thames NRA on 39 polluting incidents involving chemicals, oil and other substances in 1989-94 indicated the causes shown in Table 2.3.

Table 2.3 *Causes of incidents resulting in pollution in NRA Thames Region, 1989-94* (Source: Environment Agency)

Reason	Number of incidents
Loading and off-loading operations	6
Defective bunds	2
On-site traffic accidents	3
Defective valves, pipework and tanks	7
Spills, leaks, washings to surface water drains	5
Fires resulting in discharge of contaminated fire fighting water	5
Miscellaneous	11

A report by the EPA (1978) analysed the reasons for 84 spills from storage tanks. It was found that 60% of the incidents were due to overfilling and 31% due to tank rupture.

A relatively small proportion of incidents stem directly from failure of primary containment. Most incidents take place where materials are stored, transferred by pipework, and from loading and off-loading operations. On-site traffic movements are a regular cause of incidents involving rupture of tanks, drums, vessels, bunds or pipework.

Reasons for incidents include:

- accidents
- fires and other emergencies
- poor maintenance and supervision
- absence of bunds
- absence of interceptors
- failure to manage containment systems
- failure of alarm systems causing overtopping of storage vessels
- illegal use of surface water drains, sewers and soakaways
- inadequate drainage arrangements
- run-off of storm water from contaminated areas
- illegal waste disposal
- unsatisfactory drum storage compounds
- seepages from ground contaminated, in particular, with oil
- negligence
- vandalism
- arson
- lack of awareness
- human error.

At first glance this list may be daunting, but through improvement of management systems and procedures and implementation of the guidance in this report, the reader can greatly reduce the risk of an incident occurring.

3 UK and EU legislation relevant to containment

3.1 GENERAL FRAMEWORK

EC legislation aims to protect the aquatic environment by either:

(a) laying down water quality standards to achieve specified objectives, e.g. the protection of shell fisheries or the quality of water for abstraction for potable use, or

(b) setting emission standards or environmental quality objectives for specified dangerous substances.

Some of these standards and objectives may be used to define incident categories and for quantifying the effect of spills on the aquatic environment, an integral part of risk assessment methodology (see Section 7). The broad requirements of EU directives are enshrined in the provisions of the relevant UK legislation as summarised in this section.

The UK has ratified several international conventions including the Oslo and Paris Conventions and the North Sea Conference. The latter are relevant in that they list dangerous substances to be controlled in the marine environment and aim to control inputs of dangerous substances to the sea from land based sources. The UK 'red list', a list of those substances considered by the UK Government to be the most dangerous in the aquatic environment, was produced as an input to the 2nd North Sea Conference in 1987 (DoE, 1988).

There is a great deal of UK and EU legislation which is aimed at controlling incident risks and pollutant discharges, and for the establishment of water quality objectives and standards. The legislation is essentially aimed at controlling known, or planned, point source discharges of trade and sewage effluents to the aquatic environment, i.e. 'end-of-pipe' controls. The lack of specific legislation and related guidance on containment discourages any uniformity in approach and consistency in the application of design and construction standards.

3.2 SUMMARY OF RELEVANT LEGISLATION

There is very little in UK or EU environmental legislation directed specifically towards the design and construction of containment systems on industrial premises. Nevertheless, designers of containment systems need to be aware of the considerable body of legislation which is indirectly relevant to the containment of dangerous processes and substances since, in some of the legislation, there is a very strongly implied requirement for the containment of materials capable of causing water pollution. The legislation enables the statutory authorities to exert strong pressure on operators to improve the integrity of containment.

Much of the UK legislation is 'enabling'; i.e. the Acts themselves set out only the broad objectives and principles but enable regulations to be produced under appropriate sections of the Act to set out the detailed controls. The DoE publishes Circulars which provide definitive guidance on Statutes and Regulations.

The UK legislation most relevant to the containment of hazardous substances in the context of preventing water pollution is summarised in Box 3.1 and described in more detail below.

Box 3.1 *Summary of principal UK legislation affecting containment*

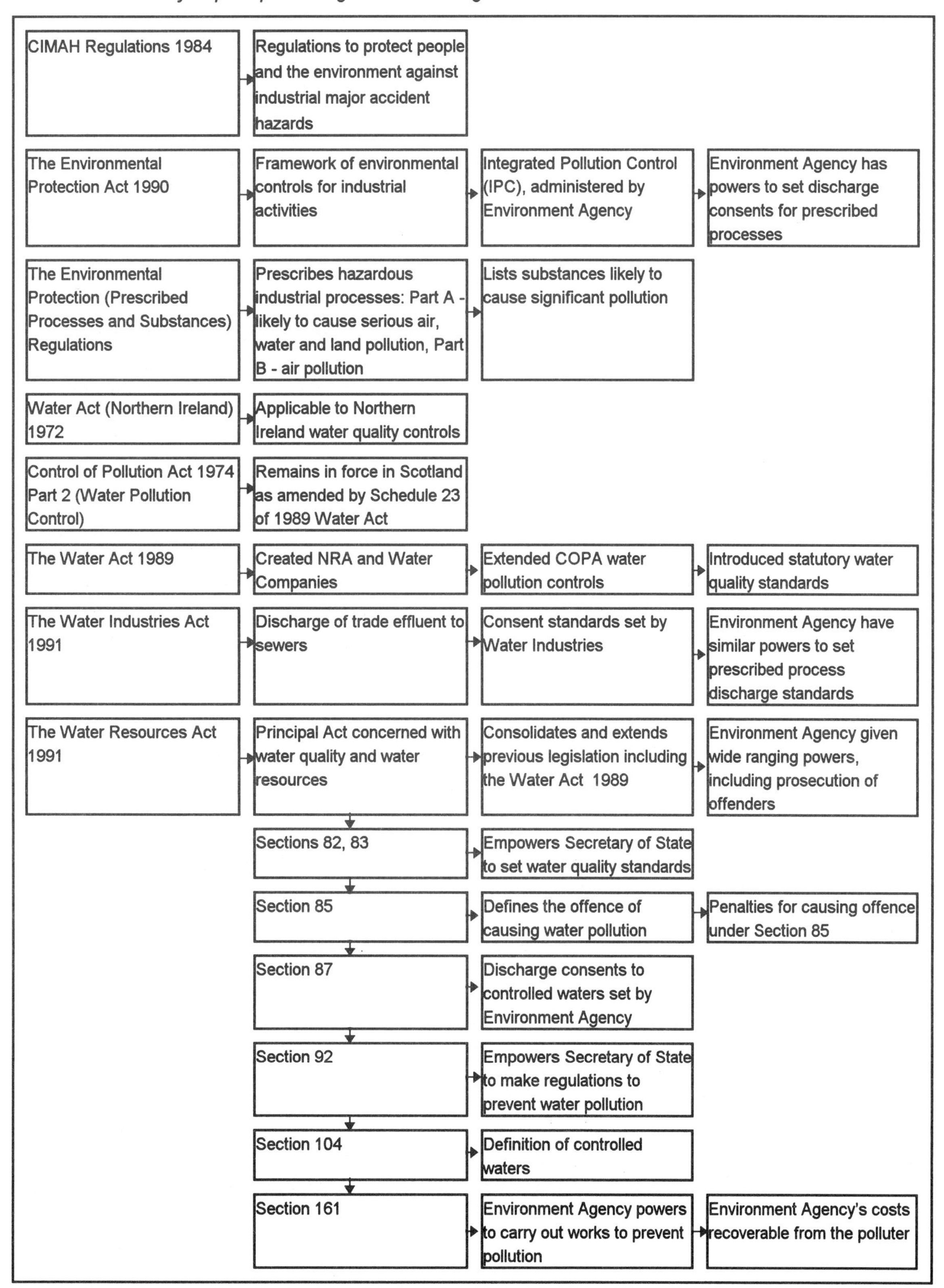

3.2.1 Water Resources Act 1991

The Water Resources Act (WRA) 1991 consolidated the 1989 Water Act and its provisions currently form the basis of most of the water law in England and Wales.

In Sections 82 and 83 the Secretary of State is given the power to prescribe systems for classifying water quality and to set statutory water quality standards. He may also specify water protection zones. The Environment Agency has applied for the first water protection zone, to be set up in the River Dee catchment as a response to a major incident in 1984. This would enable the Agency to require companies to carry out risk assessment and take precautionary measures against accidental spills and leaks.

Under Section 85 it is an offence to "cause or knowingly permit, without proper authority, any poisonous, noxious or polluting matter or any solid waste to enter controlled waters". The offence includes accidental, unplanned or deliberate acts. The words noxious, poisonous and polluting are sufficiently broad to include most of the dangerous substances which might result from an industrial incident.

The WRA is concerned with the quality of coastal, surface and groundwaters. Section 92 of the Act empowers the Secretary of State to make regulations which would:

> "prohibit any person from having custody or control of any poisonous, noxious or polluting matter unless prescribed works and prescribed precautions and other steps have been carried out or taken for the purpose of preventing or controlling the entry of the matter into any controlled waters."

Under Section 161 of the Act, the Environment Agency has extensive powers to prevent water pollution. These powers extend to carrying out work to prevent a pollutant entering controlled waters and to the cleaning up of a spillage where it may have affected water quality. The costs incurred in carrying out this work can be recovered by the Agency. In addition, the Agency has powers to serve 'Works Notices' which allow it to require a person to carry out work or operations to prevent pollution occurring where, in its opinion, it appears likely that polluting matter may enter controlled waters. The Notices can also be used when polluting matter has already entered controlled waters and the Agency wishes the responsible person to undertake a clean-up operation.

3.2.2 Water Industries Act 1991

All discharges of trade effluent to public sewers require the approval of the appropriate sewerage undertaker and will be subject to the terms and conditions of a trade effluent consent issued under Section 118 of the Act.

3.2.3 The Environmental Protection Act 1990

The Environmental Protection Act (EPA) 1990 provides a statutory framework for the control of a wide range of activities that can affect the environment. Part I of the EPA provides a mechanism, Integrated Pollution Control (IPC) (DoE/Welsh Office, 1991), whereby stringent controls can be applied to those industrial processes which have the greatest potential to pollute the environment. The Environment Agency applies these controls through a system of authorisations, which includes the consenting of aqueous effluents.

The main objectives of IPC are:

- to prevent or minimise the release of prescribed substances and to render harmless any such substances which are released, and to render harmless any other substances which might cause harm if released

- to develop an approach to pollution control that considers discharges from industrial processes to all media in the context of the effect on the environment as a whole.

The Environment Agency has powers to impose measures to prevent or minimise releases of dangerous substances to the environment by applying the principles of BATNEEC ('best available techniques not entailing excessive cost') in order to achieve the first of these objectives, and by selecting BPEO ('best practicable environmental option') in order to achieve the second objective. The IPC system does not, however, cover the large number of processes which, although inherently less hazardous than those prescribed under IPC, have considerable potential to pollute water.

3.2.4 The Environmental Protection (Prescribed Processes and Substances) Regulations 1991 (SI/1991/472)

(Amendments: SI/1992/614 and SI/1993/2405)

These Regulations, with subsequent amendments, prescribe processes for control by IPC. Schedule 1 is split into six groups of industrial processes, subdivided into two parts, A and B. Part A lists the processes likely to cause serious multi-media pollution of air, water and land. Part B is concerned with processes likely to cause significant air pollution only, which are separately controlled by Local Authorities. Each individual industry grouping has allocated sub-sections. Part A covers 105 processes and 5000 different installations. Schedules 4, 5 and 6 prescribe substances considered most dangerous in air, water and land. Schedule 5 is identical with the UK red list.

The Environment Agency has to date issued a series of over 60 Process Guidance Notes (HMIP, 1991) describing the techniques required for pollution abatement for Part A processes. Although the Notes do not provide specific advice on containment, they do broadly outline the features the Environment Agency expects to be used to prevent accidental pollution of water and soil. In order to implement IPC, the Agency has wide powers to impose conditions in authorisations relating to containment measures for the storage, transfer and use of chemicals, products, wastes and contaminated fire fighting water.

3.2.5 The Waste Management Licensing Regulations 1994 (SI/1994/1056)

Waste management is controlled by regulations made under Part II of the EPA. The Waste Regulation Authorities issue licences for waste treatment, storage and disposal activities which generally include conditions on the storage of waste materials.

Regulation 15 concerns the disposal of wastes containing List I and List II substances into landfill sites and effectively implements the EU Groundwater Directive.

3.2.6 The Control of Industrial Major Accident Hazards Regulations(CIMAH) 1984 (SI/1984/1902)

(Amendments: SI/1988/1462 and SI/1990/2325, and SR/1985/175 (Northern Ireland))

The main purpose of the CIMAH Regulations is to prevent major chemical industrial accidents and to provide safeguards to protect people and the environment. CIMAH is concerned with hazardous activities such as the processing and storage of dangerous substances. Flammable, reactive, explosive or toxic materials, liquefied gases and activities such as warehousing are controlled and materials and activities falling within the scope of the Regulations are listed. Manufacturers have to demonstrate to the HSE that they have identified potential major hazards and have taken adequate steps to prevent or limit the consequences of any accidents which do occur.

The form of containment systems used to arrest the escape of dangerous substances, including contaminated fire fighting water, would seem to be a material consideration in drawing up the emergency plans required by the Regulations. CIMAH therefore provides a mechanism for imposing wide ranging conditions on the storage of hazardous chemicals.

A guide to the Regulations has been published by the HSE (1984).

3.2.7 The Town and Country Planning Act 1990

Various Acts and Regulations control the use of land and have an important influence on the design of the built environment. A planning application must be made for any significant new development or change of use. The Local Planning Authority decides whether or not to grant planning permission based on planning policy guidelines, development plans, central Government policies and any other material considerations.

Proposals to build new plants and storage facilities, or to modify existing plant significantly, must be submitted to the local authorities for approval. This gives the Environment Agency the opportunity to request containment provisions for the storage of materials representing a hazard to surface or groundwaters. After statutory consultations have taken place, any planning permission granted may include conditions to meet environmental standards.

3.2.8 The Town and Country Planning Act (Assessment of Environmental Effects) Regulations 1988 (SI/1988/1199)

(Amendments: SI/1992/1494 and SI/1987/37)

The Regulations classify major developments into two types, referred to as Schedule One and Schedule Two projects. An environmental assessment must be submitted with the planning application for a Schedule One project. The Planning Authority may require an assessment for a Schedule Two application where it is likely to have a significant environmental effect. DoE Circular 15/88 (DoE/Welsh Office, 1988) sets out three main criteria of 'significance', the third of which relates to the adverse effects of the discharge of pollutants. Further guidance is given in a guidance note issued by the DoE (1989).

These Regulations are significant in that they give planning authorities powers to ensure that adequate measures are taken in the design of major new developments to protect the environment. An application must satisfy the Planning Authority that provisions have been made for the containment of chemicals or other materials with the potential to damage surface or groundwaters in the event of loss of containment, fire or other emergencies.

3.2.9 The Control of Pollution (Silage, Slurry and Agricultural Fuel Oil) Regulations 1991 (SI/1991/324; Scotland, SI/1991/346)

These Regulations were made under Section 110 of the Water Act 1989 (which was transposed into Section 92 of the WRA 1991). Although farm wastes are outside the scope of this report, these Regulations are significant because they are the only UK Regulations which specify design criteria for containment. Further guidance has been published by the DoE (DoE, 1991b).

3.2.10 The Petroleum Consolidation Act 1928, The Petroleum Bulk Stores Regulations 1979 (SI/1979/313) and the Retail and Private Petroleum Stores Regulations 1979 (SI/1979/311)

These three Acts and Regulations provide a legal requirement for containment of petroleum, including the bunding of above-ground tanks and encasement of below-ground petrol tanks together with notional restrictive zones. Their primary purpose is for fire control.

3.2.11 The Highly Flammable Liquids and LPG Regulations 1972 (SI/1972/917)

These Regulations specify requirements for the storage (including bunding), labelling, handling, etc. of flammable liquids and liquefied gas.

3.2.12 The Dangerous Substances in Harbour Areas Regulations 1987 (SI/1987/37)

These Regulations control the carriage, loading, unloading and storage of dangerous substances in harbour areas. Regulation 30 concerns the construction, maintenance and siting of tanks used and the storage of dangerous substances in bulk. An approved Code of Practice (HSE, 1987) relating to the Regulations was published in 1987 in which the requirements for storage tanks were outlined, but only in very broad terms.

3.2.13 Planning (Hazardous Substances) Regulations 1992 (SI/1992/ 656; Northern Ireland SR/1993/275)

The hazardous substances and their threshold quantities are listed in these Regulations. The Regulations enable local authorities to control the storage and use of hazardous substances by a system of consents. The HSE are statutory consultees.

3.2.14 Scotland: The Control of Pollution Act (COPA)1974

In Scotland, water pollution control is the responsibility of the Scottish Environmental Protection Agency (SEPA). The Water Act 1989 made no changes to the functions and roles of the statutory organisations in Scotland. The Control of Pollution Act 1974, as amended by the Water Act 1989, remains in force.

3.2.15 Northern Ireland: Water Act (Northern Ireland) 1972

It has been the tendency in recent years for environmental legislation in Northern Ireland to lag behind that in Great Britain by several years. The Government has undertaken to deal with the backlog of legislation and set specific deadlines, but it seems that these commitments are running behind schedule.

The Environmental Protection Act covers Great Britain. It appears to be the intention to cover Northern Ireland only by that part of the EPA relating to air pollution control, rather than by the full controls of IPC. This is perhaps a reflection of the fact that there are very few processes which would require control under IPC. Water pollution control is exerted through the Water Act (Northern Ireland) 1972 which is currently under review; the WRA does not extend to the Province. Controls are exercised by the DoE (NI) Environment Service.

4 Existing technical guidance

4.1 INTRODUCTION

The design of containment systems may be separated into two parts.

1. The selection of the most appropriate type, form and size of system, taking into account factors such as hazards, siting, cost and operability.

2. The detailed design and specification of the chosen system and the system's components.

There is a considerable amount of information on the first part, though not all related specifically to containment, but very little on the second.

Section 4.2 summarises the more important publications which are concerned directly with one or more aspects of secondary containment design. Section 4.3 draws attention to publications which, although not concerned specifically with containment design, provide guidance on related engineering topics.

Appendix A1 provides a quick and convenient guide to the sources of information in each topic area.

It should be noted that the recommendations made later in this document diverge from existing guidance in several areas.

4.2 SUMMARY OF EXISTING GUIDANCE

4.2.1 The Environment Agency

The Environment Agency (through the NRA) has published a series entitled Pollution Prevention Guidance Notes (PPGs) (NRA, 1994a - 1994i) which set out the principles of good practice for the prevention and mitigation of water pollution.

The PPGs cover a range of operations where precautions to prevent water pollution are necessary. They include: fuelling stations and depots, vehicle parks, demolition and construction sites, non-mains drainage sewage disposal and sewage works, storage of chemicals, foodstores, highway depots, dairies, commerce and industry. Guidance is given on matters such as siting, bunds and bund capacity, oil separators, drainage, catchments and facilities for the retention of contaminated storm water and fire fighting water.

PPG18, *Pollution prevention measures for the control of spillages and fire fighting runoff* (NRA, 1994i), covers the control of chemical spillages and fire fighting water runoff resulting from incidents at industrial and commercial sites. The need for a risk assessment of the environmental hazards is emphasised and factors influencing site sensitivity and site control are discussed. It makes clear the site owner's responsibility for developing and implementing detailed emergency plans which should include:

- full operational and control procedures for dealing with incidents

- up-dated records of all hazardous and non hazardous substances kept on site
- detailed drawings of the site, its drainage and all environmental protection facilities.

Specific guidance is given on chemical stores, tanks, silos, bunds, external storage areas, drainage, and provisions for on-site delivery, and on the handling of polluting substances generally. Although requirements for secondary containment systems and temporary measures for containing pollutants are covered, PPG18 does not provide detailed guidance on system design or specification.

4.2.2 Health and Safety Executive (HSE)

Although the HSE guidance is aimed primarily at safety, the publications summarised below contain information which is also relevant to pollution control.

The HSE has published two guidance documents (HSE, 1984) (HSE, 1992) on the CIMAH Regulations, intended to explain and set out the practical implications of the Regulations for operators of CIMAH sites. Although they do not give any detailed guidance on the design or construction of secondary containment, both publications provide useful background to help the designer understand how secondary containment fits in the context of the overall Emergency Plan required for each CIMAH site.

More specific guidance on design, construction and operation of a range of storage facilities, including the need for secondary containment, is given in the following HSE publications.

HS(G)50 (HSE, 1990a) is concerned with the storage of flammable liquids in fixed tanks of up to 10 000 m^3 capacity and includes advice on the size and construction of bunds. HS(G)52 (HSE, 1977) deals similarly with storage tanks exceeding 10 000 m^3. The guidance is concerned with the size and layout of bunds rather than with the detailed construction. It recommends, amongst other things, a maximum height of 1.5 m.

Specialist Inspector Report 39 (Bugler, 1993) provides guidance on the bunding of bulk chemical storage vessels. It includes recommendations on the situations in which bunds are required, bund capacity (recommending at least 110% of the capacity of the largest storage vessel in the bund) and some prescriptive general advice on the construction of bunds. The report includes comments on reasons for bund failures, using data extracted from the HSE incident databases.

The HSE has published draft guidance (HSE, 1995) dealing specifically with the control of fire fighting water runoff from CIMAH sites. The guidance covers very broadly issues such as environmental hazard identification and fire precautions, before moving on to deal with containment systems, including bunds, lagoons, tanks, temporary containment, etc. There is no detailed constructional advice.

4.2.3 Loss Prevention Council (LPC) / Fire Protection Association (FPA)

The five-volume *Compendium of Fire Safety* provides advice and guidance on a wide range of issues concerned with preventing, fighting and mitigating the effects of fires on industrial premises.

Volume 4 of the Compendium (FPA, 1989) contains information sheets on over 150 commodity chemicals used to produce other chemicals. These include many 'red' and 'black' list substances. For each chemical, information is provided on use, hazards, fire fighting, characteristics and any necessary safety precautions. The Compendium advises on the facilities needed for safe storage, including buildings, plant installations, tanks, drums and other containers. Requirements for containment are discussed and appropriate systems are described. Measures to prevent pollution of waters are outlined. These include brief recommendations on any need for impermeable construction and other special provisions, including drainage.

Volume 2 of the Compendium (FPA, 1995) includes data sheets on oil-fired installations, containing advice on bunding oil tanks, and on bund walls for flammable liquid storage tanks. The latter data sheet gives recommendations on height of bund walls, clearance between bund walls and storage tanks, construction and drainage. It recommends that there should be no openings in bund walls and that all service pipes should be carried either over the wall, or under the wall through liquid-tight seals.

Report SHE 8 (LPC, 1992) provides an overview of water pollution caused by substances arising from industrial incidents. It lists the 25 highest risk industries, their associated products and hazardous substances. Case histories of six of these industries are discussed in detail with comments on the processes, hazard and risk assessments, pollution incidents, and management controls.

4.2.4 Home Office

The Home Office Fire and Emergency Department (HOFD) *Manual of Firemanship* (Home Office, 1991) is concerned with preparations, provisions and procedures for fighting fires on industrial premises. In common with BS 5306 (BSI, 1990a) it ranks the fire risks for premises into six categories and processes into four classes, ranging from low to high hazard. Rules are set out for the compartmentation of warehouses. The concept of 'area of assumed maximum operation' (AAMO) allows estimates of design densities of water/extinguishing substance application to be determined according to hazard classification and process category. The Manual includes tables giving minimum fire fighting water storage volumes for the various hazard categories and rules for hydrant and other appliance supplies.

Although not directly relevant to the design of secondary containment, the HOFD Manual provides basic data on volumes of fire fighting and cooling water, and foam, likely to be required in various situations. Making appropriate allowances for fire fighting agents is often the most critical part of secondary containment design.

4.2.5 BASIS

The British Agrochemical Standards Inspection Scheme (BASIS) is a system of self-regulation by the agrochemical industry for the safe storage and transport of agrochemicals. BASIS have published guidance on the design and construction of agrochemical stores (BASIS, 1992), compliance with which is part of the requirement for registration under the scheme. The guidance covers the design of storage buildings, including fire and other health and safety precautions, and measures to prevent the escape of material or contaminated fire fighting water to the environment. Recommendations are given on several aspects of bund design, including minimum heights of bund walls and drainage from bunded areas to containment tanks. Piped drains leading from bunded areas to containment facilities are not recommended since, the BASIS guidance says, experience has shown that these systems rapidly become blocked under fire fighting conditions, and there is a continuing risk of ground movement allowing hidden leaks to develop. Open channels are the preferred option.

The role of BASIS is highlighted in the MAFF Code of Practice described below.

4.2.6 Ministry of Agriculture, Fisheries and Food (MAFF)

The Code of Practice for suppliers of pesticides to agriculture, horticulture and forestry published by the Ministry of Agriculture, Fisheries and Food (MAFF, 1990) sets out basic criteria for agrochemical stores and makes a number of specific recommendations on construction. The Code requires stores to be designed and constructed to provide containment of any spillage or contaminated water resulting, for example, from fire fighting. It suggests a retention capacity equal to 110% of stored product plus an additional 75% in environmentally sensitive situations.

4.2.7 Institute of Petroleum (IP)

The Institute of Petroleum has published a large number of guidance documents concerned with petroleum storage at refineries and distribution facilities.

Model Code of Practice Part 19 (IP, 1993) is concerned with fire precautions and fire fighting. It provides information on fire fighting water requirements which the containment system designer will find useful in estimating retention capacity.

The Environmental Guidelines (IP, 1996a) provide broad guidance on a wide range of operational issues including containment and the separation and treatment of contaminated water streams.

The Marketing Safety Code (IP, 1981) provides recommendations on the layout and design of petroleum products installations and depots.

4.2.8 United Kingdom Atomic Energy Authority (UKAEA)

The UKAEA's Safety and Reliability Directorate (SRD), HSE and LPC have published two documents relating specifically to the design of bunds around large liquid bulk storage facilities.

Barnes (1990) examines various Codes of Practice, mainly from the UK and the US, on the bulk storage of liquid chemicals and industrial materials, and compares the precautions taken to restrict the loss of such material. The report discusses the philosophy of bunding and describes the principal design features of bunds, such as materials of construction, wall height, capacity and the drainage of surface water. It concludes that there is no unified approach to bund design, while there are anomalies in existing Codes. Barnes recommended further work to compare the costs of the complex engineered containment structures that would be required to protect against complete and sudden failure of a large storage tank, against the losses associated with such an incident.

Wilkinson (1991) progressed Barnes' work by examining the consequences of the instantaneous release of a large volume of liquid resulting from the catastrophic failure of a storage vessel. The emphasis of the study is on safety rather than pollution control. The findings indicate that pressures behind bund walls can peak at up to six times normal hydrostatic pressure in certain circumstances. Advice is given on bund size and geometry in order to minimise the risk of failure by spigot flow, 'sloshing' or overtopping.

The findings and recommendations of Barnes and Wilkinson are considered further in Section 10.

4.2.9 Other relevant guidance

This section summarises other published guidance that is relevant to containment design.

German Regulations

German legislation, e.g. State of Hesse Installation Regulations (State of Hesse, 1982) sets mandatory standards for installations where water-endangering substances are stored or handled. Specific requirements are set out for containment, including the retention of contaminated fire water.

The State of Hesse Regulations are based on six basic principles which include a 'duty of containment' in which 'escaping water-endangering substances must be detected quickly and reliably, retained and properly disposed of. As a rule the plant must be equipped with a leak-tight and durable catchment space (bund) unless it is double-walled and provided with a leak indication instrument.' Another basic principle concerns 'outlets' where it is stated that 'as a fundamental principle, catchment spaces (bunds) shall not have any outlets.'

The German Regulations refer to VCI recommendations (VCI, 1988a) for the calculation of fire water retention volumes for warehouses. Tables are given specifying a range of retention capacities according to the fire protection equipment in place at the site, and the quantities and types of chemicals stored.

The VCI Stoffliste (VCI, 1988b) classifies a wide range of chemicals according to their water pollution potential, enabling graduated safety precautions to be formulated. Four classes are defined ranging from WGK 0 (no hazard) to WGK 3 (severe hazard). There are approximately 400 chemicals in class WGK 3.

National Water Council (NWC) - rainwater catchment and drainage systems

The NWC five-volume manual *The Wallingford Procedure* (HR Wallingford, 1986) describes the procedures for the design of water catchment zones. It includes meteorological and soil data maps of Britain. Volume 1 of the manual is particularly relevant as it covers the modified Rational (Lloyd Davies) method which is suitable for the hydraulic design of small catchments up to 150 ha and pipe sizes up to 1 000 mm diameter.

Drainage must be designed and constructed so as not to permit pollutants to infiltrate into groundwaters. Tables prepared by Wallingford and Barr (1994) are well established aids for designing pipe networks and channels.

The Water Research Centre (WRC) publishes a number of manuals for the specification and construction of sewers. Useful information for the design of small sewerage systems is contained in *Sewers for Adoption* (WRC, 1990).

Pollutant characteristics

The range of substances with the potential to cause water pollution is very wide. The following references provide sources of information on the chemical, physical, biological and ecotoxicological properties of various chemicals and substances.

Edwards (1992) presents lists of substances considered dangerous in the aquatic environment by the UK Government, the EC, the Oslo and Paris Commissions and the North Sea Conference. Practical implications, including the need for containment, are discussed.

The European Centre for Toxicology and Ecotoxicology of Chemicals (ECTEC) provides environmental hazard assessments for substances on the basis of ecotoxicology and toxicology (ECTEC,1991).

The Royal Society of Chemistry publication, *Dictionary of Substances and their Effects* (*DOSE*) (Richardson and Gangolli, 1995) is a comprehensive reference manual providing physico-chemical and ecotoxicity data on some 6 000 substances, such as their impact on a variety of life forms including human and aquatic. The substances are selected from lists, including The EC Classification Packaging and Labelling Regulations, The EC 'Black' and 'Grey' substances, the UK DOE 'Red' list and the USA and Canada Priority Pollutants List.

The Royal Society of Chemistry publication, *Toxic Hazard Assessment of Chemicals and Risk Assessment of Chemicals in the Environment* (Richardson, 1988) covers a wide range of topics including chemical disasters, accidental emissions, water pollution control, contribution of toxicity to risk assessment and many other aspects relating to the safety of chemicals in the environment.

4.3 PUBLICATIONS RELEVANT TO THE DESIGN OF CONTAINMENT SYSTEMS

Technical references and standards which are applicable to the detailed design, specification, structure and construction of containment systems are included in the following sub-sections. Although none of the publications deals specifically with containment design, the guidance and recommendations contained are relevant to the design process.

4.3.1 British Standards

British Standards and Codes of Practice provide guidance and recommendations on a wide range of components and structures which may be relevant in parts to the design of containment systems. The most important, and relevant, of these are included in the list of references at the end of this report. Although some of the Standards and Codes deal with storage tanks and retaining structures, none of them deals specifically with secondary containment design. Issues such as capacity requirements, drainage arrangements, design life, testing, load factors, etc. are not therefore codified.

4.3.2 Site and geotechnical investigation

Guidelines for site investigation prepared by the Institution of Civil Engineers' Site Investigation Steering Group (ICE, 1993) describe geotechnical tests to determine the characteristics and suitability of a soil for many forms of construction, including lagoons and embankments.

Once the physical characteristics of sub-soils have been determined, computer software packages such as OASYS (New Civil Engineer, 1994) may be used in soil stability calculations as part of earthwork design.

4.3.3 Lagoon design

Earthwork lagoons are used for the containment of contaminated waters at many industrial sites. CIRIA Report 126 (Mason, 1992) deals with shallow to medium depth lagoons, giving advice on site selection, soil suitability, design, construction, impermeable liners, leakage detection and maintenance. Twort *et al* (1993) and Linsley and Franzini (1992) provide guidance on hydraulic engineering, flow nets and other aspects of design which are relevant to larger earthwork structures.

Geotextiles and geomembranes are used increasingly in earthworks for reservoirs, lagoons and waste storage. Ingold and Miller (1990) describe the properties and applications of these products and techniques for using them in earthworks' construction. The *International Directory of Geotextiles* (Rankilor, 1986) provides a comprehensive list of geotextiles and their properties. Geotextile manufacturers provide design literature and many also offer a design service for the use of their products.

A number of artificial impermeable lining systems have been developed for use in water retaining structures and for landfill sites. Many manufacturers, e.g. the OHV Group (Polyfelt), Terram, MMG and Butyl Rubber provide design manuals and other technical information.

Further information on the design and evaluation of geosynthetics and earth structures for the containment of pollutants is provided by Sharma and Lewis (1994), and in two CIRIA Special Publications (Jewell, 1996) and (Privett *et al*, 1996).

4.3.4 Tank design

The design of prefabricated storage tanks in steel, glass reinforced plastic and other structural materials is covered in a number of British Standards and Codes of Practice (see Reference list). Some of these tanks may be used for secondary containment.

In-situ reinforced concrete is the most common construction method for large below-ground tanks. CIRIA Technical Note 144 (Aubrey, 1992) provides guidance on sealants for incorporation into the design of watertight joints in concrete, one of the critical aspects in the design of *in-situ* concrete tanks. Guidance on the structural analysis of large circular and rectangular *in-situ* tanks is given by the Portland Cement Association, USA (1969). The British Cement Association (BCA) also publishes information on this form of construction.

The *Manual of Reinforced Concrete* (Institution of Structural Engineers, 1985) and *The Reinforced Concrete Designer's Handbook* (Reynolds and Steadman, 1991) both provide detailed guidance on the design and construction of reinforced concrete retaining structures.

Smaller tanks and retaining walls may be built in reinforced brickwork. Two publications by The Brick Development Association, (Haseltine and Tutt, 1991) and (Curtin *et al*, 1983) provide useful guidance.

4.3.5 Diaphragm/ cut-off/ curtain walls and sheet piling

A developing technology is the use of below-ground vertical barrier systems to prevent the migration of contaminants within the ground and groundwaters. A number of techniques, including impermeable cut-off walls and injection grouting, used to isolate sites from sensitive aquifers are described in Johnson *et al* (1991). BRE Digest 395 (1994) describes the cement-bentonite slurry trench technique for cut-off walls, a system which is widely used in the UK. CIRIA Special Report SP124 (Privett *et al*, 1996) gives further general guidance.

The *Steel Designers Manual* (Steel Construction Institute, 1992) provides guidance on the design of retaining structures utilising steel sheet piling.

4.4 QUICK GUIDE TO SOURCES OF INFORMATION

Appendix A1 draws together the references cited in this section into topic area headings. Most of the references cover more than one topic so they appear more than once in the listing.

5 Current practice

5.1 INTRODUCTION

This section summarises current practice in terms of the secondary containment and other protective measures provided on industrial sites. The following comments are based largely on discussions with plant operators, and on observations made on visits to many sites, ranging from very large to very small, across all industrial sectors during the course of this study.

It should be stressed that the following is an account of current practice, which in many instances deviates from what is regarded as best, or even acceptable, practice.

5.2 EXTENT AND TYPE OF CONTAINMENT

5.2.1 Bunds

The capacity of secondary containment provided in the form of bunds normally ranges from 5% to 110% of the contents of the largest individual tank within the bund. Larger bunds are found on storage facilities in sensitive locations covered by the BASIS scheme (BASIS, 1992) in compliance with the scheme's requirement for an additional 75% on top of the normal 110%. Any bund smaller than 5% is generally regarded as a drip tray.

It is common practice for there to be several tanks in one bunded area, in which case the total capacity of the bund is likely to be much greater than the volume of the largest tank and may in some cases exceed the total volume of all the tanks. Bunds are generally designed to include stock tanks and their associated pumps and pipework, so that spills from the latter items of equipment (which are more common than from the stock tank itself) will also be contained within the bund. In most cases bunds are purpose-built, but in some cases existing structures may also be used. An example of this is the storage of drummed wastes inside old settling tanks.

Full-sized bunds (defined here as those with a capacity at least equal to the biggest tank inside the bund) built around individual tanks or small groups of tanks are now constructed almost exclusively from reinforced concrete. Very often, both the base and the walls of a bund are coated on the inner side with an epoxy resin or similar sealant to provide additional protection against attack by corrosive materials and to improve impermeability.

On older sites there are many examples of full-sized masonry bunds, but these are slowly being replaced with reinforced concrete bunds. However, the smaller capacity 'drip trays' are frequently still built using bricks or concrete blocks.

Earth or clay bunds are often found, particularly at older sites, around tanks of fuel oil and related materials or less polluting materials such as dilute acids or alkalis. At one site, large storage tanks containing organic liquids were provided with off-set containment tanks. These consist of a low retaining wall round the tanks forming a shallow bund that drains into an adjacent, much larger, sub-surface concrete tank. The tank in this situation may be regarded as a large sump.

In the petroleum and petrochemical industries and in the contract storage sector, materials are often stored in very large tank farms at ports and on the banks of estuaries. These facilities may comprise up to 50 individual tanks with a capacity of up to 100 000 m^3 of bulk liquids of various compositions. The bunding of tank farms frequently takes the form of earth embankments lined with clay. The whole of the area within the bund is lined with clay or some other impermeable membrane. The membrane may extend up the side of the retaining embankment. The area is graded so that rainwater and any spilled material collects at a number of low points or sumps for disposal elsewhere.

A perennial problem with bunds is how to remove rainwater. Methods that have been observed include:

- collection at low point, or in a sump, and removal by mobile pump, 'gully sucker' or fixed pump (either manually or automatically switched) (see Figure 5.1)
- discharge through a valved outlet in the bund wall (see Figure 5.2)
- discharge over a weir to the plant drainage system and treatment plant
- removal by contractor for off-site disposal
- pumping to an on-site incinerator.

The first two methods in the above list are by far the most common.

Figure 5.1 *Bund with sump for collection of rainwater* (Reproduced courtesy of GATX)

Figure 5.2 *Bund with valved rainwater outlet* (Reproduced courtesy of the Environment Agency)

Discharge of rainwater by means of automatically switched pumps (e.g. controlled by float switches) and weirs is unsatisfactory in so far as the bund would be effectively by-passed in the event of a major spill. At one plant, the discharge of harmful substances was controlled by a device that prevented the pump from operating if the water in the bund exceeded a pre-set concentration of contamination.

Routine sampling and analysis of the contents of bunds prior to discharge happens only on a few sites. The general view appears to be that the plant operatives will know if the liquid in the bund or sump is other than rainwater or if the rainwater is contaminated. In situations where the bund contents are pumped to a treatment plant that can deal effectively with such hazardous materials this may be a reasonable approach, but where the material passes directly to a site's surface water drains it introduces a degree of risk that is unlikely to be acceptable.

Bunding is generally not provided for liquefied, non-flammable, non-toxic gases. It is assumed that spillages of these materials will evaporate rapidly without causing damage.

There are many examples of bunds that have become ineffective as a result of inadequate maintenance. It is an extremely common experience of those involved in spillage prevention and control to find that bunds provided around stock tanks are rendered useless as a result of holes being drilled through the walls at base level, or through valved outlets being left open to allow rainwater to escape. Bunds and other protective devices can lead to a false sense of security. A bund with a hole in it may be worse than no bund at all.

The provision of valved outlets at base level is a very common practice (see Figure 5.2), but it is a poor design feature since in very many cases they are left open by operatives. During a survey of companies in the engineering sector, 75% of such valves were found to be open. Even the largest multi-nationals cannot escape criticism in this respect. Although the provision of lockable valves improves matters, it does not overcome the problem entirely.

Other common reasons for the failure of bunds include:

- permeable walls or base due to poor design, construction or maintenance
- reduction of capacity by storage of other materials and wastes within bunded area
- juxtaposition of bund and tank allowing contents of tank to jet out beyond bund walls (see Figures 5.3 and 10.8, and Box 10.5)
- increasing the capacity of vessels within a bund without increasing bund capacity accordingly
- structural weakening caused by routing pipes through walls
- removal of sections of bund wall to permit vehicular access
- poor security leading to damage, theft or vandalism.

Figure 5.3 *Example of a bund vulnerable to 'jetting' failure*

5.2.2 Separators

On older sites where relatively small amounts of materials are stored there is heavy reliance on the use of separators (sometimes referred to as interceptors, drain catchpits or drain traps), rather than bunds, to provide protection. Separators are 'flow through' devices, as distinct from bunds and sumps that do not have an outlet. They may be regarded as providing primary treatment, i.e. phase separation, for materials that may contain solids, oil or other process liquids that are immiscible with water. They are generally designed to deal with minor spills and wash waters from process areas and filling points, rather than with major incidents. They are unlikely to have sufficient capacity to retain more than a small proportion of the contents of a tank in the event of a major failure.

Separators take a number of different forms, ranging from simple traps for solids or oil, to large multi-compartmented reinforced concrete tanks. In some chemical industry applications, separators are designed to allow separate collection of solids and non-aqueous liquid phases. For separating solids, a separator may have two compartments; the first (smaller) compartment to separate the coarse solids, and the second (larger) compartment to separate the fine solids. For immiscible materials that are lighter than water, e.g. diesel, paraffins and heavy fuel oil, the separator would contain compartments connected by underflow weirs. For materials that are denser than water, e.g. chlorinated solvents, the compartments are linked by overflow weirs.

At one of the sites visited, which was in a particularly sensitive location with respect to water pollution, all new separators were constructed with double linings and were fitted with devices to detect any leak through the inner lining.

On some sites the outflow from separators is continuously or routinely sampled and monitored to ensure that no polluting materials are carried over into the drainage system. On other sites, however, there is no routine monitoring and reliance is placed on preventing overflow by periodic emptying of the retention chambers.

5.2.3 Drainage systems

Site drainage provides a potential pathway through which material can flow rapidly from the source of the escape to the receiving water (see Section 7) and it is therefore a critical consideration in the assessment of the risks posed by an existing operation and in the design of new plant. On some older sites, particularly where drainage systems have been extended piecemeal over a period of many years, operators may not be fully aware of the layout of the system and the extent to which trade effluent and surface water drains are separated or interlinked.

Modern drainage systems provide generally more effective segregation of clean water, wastes and other liquids so that it is easier to isolate hazardous materials and divert them either to appropriate treatment plants or to secondary containment prior to disposal. There is widespread appreciation that it is easier to control, treat and dispose of relatively small quantities of strong pollutant, as will normally arise in a properly segregated drainage system, than larger quantities of pollutant that have been diluted with, for example, storm water.

An analysis of pollution incidents (see Section 2.2) has shown that washing down spilled materials into road grids and drains is a significant problem. This occurs frequently through the operators' ignorance of where the various drains discharge and the extent to which they are interconnected. For the same reason, the regulators may find it difficult to determine and take action on the source of an incident. In order to help overcome this, several large companies operating on sites with complex drainage arrangements have colour coded the road grid covers; typically blue for drains discharging to surface water and red for drains leading to effluent treatment. (Note: colour coding of drainage systems is recommended by the Environment Agency, but there is at present no uniform approach.)

Drainage systems may provide some degree of secondary containment by virtue of their capacity. On some sites, there are plans and procedures for blocking off site drainage systems in the event of an incident in order to prevent, or slow down, the spread of the escaped material. In some cases this is achieved by building-in properly engineered sluices or penstocks, whilst in others it may involve simply dropping sand bags down inspection chambers, manholes or other access points. Although rudimentary, the latter approach may be the only practicable method of containing spilled materials at older sites on hillsides close to rivers or sewage treatment plants where the retention time in the drains may be very short. However, there have been reports of drainage pipes and seals being extensively damaged due to extended periods of exposure to escaped substances.

On most sites, the drains consist generally of buried ceramic pipes, or brick channels set into the ground. On some of the larger and more sensitive sites, higher specification options such as concrete lined ductile iron piping, or concrete channels lined with acid resistant tiles, fibreglass or polypropylene, have been installed; but such high specification is very uncommon.

The policy of many companies is to move away from sub-surface drains, which are difficult to monitor and maintain and may give rise to chronic leakage, and towards above-ground systems where any problems are easier to detect and remedy.

Drainage pipelines are sometimes bunded. Approaches to this vary both in form and effectiveness. At its simplest, a pipeline may be laid in a brick or blockwork channel to provide protection against physical damage and corrosion, stability, visibility and ease of maintenance. Such channels would not normally be watertight and should not, therefore, be regarded as bunds. At the other end of the scale, pipelines are sometimes fixed in reinforced concrete channels, designed and constructed to water retaining standards so that leakage or even a major discharge resulting from a pipe rupture would be safely contained (see Figure 10.4).

At one site (see Section 6.3 for detailed description) the entire effluent pipework system has been raised above ground and contained within a concrete structure lined with an impermeable membrane and graded so that any spill would drain into strategically located sumps. Part of this system is shown in Figure 5.4 where the pipeline, contained within a covered, impermeable conduit, crosses a watercourse (bottom left of the photograph). On another major site, the pipework runs over impermeable ground that is graded to a drain and sump system so that the transfer system is effectively bunded.

Figure 5.4 *Bunded pipework crossing over a watercourse* (Reproduced courtesy of FLEXYS Rubber Chemicals Ltd.)

Double-walled drainage pipes, with facility for leakage detection in the cavity between inner and outer walls, are available but their use for pollution control is uncommon. A problem reported with cavity leakage detection is that alarms have been falsely activated due to condensation in the cavity or high groundwater levels causing water penetration at joints and junction boxes.

5.2.4 Storm water tanks

Rainwater falling on some parts of a site, e.g. process areas, bunded areas around tanks, filling points and drum storage areas, is likely to become contaminated with the materials used at the site. At some sites, particularly those in sensitive locations, operators collect and treat all contaminated surface water prior to discharge (see Figure 5.5). If the capacity of the treatment plant is exceeded, the excess is retained in tanks or basins for subsequent treatment. On some sites, all rainwater falling on the site is routed through the treatment plant, whether or not it is likely to be contaminated.

Storm water basins are sometimes located off-site and a single basin may be shared between a number of operators. In one case, a single basin serves the whole of an Industrial Estate.

Figure 5.5 *Typical example of storm water tank* (Reproduced courtesy of Elf-Atochem)

5.2.5 Emergency waste basins

Several companies have systems in which spills and contaminated water from stock tanks or loading areas flow over graded hardstandings towards site drains which discharge to an emergency waste basin. The latter generally takes the form of a shallow depression in the ground, taking advantage wherever possible of the local topography. Some basins are lined with an impermeable membrane, others are not. It is necessary to make suitable provisions for the removal of rainwater.

At one major site a reinforced concrete basin has been built adjacent to a chemical plant, large enough to take the whole inventory of the plant and the anticipated volumes of fire fighting foam and water that would be applied in the event of a major incident. There are a number of reasons for adopting this type of containment system (defined as a 'remote' system in Section 8) rather than bunding, including:

- if a spilled liquid catches fire it may be transferred away from the vicinity of the tank or tank farm from which it was lost, thereby minimising the risk of fires, explosions, etc. in other tanks
- a single emergency basin may be designed to serve several primary storage or process areas
- it may be possible to install remote containment in situations where bunds around the primary operational areas could not be provided owing to space restrictions.

At another location a large basin (1 million gallons) has been built so that if a spill results in effluent of suspect quality the entire site effluent can be diverted into it and retained pending testing and a decision on disposal.

5.2.6 Kerbs and ramps around catchment areas

The construction of kerbs around areas of hardstanding as a means of containing minor spills, drips and leaks is very common across a wide range of industrial premises and processes. Ramps are installed at appropriate positions to allow vehicular access. Kerbed areas are commonly used in the following situations:

- areas used for manufacturing, formulating and warehousing
- surrounds to tanks where bulk liquids with a relatively low polluting potential are stored
- loading and off-loading areas
- storage areas for raw materials, product or waste in drums, with the kerbed area typically providing retention for several drums should these fail simultaneously.

Any spilled material, washing down water or rainwater drains to a low point or a sump within the kerbed area, from which it is removed by pumping.

At one plant, kerbs have been built into each floor of a new multi-storey facility used for manufacturing chemicals in order to route any spillages to the containment system on each floor, and to protect stairwells.

Kerbs can also be used to contain or direct the flow of the much larger volumes of water used in fire fighting operations.

5.2.7 Roofed areas

Very reactive or toxic products, raw materials or wastes are frequently stored in roofed areas to reduce the risk of contaminating rainwater. In the chemical industry, filling points are frequently roofed over (see Figure 5.6).

5.2.8 Tanks

A number of features may be built into primary storage tanks to provide additional protection. Increasingly common for process vessels and underground fuel storage tanks is the use of double-walled storage tanks, in some cases with leak detectors in the cavity between inner and outer skins. Double-kinned tanks may be built so that the outer skin provides containment for a possible overflow of up to 10% of the inner tank capacity; such tanks are sometimes referred to as 'self bunding' (see Figure 8.1(a)). Another common development is the use of 'integrally bunded' tanks (see

Figure 8.1(b)) where a separate outer tank is provided to contain any spill or leakage from the inner.

High level alarms are used extensively, particularly on larger sites, to provide warning against overfilling of tanks.

Figure 5.6 *Example of filling point with roof over* (Reproduced courtesy of Elf-Atochem)

5.2.9 Total containment

Some very hazardous materials, for example anhydrous hydrofluoric acid and oleum, are stored in primary vessels within specially designed buildings that act as the secondary containment system. All ventilation from the building passes through scrubbers and there are no drainage points, so that in the event of a spill atmospheric emissions and liquid discharges would be prevented.

Some companies in the waste management sector are introducing a policy of total containment of all liquid and solid materials imported into the site. The final waste products and liquid effluents are incinerated.

5.2.10 Drum storage

It is normal practice for drummed materials to be held in kerbed areas where there is no outlet to surface water drainage. On some sites, drum storage areas are sub-divided or segregated in order to keep different materials apart. 'Over-drumming' (the practice of placing one drum inside another) is often practised in the waste management industry for the storage of particularly hazardous wastes or when the primary drums are in poor condition.

The storage of waste on a site where it is produced requires licensing where the quantity is above a prescribed threshold, and conditions in the licence are likely to specify containment measures in some detail.

The arrangements made for storing drummed materials are extremely variable and there are many examples where drums of hazardous materials are stored inappropriately. In general, the situation tends to be managed better within large operations than at small sites. Small engineering companies where materials such as diesel oil, cutting oils, lubricants, paints, thinners and waste oils are traditionally held in drums have been identified as having a poor record in this respect. An extreme, but not uncommon, example of bad practice is shown at Figure 5.7.

The Environment Agency is currently drafting guidance on drum storage.

Figure 5.7 *Drum store - example of poor practice* (Reproduced courtesy of the Environment Agency)

5.2.11 Effluent treatment plants

Large industrial sites are often equipped with one or more effluent treatment plants to treat aqueous effluents from either the whole or part of the site. It is common practice for the drainage from hardstandings in and around storage and process areas to be routed direct to the treatment plant. The effluent treatment plant itself may be bunded.

Treatment plants can be damaged and put out of action for a long period of time if they are overloaded, as might happen for example in the event of the sudden entry of large quantities of contaminated material during an incident. On many sites, but not all, storage facilities are provided to buffer the treatment plant against such overload. The type of material reaching the plant, as well as quantity, must be taken into account. For example, a major spill of acid or alkali is likely to overwhelm completely the ability of a treatment plant in which neutralisation takes place. Similarly, a major spill of an organic solvent could not be dealt with in a biological treatment process. Small spills of organic solvents are also a hazard since they can pass through treatment works largely unaffected and cause problems with the public water supply.

5.3 RETENTION OF CONTAMINATED FIRE FIGHTING WATER

This section should be read in conjunction with Sections 5.2.4 and 5.2.5 since retention of fire fighting water is usually combined with measures to collect major spills and storm water.

Until recently, the provision of facilities such as basins, lagoons and tanks for the purpose of retaining contaminated fire water was uncommon at industrial sites in the UK. In the aftermath of the Sandoz incident in 1986, many companies in the chemical, agrochemical, pharmaceutical, petroleum and waste management sectors have installed retention systems. The reasons include a desire by companies to improve their image in the eyes of the general public, to move towards a general policy of total control of emissions and to help ease planning problems when they apply for new plants or modifications to existing facilities.

The installation of contaminated fire water retention basins is invariably a retrospective action at existing facilities. Most large UK chemical companies make their decisions on whether to install fire water retention systems based on risk assessment procedures (see Section 7) although some, reacting perhaps to pressure from non-UK parents, intend introducing retention basins at all sites where practicable.

Site operators calculate the volumes required for fire fighting water containment in a number of different ways. In some cases, calculations are based on the assumption that any fire would be fought using only foam or water applied through sprinkler systems. In others, additional allowance is made for water from hydrants and fire fighting appliances, damping down water and, perhaps, the simultaneous occurrence of a heavy rainstorm.

5.3.1 Examples of existing facilities

Box 5.1 lists a number of major companies and sites where retention arrangements for contaminated fire fighting water are either in place, under construction or are being planned. It is presented to indicate the significance which major companies now attach to the issue.

Typical examples of fire fighting water basins are shown at Figures 5.8 and 6.3.

Box 5.1 *Examples of major sites where fire fighting water retention facilities have been, or are being, constructed*

Company	Site	Company	Site
Monsanto	Ruabon	Beck and Politzer	Manchester
Zeneca	Huddersfield, Grangemouth and Yalding	A H Marks	Bradford
Albright & Wilson	Avonmouth	Elf Atochem	Widnes and Manchester
Courtaulds	Derby, Leek and Stretford	Fine Organics	Teesside
Du Pont	Maydown	Cleanaway	Ellesmere Port
Sandoz	Leeds	Rechem	Fawley and Pontypool
Burroughs Wellcome	Dartford	Lanstar	Carrington
Fermin-Coates	Bury St Edmunds	Texaco	Milford Haven
Bush, Boake, Allen	Widnes		

The facilities used for containing contaminated fire fighting water range from the rudimentary, to fully engineered structures.

The following list provides a very brief indication of some of the types of containment recently seen used for this purpose:

- purpose-built lagoons, lined with clay or other impermeable membrane
- adaptations of natural depressions in the land, with or without a liner
- redundant effluent treatment settling basins
- catchment area formed by constructing a kerb along a low lying area of site alongside a river bank
- redundant primary storage tanks
- storm water tanks at an adjacent sewage treatment works
- a large capacity bund around a tank farm
- land drainage ditches adjacent to the plant, with facility for sealing off with penstocks
- a low-lying car park doubling as an emergency basin.

Figure 5.8 *Example of fire fighting water retention basin* (Reproduced courtesy of Du Pont (U.K.) Ltd.)

5.3.2 Warehouses

The Sandoz incident promoted a great deal of activity in the industrial warehousing and storage sectors. Before that, the Woodkirk fire in 1982 (Stansfield, 1983) involving an agrochemical warehouse brought about the BASIS scheme. It is now common practice in the UK for stores and warehouses containing hazardous materials to be bunded to provide containment for the full inventory of material stored, washing down water and fire fighting water delivered from fixed installations. Warehouses may be sited on primary production sites or at formulation sites or at isolated off-site locations.

Bunding for water delivered through fixed fire fighting installations is usually provided by the external retaining walls and the internal compartments of the building, combined with ramped access at doors and vehicular access points.

In recently constructed warehouses designed to store toxic materials, there is unlikely to be any internal drainage provision other than for the floors to fall to a sump from which any spills may be pumped to effluent treatment or other means of disposal or storage.

A potential weakness in some existing warehouses is where internal plastic rainwater pipes pass through ground slabs. In the event of a fire these pipes may melt, allowing pollutants to enter the surface water drainage system. A solution is to encase the pipework in concrete to not less than the height of the bund surrounding the warehouse.

5.4 OVERALL POSITION

5.4.1 Containment of hazardous materials

In situations where particularly hazardous or toxic materials, e.g. sodium cyanide liquor, pesticides and herbicides, or liquefied gases are handled or stored, it is normal practice to protect against loss of substance to the environment by bunding. Bunding for flammable liquids is a statutory requirement.

Bunds are less common where other materials are concerned, e.g. chemicals such as acids and alkalis, sulphides, hypochlorite, ammonia, raw materials used in effluent treatment, aliphatic and aromatic solvents, halogenated solvents and organic compounds. The uncontrolled release of any of these materials could be extremely damaging to the aquatic environment.

It is common practice for materials of low toxicity to be held in un-bunded tanks or for bunding to consist of little more than a low kerb to collect minor spills.

The storage of such products and raw materials as sugar, salt, oils, fats, vinegar and milk is not normally bunded, often because the site operators do not regard them as having potential to cause damage to the environment.

5.4.2 Factors influencing current practice

Secondary containment systems may be provided on industrial premises for a wide range of reasons, of which one or more of the following is likely to have provided the key driving force:

- company policy
- the consequences of previous incidents
- pressure from the statutory authorities
- use of highly toxic materials in significant quantities
- high environmental sensitivity of the site
- the storage of high value products or raw materials
- health and safety issues
- achievement of certification under BS 7750 or the EU EMAS scheme
- a high level of environmental awareness
- agreements reached during the planning process for new plant
- multi-national standards of large organisations.

As a corollary, if none of the above driving forces is present it is unlikely that secondary containment measures will have been provided. The following negative influences are also important:

- insufficient space to install physical structures
- old, non-segregated drainage systems
- ignorance of the potential impact of uncontrolled releases
- lack of expertise and resources.

5.4.3 Developments and trends

Technological developments, new regulations and more stringent conditions in authorisations and licences are leading to changes in process design, plant, equipment and operating procedures. The following future trends were identified by the oil industry as part of a study of soil and groundwater protection (CONCAWE, 1990) but they apply equally to the storage and use of hazardous materials in general. Some of the measures are now in place and others are likely to come into operation in the near future:

- above-ground storage tanks will have double bottoms, which may be provided with instrumentation to detect leaks, and/or be sited on impermeable bases
- pipework is increasingly being contained within a second sleeve or, if underground, within impervious pipe chases
- there is an increasing requirement for tanks and pipework to be pressure tested
- corrosion resistant fibre-glass tanks and interceptors are being introduced for some underground applications
- underground drains, pipes and storage tanks are being replaced by overground alternatives
- there will be increasing use of continuously welded pipework
- for storage tanks, the use of impermeable linings, high level alarms and leak detector/prevention devices will increase
- non-destructive inspection and testing techniques for vessels and pipework are being continually improved
- the increase in extent of impermeable surfaces on sites will necessitate increased holding and treatment capacity
- managerial practices are being developed and improved, e.g. increased safety and environmental auditing and improved operational procedures.

In addition the present study has identified the following trends:

- moves towards the collection of site effluents, surface water and contaminated fire fighting water will increase as sites move towards a policy of controlled containment, testing and treatment prior to discharge
- risk assessment techniques will be increasingly used to identify those sites where containment is most needed (see Section 7)
- increased auditing will result in the identification of potential impacts from uncontrolled releases, resulting in improvements to containment systems generally
- improved containment will be provided at all locations where hazardous materials are used and the containment of materials of moderate or low damage potential will increase, particularly at environmentally sensitive locations
- companies generally accept that 110% bunding of vessels and tanks holding hazardous substances is necessary for new plant almost irrespective of the environmental sensitivity of the site.

6 Case studies

6.1 INTRODUCTION

As part of the review of current practice regarding containment a number of industrial premises were visited. At many sites there were examples of good practice and a small selection of these was studied in some detail in order to illustrate various general and specific features. The operators at the sites described in Sections 6.2 to 6.7 below fully co-operated with and have expressed agreement with the contents of these case studies.

The sites of the case studies were chosen:

- as examples where loss of containment and environmental incidents have occurred and where the corrective actions taken illustrate good containment practice
- to represent a cross-section of industrial sites in terms of the size of the organisation, the nature of the operations and the range of potential pollutants.

6.2 ALLIED COLLOIDS

Company Allied Colloids, Low Moor Works, Bradford, West Yorkshire

6.2.1 Introduction

Allied Colloids manufactures a wide range of speciality chemicals, mainly polymers for use in effluent and water treatment and in the manufacture of paper, paint, textiles and agrochemicals. A wide range of raw materials is used, bought in from other producers. The Bradford site is a CIMAH top tier site by virtue of the storage of 600 tons of acrylonitrile. Total site area is approximately 60 acres.

A major incident took place on 21 July 1992 when there was a fire in the raw materials warehouse that spread to an external drum storage compound. It is estimated that 36 fire tenders supplied four million gallons of water, most being delivered over an initial three-hour period with the remainder used to damp down the fire for up to 48 hours after it started.

The site drains and pumping systems became blocked with viscous polymeric solids. Contaminated fire water entered the public sewer and passed to a sewage works where it was initially contained in stormwater tanks. These overflowed and the water entered Spen Beck and then the rivers Aire and Calder. About 10 000 fish are thought to have been killed over a period of several days in a 50-km stretch of water. The incident was reported to the European Commission as a major accident under the terms of the Seveso directive. It provoked local and national concern.

6.2.2 Fire fighting water project

Since the fire Allied Colloids has designed a water supply, drainage and water retention system so that contaminated rain water, effluents or fire water produced in any future incident will be retained entirely on site and tested before release.

Consultants were used to (a) carry out a risk assessment, (b) calculate fire fighting water capacity and flow rates and (c) design the drainage and fire water retention basin. The project design was accepted by the Regulators, fire services and insurers.

The major components and features of the project are described below.

Fire water supply reservoir

The 27 m x 27 m x 7 m reservoir was constructed in reinforced concrete to potable water service reservoir standards (BS 8007). It was compartmented and located at the highest point of the Bradford site. The reservoir was designed to hold one million gallons of water which was considered sufficient for the worst case disaster scenario. This volume of water is much lower than that used in the incident itself due to a combination of (a) a much more effective and faster method of applying water to a fire, (b) a sprinkler system in the new, compartmented warehouse and (c) better segregation of chemicals throughout the site. Filled ground beneath the reservoir was stabilised and its load bearing capacity improved by pressure injection grouting of the sub-strata.

Ring main

A 3 km ring main, consisting of 300 mm diameter HDPE pipe, was installed around those areas of the site considered most at risk, with American type hydrants having four outlets to British Standard design (Figure 6.1) at appropriate points. Water can be pumped into the ring main using two diesel pumps, rated at 2 000 gallons per minute (8 to 9 hours supply time) at 7 bar pressure. Additional water could be provided, if necessary, from fire tenders.

Figure 6.1 *Fire hydrants being installed in ring main* (Reproduced courtesy of Allied Colloids)

New drainage system

The site has effluent treatment plants for solids' separation, pH adjustment and biological treatment and numerous effluent sumps and pumping installations. The new plant is designed to keep the rain water and process effluents separate.

Process effluents pass through treatment plants to a consented outfall to foul sewer. Rainwater is dealt with in one of three ways:

- rainwater from road and roof areas where there is no risk of contamination passes directly to a surface water sewer
- rainwater from areas where there is some risk of contamination is collected, tested and passed to effluent treatment or direct to sewer as appropriate (automatic TOC, conductivity and pH meters are provided at key locations)
- rainwater from areas where there is a high risk of contamination passes to effluent treatment. Rainwater is collected in sub-surface, open collection channels consisting of pre-cast concrete units covered with grids (Figure 6.2).

Figure 6.2 *Sub-surface, open collection channels for rainwater* (Reproduced courtesy of Allied Colloids)

Spills and wash down waters from process areas, filling points, etc. normally pass to effluent treatment, but in the event that they are too contaminated they can be diverted to the fire water basin pending a decision on the disposal option.

In the 1992 incident burning organic materials floated on the fire water. The new system incorporates underflow weir arrangements to extinguish such fires by depriving any burning materials of an air supply.

All drains are in corrosion-resistant, concrete-lined ductile iron pipes with a high integrity jointing system. As the drainage network passes from the highest to the lowest points of the site the drain diameter increases from 300 mm, through 450 mm to 800 mm. The drainage system has been designed to allow internal CCTV inspection for checks of integrity.

Retention basin

The retention basin is constructed in reinforced concrete, designed to BS 8007, and located at the lowest point of the site. It can hold up to one million gallons of contaminated fire water to a depth of 4.5 metres flowing into it by gravity (Figure 6.3). The tank is divided into seven segments, two of which are epoxy lined. Under normal circumstances only two of the tanks contain effluents at any one time. There are facilities to pump effluents between the compartments to allow flexibility in the storage, treatment and disposal options.

Figure 6.3 *Retention basin in course of construction* (Reproduced courtesy of Allied Colloids)

In the event of a fire, all water used for fire fighting will be diverted to the retention basin where it will be held pending disposal. The volume of the retention basin was calculated on the basis that it would never be more than half full - i.e. in the event of a 1 in 25 year 8-hour storm occurring at about the same time as a fire, the storm water would not account for more than 0.5 million gallons.

Costs

The cost of the complete fire water project described above was approximately £4 million. The major cost items were: fire fighting water supply reservoir £0.5 million, fire fighting water retention basin and drainage system £2.0 million, and ring main and hydrants £0.5 million. The Civils work was completed in mid-1995.

6.2.3 Other containment measures

Various other containment measures are in place at Low Moor, some of which were in existence before the incident, with others having been installed in conjunction with the fire water project.

Storage tanks

All new plant storing bulk liquids is protected by fully engineered bunds, each with a capacity of 110% of the largest tank within the bund. Rainwater is removed by pumping to the on-site effluent treatment plant; there are no valved outlets. All new product tanks have concrete bunds lined with epoxy resin.

Loading/off-loading points

Filling points are all sited on hard standings, graded and drained to sumps from which spillages can be collected, removed and subsequently dealt with. Roads and hard standings are kerbed.

Warehouse

A new warehouse has been built close to the site of the one destroyed by fire. It is designed to high fire prevention standards, with emergency facilities incorporated into the design. It is segregated into various bays and is equipped with a full sprinkler system.

6.3 MONSANTO

Company Monsanto[1], Ruabon Works, Wrexham, Clwyd

[1]*In 1995 the company's name was changed to FLEXYS Ltd.*

6.3.1 Introduction

Monsanto, an international chemical company, has been manufacturing a range of chemicals at its site at Ruabon since 1867. Major products are rubber chemicals, carboxylates and water treatment chemicals, whose production involves the use of a wide range of raw materials, many of which are toxic in the aquatic environment (e.g. chlorine, ammonia, cyanides, acids, alkalis, phenols, aniline, carbon disulphide). The Ruabon site is a CIMAH site by virtue of the storage of 40 tons of chlorine. The works occupies 68 acres and is situated on a steep hillside. The site is in three geographically separate parts: production areas, warehouse and waste water treatment plant (WWTP).

The River Dee, which is class 1A, flows in the valley bottom close to the site. Two minor tributary streams of the Dee pass through the works, culverted in places. Dee water is abstracted downstream of Ruabon by North West Water and Welsh Water and by the Chester and Wrexham Water Companies for the potable water supply of about two million consumers. Old mine workings beneath the site drain directly into the river via adits which are still extant.

6.3.2 Incidents in 1984/5

A major polluting incident took place in January 1984 when a quantity of phenol was spilled into the River Dee from the premises of a small company at Clwyd. Chlorination of the water in the downstream treatment works produced chlorophenols, which have a very low taste threshold. The incident provoked local and national concern.

In January 1985 effluent was released from Monsanto's site into a mine shaft from which it flowed into the River Dee. There were three other incidents of pollution of the River Dee, two from industrial sources, between January and March 1985. (Monsanto was involved in only the second of these five incidents.)

6.3.3 Water protection zone

Under section 93 of the Water Resources Act 1991 the NRA asked the DoE and the Welsh Office to designate the River Dee catchment as a water protection zone, the first such request to be made in the UK. Under the proposed scheme, consent will be required for the storage and use of chemicals in the Dee catchment. The Regulators may require a quantitative risk assessment to show that the risk of water pollution is as low as reasonably practicable. Where this is not so, firms will be expected to take precautionary measures such as bunding, improving site drainage, reducing the quantities of chemicals stored or changing operational procedures.

6.3.4 The project

Realising the sensitivity and vulnerability of the Ruabon works, Monsanto initiated a series of schemes to increase the integrity of its containment measures. Potential hazards and problem areas were identified and an overall site strategy formulated. This translated into a project consisting of several separate, but inter-related, schemes which were progressed in phases during the period 1986-92. There was extensive consultation with the Regulators. The major components of the project are described below.

Effluents and drainage

Mains collecting strong process effluents were raised above ground. Underground mains taking weak effluents from non-process areas were partly refurbished.

The weak and strong effluent mains pass to a central point from which the effluents flow through a pipeline culvert to the WWTP. The pipeline culvert consists of a square cross-section structure, concreted at the base and part way up the sides which acts as a bund, the rest being an open steel framework with a roof. The concrete base and sides of the culvert are lined with butyl rubber for protection. There are three effluent mains in the channel, carrying strong and weak effluent, and one spare. The effluents flow in separate pipelines for operational reasons and to aid spill segregation. The mains are constructed of GRP/polypropylene.

At its lowest point the culvert is provided with a reinforced concrete sump to collect spills from the pipelines. Spills are pumped to one of the large tanks at the WWTP. The sump is provided with an alarmed level indicator to detect leaks from the effluent transfer pipelines into the culvert.

Storm water

There are two inter-linked storm water retention basins with a total capacity of 3 000 m^3. Basin B1 can overflow in severe weather into basin B2 (Figure 6.4).

Figure 6.4 *Basin B1 (foreground) linked to basin B2 (centre left)* (Reproduced courtesy of FLEXYS Rubber Chemicals Ltd.)

Rainwater falling on roads, roofs, processing and storage areas and all hard standing flows via road grids or sub-surface road channels to the weak effluent drains. In many areas the site has been regraded and kerbed to ensure that all storm water is collected in this way (Figure 6.5). If the flow of storm water exceeds the rate at which it can be accepted at the WWTP, the weak effluent overflows into basin B1. This effluent is subsequently pumped out of B1 back into the weak effluent drain.

Figure 6.5 *Site roads graded, kerbed and provided with sub-surface drainage channels* (Reproduced courtesy of FLEXYS Rubber Chemicals Ltd.)

The storm water collection basins B1 and B2 are designed to hold the water falling on the site in an eight-hour one-year storm (600 m^3) and the volume generated by a major fire (2400 m^3) if these events were to occur at about the same time. Since a major fire is unlikely to occur simultaneously with a major storm, in practice basins B1 and B2 provide capacity for an eight-hour 50-year storm. Basin B2 will only be used in the event of a major storm or a fire.

Fire fighting water (process areas)

Fire fighting water (FFW) is supplied from two steel tanks with total capacity 2400 m^3, sufficient for about two hours supply. The volume of the FFW supply determined the volume chosen for the contaminated FFW retention basins. The tanks are connected to an underground ring main which supplies fire hydrants, deluge or sprinkler systems to protect all the most vulnerable storage and process areas across the site.

In the event of a fire, contaminated FFW from one part of the processing area of the site would flow into B1 (overflow to B2) by the same route as the storm water. Contaminated FFW from the rest of the processing area would flow into a separate basin B3 which has a capacity of 2400 m^3 (Figure 6.6). FFW which collected in basin B2 and B3 would be later pumped out into the weak effluent stream using a mobile submersible pump.

Fire fighting water (warehouse)

The warehouse is not equipped with a sprinkler system and has no internal bunding to contain contaminated fire fighting water. FFW emerging from the warehouse would flow by gravity to a metalled underpass beneath a public road. In normal circumstances rainwater from the underpass drains through a dedicated pipeline to the WWTP main storage tanks. In a fire situation the underpass (approximate capacity 1600 m^3) would act as an emergency basin and contaminated FFW would be lagooned since (a) the fire water pumping rate would exceed the flow rate to the WWTP and (b) the pipeline can be sealed off so that the contaminated FFW can be retained in the basin if necessary. Since the road verges and adjacent embankments are not protected, some soil would become contaminated in this scenario.

Figure 6.6 *Open concrete channel to convey FFW into basin B3* (Reproduced courtesy of FLEXYS Rubber Chemicals Ltd.)

Construction of retention basins

The basins are dug out in the form of sub-surface lagoons with bank slopes generally of 1 in 3. The retention system consists of:

- a base layer of 150 mm of sand and gravel
- a 1.5 mm HDPE sheet geomembrane (Note: the underlying ground is not impermeable)
- a woven geotextile
- a blanket of gravel bound in bitumen.

The liners in basins B2 and B3 are overlain with a honeycomb geotextile which is filled with soil and grassed over to provide stability. The piped inlets/outlets are set in concrete abutments. The basins are not provided with underground leak detection systems, but regular analysis of water from nearby boreholes would detect leaks.

Secondary containment

All storage of bulk liquids in new plant is protected by fully engineered reinforced concrete bunds giving 110% protection to the largest tank within the bund. Most tanks are provided with high level alarms to prevent over-filling. Rainwater is removed from low points or sumps in the bunded areas either by steam ejectors or pumps discharging into the weak effluent mains; there are no valved outlets.

Monsanto aims to design its future bunds on the design principles advocated by Barnes (1990). The diameter of the bund will be at least twice the tank diameter in order to overcome the potential problem of liquids jetting over the bund wall (see Section 10).

Filling points are sited on hard standing graded to channels in the road which lead to sumps where spillages can be collected and subsequently dealt with by pumping into the site drains or by removal off-site.

The area where waste drums are stored is surrounded by a drain to pick up rainwater and convey it to the weak effluent drains. The same system would collect any spills or leaks of waste chemicals from the drums.

Waste water treatment plant

The plant was designed to treat an effluent flow of up to 7000 m^3 per day containing about 8500 kg per day COD. The operation of the WWTP is outside the scope of this report but there are several features of interest regarding containment.

Storage and balancing in three steel tanks each of 1800 m^3 to smooth out variations in the flow and composition of the influent. A fourth steel tank (1000 m^3) can be used for emergency storage pending treatment in the event of a major spill.

1. The sides and base of the holding tanks are constructed of bolted steel plates with enamel lining for corrosion resistance and high overall integrity. Each tank sits on a bed of sand/bitumen laid on a low concrete plinth so that leaks can be easily detected.

2. Further storage capacity for weak effluents in the event of a major spill is provided by two underground concrete storage tanks close to the WWTP (total 4500 m^3).

3. The mixed effluent passes into a clarifier the base of which is raised above the level of the bunded compound for ease of inspection and maintenance.

4. The whole of the WWTP is bunded.

Costs and timescale

The capital cost of the scheme was about £5 million, including the cost of the WWTP, new pipelines and culvert, kerbs and drainage channels, raising mains above ground and emergency FFW storage basins, but excluding the fire water supply system which was already in place.

The WWTP was built in 1988-90. The storm water and contaminated FFW retention basins were finished in 1992.

6.4 A H BALL

Company A H Ball, Tilford Road, Farnham, Surrey

6.4.1 Introduction

A H Ball is a firm of civil engineering contractors whose main business is the installation and maintenance of water supply pipelines for the water industry in Surrey and Hampshire. It has a turnover of about £8 million per annum and employs about 130 staff at its main base at Farnham. The site is used for the parking of vans and the storage of equipment and fuels. The potential contaminants present on site are diesel, gas oil, petrol, paraffin and waste oil, stored in maximum quantities of about 12 000 litres. The two acre site is located in a wooded valley through which flows the Bourne Stream, approximately 25 yards from the fuel storage area. The Bourne is a small stream which flows into the River Wey.

In 1990 there was an incident at the site when vandals interfered with the sight glasses on a fuel tank, resulting in a loss of several gallons of gas oil to the Bourne Stream.

6.4.2 Containment measures

As a result, the NRA recommended ways in which the company could improve its storage facilities to reduce the likelihood of similar events. These were implemented by A H Ball in the 12 months following the incident, at a total cost of approximately £18 000, as described below.

1. Soil contaminated with oil was removed from the site. The area beneath the fuel tanks and between the tanks and the Bourne was then concreted, with falls graded away from the stream and into a treatment system (prior to the incident it had consisted of unprotected ground).

2. The bund present prior to the incident was left in place and a second bunded area was built. Each bund is capable of holding significantly more than the total quantity of fuels and waste oils contained within it (Figure 6.7). The bunds incorporate sumps from which spilled fuels are pumped into a waste oil tank, which also takes waste oil from vehicle sumps. There are no valved outlets from the bunds.

Figure 6.7 *Bunding arrangement for fuel tanks* (Reproduced courtesy of A H Ball & Co. Ltd.)

3. An improved method of fixing the sight glasses to the tanks was adopted, which is said to be much more secure against possible vandalism (Figure 6.8).

4. Water falling on the concreted area and water used for washing vehicles drains to an underground grit bin of approximately 500 gallon capacity to remove grit.

5. From the grit bin, contaminated water flows into a compartmented, underground interceptor in the form of a fibreglass tank. Oil is separated in the tank and water overflows to the public sewer. Oil and oily water from the bund sumps, the waste oil tank and the interceptor is removed at intervals by a contractor.

6. Some general drainage work was undertaken in parallel with the above measures, e.g. roof drainage diverted from sewer to surface water and domestic effluents from surface water to sewer.

Figure 6.8 *Sight glass enclosed in metal sheath to reduce risk of breakage through vandalism* (Reproduced courtesy of A H Ball & Co. Ltd.)

6.5 MHH ENGINEERING

Company MHH Engineering, Bramley, Guildford, Surrey

6.5.1 Introduction

MHH Engineering is a light engineering company specialising in the manufacture of torque wrenches and related equipment. It employs 75 staff and occupies a six-acre site. The company uses proprietary, non-chlorinated oils during metal cutting operations and various oils for the machines, such as gear box oil. No other potentially polluting substances are stored or used on site.

The metal cutting process produces metal swarf coated with cutting oils. The swarf is placed into skips at the site boundary. The skips are sited within 20 to 30 yards of Cranleigh Waters which is a tributary of the River Wey. Over the years oil ran out of the bottom of the skips and soaked into the ground beneath them, threatening to pollute the river. The waste storage area is also used to store general building rubble and scrap in skips.

6.5.2 Containment measures

In 1992 the company decided that the condition of the area around the skips was environmentally unsatisfactory and carried out several improvements. Routine visits to the site are made by the Regulator in view of its proximity to the river and in 1993 the following additional measures were suggested by the NRA and implemented by the company on a voluntary basis.

1. The oil soaked ground in the storage area was removed from site and the area was concreted.

2. The concrete base was divided into two areas, each of which was graded so that all rainwater drains towards the back of the area. The two areas are separated by a slight lip which prevents rain flowing from either area into the other. A rectangular, underground, fibreglass drainage sump of 65-gallon capacity is located at the back of each area in which rainwater collects. The skips containing oily swarf are located in one area and the general rubbish skips in the other.

3. The risk of a spill to the river and of ground contamination was reduced further by building a 400 mm high retaining wall along the edge of the storage area closest to the river.

4. Contaminated water from the two drainage sumps is emptied by MHH into 45-gallon drums using a mobile pump. Waste oils from machine sumps etc. are also deposited into these drums. At intervals the tanks and drums are emptied by a contractor for off-site disposal. The skips containing oily swarf are sheeted over out of normal working hours to minimise the amount and extent of oil contaminated rainwater (Figure 6.9).

Figure 6.9 *Metal swarf and cutting oils in skips sited on graded and drained concrete standing, showing sheeting to reduce ingress of rainwater* (Reproduced courtesy of MHH Engineering)

5. Two waste holding tanks with a total volume of 1200 gallons have been placed adjacent the concreted area. The tanks are protected by a bund wall which has a capacity of 110% of the contents of both tanks. A sump in the corner of the bund enables rainwater and potential spills from the tanks to be removed by pumping. The intention is to install pipework and a pump between the drainage sumps and the waste holding tanks so that oily waste water can be pumped from the sumps into the tanks, thereby eliminating the use of 45-gallon drums.

The total cost of the measures taken to date is about £4000. The proposed measures will cost a further £1000 approximately.

6.6 HARCROS

Company Harcros Timber and Building Supplies, East Moors Rd, Cardiff

6.6.1 Introduction

Timber treatment prevents deterioration of wood from fungal and insect attack. One of the commonly used industrial timber treatment processes uses a British Standard formulation type which is a mixture of copper, chrome and arsenic (CCA) dissolved in water at a total metal concentration of about 3%. When impregnated into the timber the chemicals react and become fixed in the wood structure giving the treated timber its long service life.

The chemicals used in timber treatment have the potential to cause severe damage if released to the aquatic environment.

At its Cardiff site, occupying approximately three acres, Harcros has operated a CCA plant for many years for the treatment of softwoods. The only aqueous discharges are of sewage to the foul sewer and of rainwater which eventually discharges into the Bristol Channel.

In late 1994 the company decided to move the plant from an outdoor location to a new, undercover area provided with a range of improved containment measures. These are described below.

6.6.2 The CCA timber treatment process

The process of treating wood with the CCA wood preserving fluid consists of the following fully automated stages (Figure 6.10).

1. Timber is loaded and secured onto a bogey mounted on a rail track.
2. The bogey is pulled into a cylindrical treatment vessel (approximately 1.8 m diameter by 15 m long) by a winch operated rope and pulley system.
3. The door of the vessel is sealed and a vacuum is applied to draw water and air out of the timber (approximately 30 mins).
4. The vessel is then flooded with about 25 000 litres of treatment fluid at ambient temperature from an adjacent storage tank.
5. An excess pressure of about 10-14 kg/cm^2 is applied for about 90 minutes to aid deeper penetration of the fluid into the wood.
6. The residual fluid in the vessel is pumped back into the holding tank and a further vacuum is applied to withdraw excess fluid.
7. The door of the vessel is opened via interlocks which can only be activated when there is no excess pressure in the vessel.

8. The bogey is withdrawn from the vessel by the winch.
9. The treated wood is removed from the bogey and stacked on racks to dry.

In the CCA process the metals become strongly bound chemically to the wood and are not leached out. Treated timber, when dry, can therefore be stored safely in the open and used for construction purposes in areas of high environmental sensitivity.

Figure 6.10 *A typical bunding arrangement at a CCA timber treatment plant* (©Reproduced courtesy of Hickson Timber Products Ltd.)

6.6.3 Containment measures

Treatment fluid could be released at any of the following stages:

- during delivery, storage or pumped transfer
- during pressurisation in the treatment vessel
- when opening the vessel at the end of each batch
- when the timber is loaded on to racks to dry, following treatment, during which time it may drip for several hours.

The plant uses the following techniques and design features to contain the treatment fluid:

1. The treatment vessel and storage tank are located in a bund with the capacity to hold 110% of the total volume of the treatment fluid. The area around the rail track is bunded and drained to a sump. The inner faces of the bunds are painted with bitumen, which is resistant to attack by the fluid.

2. The drying racks are in two tiers. Drips of fluid at each of the two levels are collected by corrugated, plastic drip trays draining to a gutter which in turn drains to a collecting tank (Figure 6.11). This arrangement prevents nearly dry, racked timber being re-wetted by drips of fluid from newly treated timber placed on the tier above and maximises recovery of the treatment fluid. Each racking area is additionally protected by a bund and a sump.

Figure 6.11 *Treated timber on drying racks - drips of CCA fluid drain to the collecting tank (bottom right), and the whole area is bunded* (©Reproduced courtesy of Harcros)

3. Treatment fluid collecting in sumps is pumped back into the storage tank on a regular basis. Treatment fluid collecting in the collection tanks is emptied on a weekly basis using an electrically operated submersible pump and returned to the storage tank for re-use.

4. The whole plant area, approximately 13 m by 50 m, is bunded. Access points into the plant are bunded using 'sleeping policemen'. The whole area is graded to sumps. Thus any leakage through the primary bunds round the treatment vessel and storage tank, or from the rack bunds, or jetting from the storage tank would pass into this area and be contained. All bunds are built in reinforced concrete and joints are formed with waterbars. The base of the bund is built of reinforced concrete varying from 0.2 m to 0.5 m thick, beneath which is an impermeable geomembrane. No services or drains pass through the bund walls or floors.

5. The plant is located in an enclosed building to eliminate the problem of dealing with contaminated rainwater.

6.6.4 Other features

Harcros estimates that the cost of incorporating all the containment features described above into the construction of a new plant, but excluding the cost of treatment plant hardware, would be of the order of £150 000.

Since the plant was relocated in a roofed building it has been operated on a total containment principle with all treatment fluid that is not absorbed into the wood being recovered and recycled. Preventing the contamination of rainwater, by the enclosure of the plant inside a building, is a major environmental improvement and should be considered for other potentially polluting processes where appropriate.

A borehole adjacent to the plant enables groundwater to be monitored. Regular analysis of the surface water drains around the site for the CCA metals indicates that the protective measures are effective.

Harcros considers that it is essential to build bund walls in reinforced concrete, rather than masonry, if fork lift truck or similar vehicles are in regular use close to them. Impact by such vehicles can easily damage unreinforced masonry.

Harcros also point out that management procedures, such as auditing, play a key role in improving the environmental performance of timber treatment plants.

6.7 A MAJOR FOOD PROCESSOR

6.7.1 Introduction

The company is a leading producer and distributor of fresh milk and fruit juices at a number of sites in southern England and south Wales. The company employs about 600 staff and the site, which has been in use for this purpose for about 50 years, occupies 20 acres in an area of light industry and business parks.

6.7.2 Products and processes

The site is divided into two areas, the dairy and the juice plant, sharing a common sewerage and effluent treatment system. The dairy produces about three million litres per week of liquid milk products and the juice plant produces 1.5 - 2 million litres per week of fruit juices (80% orange juice). The main processes, which are fully automated, involve the pasteurising of milk and juice, dilution of concentrates, and the addition of sugar or artificial sweeteners.

Bulk liquid milk is stored chilled in five insulated stainless steel silos (two of 115 m^3; three of 180 m^3). Bulk orange juices are stored in two stainless tanks (30 m^3 each). The management takes the view that bunding of these tanks is not necessary since (a) the contents are not hazardous to humans and (b) the aquatic environment is protected by the drainage and effluent treatment system (see Sections 6.7.3 and 6.7.4 below).

Given the need for strict hygiene and environmental controls, the vessels, pipelines, pumps, silos, etc. are on a regular cleaning schedule. Equipment in the food processing areas is fabricated in stainless steel, which is cleaned and disinfected in place using caustic soda solution (cleaning) and peracetic acid or sodium hypochlorite (sterilisation). On-site laboratories control raw material and product quality and test the various effluent streams.

In order to minimise waste and meet the terms of the consent, considerable efforts are made to reduce losses and recover products.

6.7.3 Effluents and drainage

The processes result in aqueous effluents which are very high in chemical oxygen demand (COD). In the milk and juice storage and processing areas, effluents flow over impervious concrete floors into a network of drains and sumps before finally mixing with domestic effluents in the site sewerage system. The effluents are pumped separately to the site effluent treatment plant (ETP) in single pipelines from the two plant areas.

Any spill of milk to the drains is detected by automatic sensors linked to an alarm system. Above a pre-set value, the drainage from the dairy is automatically or manually diverted at the ETP to one of the divert tanks (see below). In the event of a major spill of milk there is about 100 m^3 of capacity in the drains, in addition to holding capacity at the ETP, to permit it to be contained while off-site disposal is arranged.

Drains in the juice area are monitored in similar fashion. If the concentration of sugar exceeds a pre-set level the effluent is diverted automatically into a separate pipeline which flows to a divert tank at the ETP.

The external areas of the site are of concrete hard standing which is graded to ensure that storm water flows into the site sewerage system and thereby passes to the ETP. All drains are colour coded. Thus all trade and domestic effluents and rainwater (other than that falling on to roofs or car parks) are pumped to the on-site ETP via one of three rising mains. The pumping stations have high level alarms in order to detect any failure rapidly.

6.7.4 Effluent treatment plant

Normal effluents from the dairy and juice plants are collected in one of two treatment tanks (capacity 110 m^3 each). Diverted dairy and juice plant effluents are pumped into one of three divert tanks (capacity 40 m^3 each). The treatment tanks operate in duplex - one filling while the other is being treated/emptied. Treatment consists of (a) holding/balancing/mixing, followed by (b) pH adjustment using hydrochloric acid injected through sparge pipes.

The ETP receives an average effluent flow of 1300 m^3 per day containing about 1300 kg per day COD. This amount of COD represents about 0.7% of the total amount of COD entering the factory daily in the form of milk and fruit juices. Treated effluent from the ETP passes to a local sewage treatment works. The COD load from the factory amounts to about 80% of the COD load to the sewage works from all other sources. In view of this very high proportion, the COD load specified in the consent has to be strictly adhered to and there is a need for very strict control over wastage. Dairy and juice effluents held in the divert tanks are disposed of by various routes.

Proportional flow samples are collected from the three mains to the ETP by automatic samplers linked to flow meters. Samples are also taken from the treatment tanks and divert tanks for analysis. The site operates a positive release system - i.e. nothing is released to drain until tested and found satisfactory.

The capacity of the treatment and divert tanks is such that they provide a maximum of about six-hours holding time for the total site effluent. This is a potential bottleneck and would mean closing the dairy or juice plants in the event of problems at the ETP which could not be quickly resolved. The company therefore plans to replace the two existing treatment tanks with three new tanks (capacity 200 m^3 each). This will increase the holding capacity to about 720 m^3, increasing the holding time to 12 hours. It is estimated that the cost of the new tanks and ancillary equipment will be of the order of £275 000.

6.8 CONCLUSIONS

The six site visits and case studies illustrate a number of points.

1. The construction of containment facilities has generally been stimulated by an incident and/or by pressure from the Regulators.

2. At large sites a variety of containment system components have been incorporated into the overall containment option for the site (see section 8). At such sites a strategy for containment needs to be developed, which is likely to involve a risk assessment.

3. The cost of containment measures, in broad terms, does not appear to have been excessive and none of the companies concerned was critical of the costs involved.

4. The site operators felt more comfortable knowing that adequate environmental protection measures were in place.

5. Site operators felt that the Regulators had been co-operative and helpful.

6. Good management practices need to be in place irrespective of the containment hardware.

The case studies contain examples of many of the containment measures discussed later in this report.

7 Classification of secondary containment systems according to site hazard or risk

7.1 INTRODUCTION

It is not possible to design and build a containment system that is 100% safe, i.e. that can be guaranteed not to permit the escape of pollutant in every conceivable circumstance. No matter how much care is taken there is always a probability that, for example, a particular hazard has not been recognised, or that structural elements or materials do not behave as predicted, or that the designer or contractor has simply made a mistake. Additional risks and uncertainties are introduced throughout the service life of a containment system depending on the extent to which repairs and maintenance are undertaken and the quality of that work. With some existing containment systems, for example underground storage tanks and associated drainage systems, it may not be practicable to carry out any maintenance or to monitor performance.

Although it is not possible at this stage to provide a fully quantitative methodology which links containment design with hazard or risk assessment, this section is intended to help designers identify situations where particular care is needed and to undertake the system (or concept) and detail design accordingly. This should be done in consultation with the Regulator.

This section of the report provides guidance to help the risk assessor or containment system designer to understand the hazards and/or risks posed to the aquatic environment by the particular process and the site on which it is carried out, in the context specifically of secondary containment design. It then introduces the concept of containment system *classification* which allows the designer to link the design of containment system to these hazards or risks. Three classes of containment are defined in terms of design, construction and operational integrity:

- *Class 1* - appropriate to low hazard or risk situations
- *Class 2* - for moderate hazard or risk situations
- *Class 3* - for high hazard or risk situations.

In many companies, risk assessment techniques are applied routinely as part of the planning, design and operation of new and existing sites. They may involve a detailed examination of every part of the plant and process including the materials stored, used or produced on the site (the *sources*), the routes by which potential pollutants could be transferred to the environment (*pathways*), the sensitivity of the environment to these pollutants (*receptors* or *targets*), and operating procedures. The techniques are constantly being refined and developed as understanding of the environment increases and more information and data become available. The DoE and Environment Agency are particularly active in this area and plan to publish shortly a comprehensive risk assessment methodology. Understanding and applying many of these techniques requires specialist skills, knowledge and experience.

In the context of this report, it is assumed that any strategic decisions about a site or plant (including whether or not to provide secondary containment rather than some other protective measure) have already been taken by appropriately qualified specialists acting on the advice of any relevant consultees. These decisions may have been arrived at using one of the risk assessment techniques mentioned above. In Section 7.4.4, the best known of these hazard and risk assessment techniques are reviewed and summarised in order to give the containment designer an insight into the way in which the Regulators and others have made their assessment. With that understanding, the containment system designer will be able to see the task of designing and specifying works within the wider context of hazard and risk management as a whole.

The section then goes on to develop a hazard and risk assessment methodology intended specifically to help the design engineer with his part of the work and, to that extent, it is purposely restricted in context and scope. The aim is to help engineers to select and design appropriate containment systems according to the assessed hazards and/or risks, where it has already been decided by others that some form of secondary containment is necessary.

The relationship between risk, environmental impact and frequency of loss of containment may be represented diagramatically as shown in Figure 7.1, where different levels of risk are indicated by different regions of the chart. The overall purpose of containment is to reduce the frequency of the escape of pollutant and/or the resulting environmental impact.

It should be stressed that this diagram necessarily simplifies what is a complex set of relationships. In particular, the boundaries (represented by the shaded areas) between the regions separating the three risk levels are not in practice clearly defined.

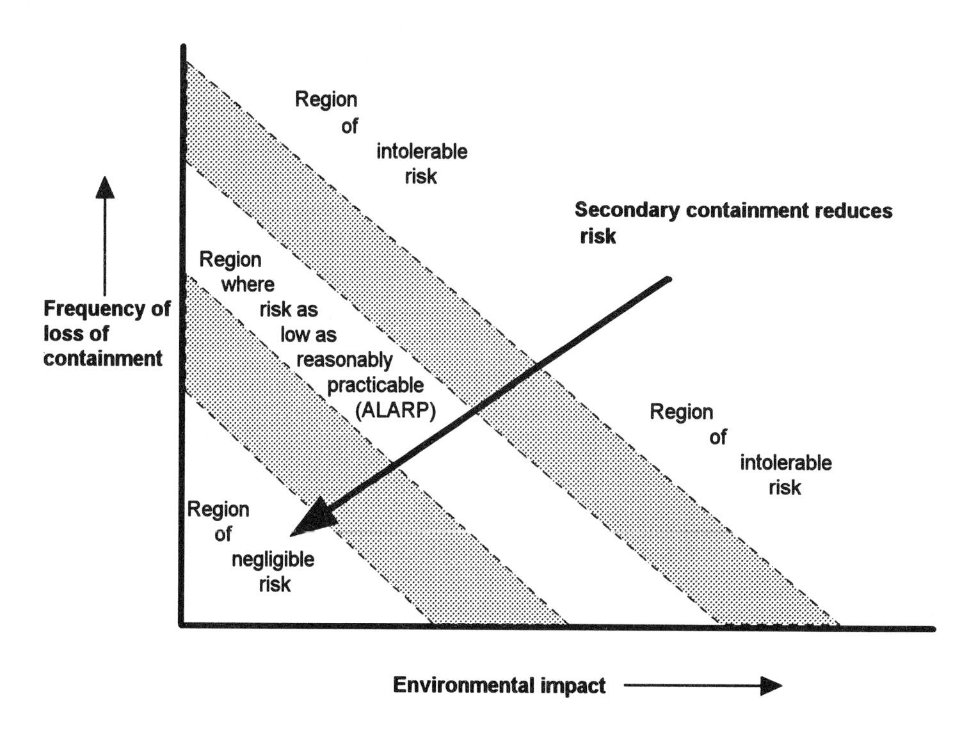

Figure 7.1 *Relationship between risk, environmental impact and frequency*

7.1.1 Layout of the section

This section of the report provides:

- a brief introduction to risk assessment (7.1 and 7.2)
- a description of the basic features of risk assessment and the relationship between hazard and risk (7.3)
- an indication of the general circumstances in which a risk assessment is likely to be beneficial and the quantitative information that will be needed (7.4)
- a framework around which risks and hazards may be assessed in the context of containment design (7.5).

Worked examples of existing risk and hazard assessment methodologies are given at Appendix A2.

The framework section (7.5) sets out in some detail a method by which the containment system designer can approach hazard and risk assessment. It provides:

- a detailed examination of the key factors - *source*, *pathway* and *receptor* - which have to be considered, and the circumstances which influence whether the hazard rating (associated with the *source*), transport potential (*pathway*) or impact potential (*receptor*) is *high*, *moderate* or *low* (7.5.1, 7.5.2 and 7.5.3)
- information on quantifying and categorising potential environmental impacts at receptors (i.e. incidents) which may result from loss of containment (7.5.4)
- a methodology for assessing the overall *site hazard rating* (categorised as *high*, *moderate* or *low*) based on combining the above factors (7.5.5)
- a description of events which may lead to loss of containment (7.5.6)
- a rationale for combining (a) the quantified *site hazard rating* with (b) the frequency of events that could lead to loss of containment, to arrive at a *site risk rating* (7.5.7)
- a means of relating the quality of containment system to be provided to the site hazard or site risk rating. Quality is defined in terms of structural integrity and degrees of redundancy built into the system: three classes are defined, corresponding to low, moderate or high hazard, or risk, situations (7.6).

Further relevant information on sources and receptors to be used within the framework is given in Appendices A3 - A5.

7.2 HAZARD AND RISK ASSESSMENT IN THE CONTEXT OF REGULATORY AND PUBLIC PRESSURE

The Regulators and the general public expect the most appropriate technology (BATNEEC) to be used by industry to prevent spills and other incidents that could harm the environment. Industries are becoming increasingly concerned about their public image and appreciate the damage that is done to their reputation by reports of spills of hazardous materials to watercourses, even when such incidents appear to have no immediate or obvious effects. As a result, many companies now aim for the total containment of potentially polluting materials. It is likely, therefore, that risk assessment will be used increasingly, not as an absolute technique to decide whether additional protection is or is not needed, but rather as a method of ranking risk so that money can be spent effectively whilst progressing towards a target of total containment and to help designers decide the safety factors to be built into the containment systems.

For new plant, or major modifications to existing plant, the regulatory bodies are likely to require containment for all but the most innocuous materials regardless of the environmental setting. Containment is likely to be the rule rather than the exception. There is a view among the UK authorities that well designed, passive containment systems, such as bunds, that are regularly inspected and properly maintained, are more effective than protective measures requiring positive action, such as a response by operatives to alarms or warnings. A site operator wishing to adopt the latter approach may have to supply clear justification to the Regulators that physical containment is impracticable or unnecessary given the particular circumstances.

For existing plant, it is likely to be more difficult to install new containment systems. There may be situations, as stated earlier in the report, where the retro-fitting of physical systems is impracticable or where the cost would be prohibitive and where other techniques or procedures may have to be adopted.

In addition to providing environmental protection, high integrity containment measures may also be required for other reasons, such as:

- bunding of flammable liquids
- bunding of toxic, volatile liquids to minimise vapour release
- preventing exposure of site operators to toxic materials
- recovery and recycling of materials to meet waste minimisation targets.

Each of these considerations may point to a different solution and it is the responsibility of the designer to balance the competing (and in some cases conflicting) requirements. It should be stressed that this report is concerned exclusively with the need to protect the aquatic environment.

7.3 BASIC FEATURES OF ENVIRONMENTAL HAZARD AND RISK ASSESSMENT

There are numerous definitions of hazard and risk and related terms but none has been universally accepted. The Royal Society Study Group on Risk (1992) used the following:

> *Risk* is the probability that a particular adverse event occurs during a stated period of time or results from a particular challenge.
>
> *Hazard* is a situation that in particular circumstances could lead to harm.

A number of risk assessment procedures have been described in the literature (e.g. DoE, 1991a; DoE, 1993; Pritchard, 1994). Although the approaches vary from the very simple through to the complex, sophisticated and fully quantitative, they each involve a number of essential components as described below.

Hazard assessment

1. An estimate of the quantity and composition of material which could escape (*sources*), the amount which could reach surface or groundwater and the timescales involved (*pathways*) and the environmental sensitivity of the receiving water (*receptors*), taking on-site mitigating factors into account.

2. Quantification of site hazard rating in terms of the possible concentration of the pollutant in the receiving water at a particular location or target, taking into account dilution in the receiving water and other off-site mitigating factors.

3. A comparison of the potential pollutant concentrations in the receiving water, following an incident, with any tolerable concentrations agreed with the appropriate Regulators.

Risk assessment

1. Identification of the events which could lead to the release of polluting materials from primary containment and a quantitative assessment of the frequency with which such events could occur.
2. An assessment of whether or not the risk is tolerable, taking into account the seriousness of the potential environmental damage and the likely frequency of occurrence. Views on what is tolerable may vary significantly between dischargers, the regulatory authorities and the public.

Action

Where the risk assessment indicates that an event could result in significant environmental damage, at an intolerable frequency, the site operator would need to consider one or more of the following actions:

- installing new, or improving existing, containment systems - **the subject of this report**
- reducing the inventory of hazardous materials
- modifying on-site pathways to minimise the likelihood of escape of pollutant
- changing or relocating the process or activity
- changing operational and/or management practices.

The effects of changes in the nature, size or frequency of potential releases as a result of actions taken may be modelled using risk assessment techniques.

The hazards relating to *source*, *pathway* and *receptor* may be combined to give an overall *site hazard rating*. On completion of the hazard assessment the risk assessor may decide that there is sufficient information to permit decisions to be taken regarding containment measures, and that a risk assessment is not warranted. Alternatively, it may be decided to extend the hazard assessment to include a study of the probability of occurrence of events that could result in the loss of containment. The combination of hazard rating and the probability of an incident occurring gives a measure of the risk.

7.4 APPLICATION OF HAZARD AND RISK ASSESSMENT

7.4.1 The need for hazard or risk assessment

It is recommended that a site operator should carry out a hazard or risk assessment if:

- quantities of substances dangerous to aquatic life (in particular any of the substances listed in legislation as shown in Appendix A3) are present on the site
- pathways exist for the entry of spilled materials into surface or groundwaters
- the site is situated close to surface or groundwater.

7.4.2 Information required for risk assessment

In assessing risk it is necessary to take account of a wide range of factors (involving many uncertainties) relating to the environmental consequences of unplanned and uncontrolled events. Wherever possible, these factors should be defined and quantified in order to minimise the subjectivity of qualitative assessments. The information likely to be needed in any environmental risk assessment includes:

- the impact that a sudden, major discharge of pollutant would have on the target
- the flow characteristics of the surface or groundwater (rates, stratification, mixing, etc.)
- possible duration of any incident
- the likely attenuation of the effects of the pollutant by chemical reactions, evaporation, absorption etc., both on site and in the receiving waters
- the specific gravity of the potential pollutant - will it float (e.g. oil) or sink (e.g. carbon tetrachloride)
- the likely public reaction to aesthetic considerations such as colour, taste and odour
- migration times through pathways such as drains, treatment works, soils and aquifers
- the likely frequency of failure of the primary containment (in practice this can be expressed only in very broad bands).

Risk assessment requires a quantitative appraisal of the likely frequency of failure of plant and equipment and a decision (in agreement with the appropriate authorities) on the frequency at which specified environmental impacts can be tolerated. The detailed investigation of the reasons for the potential failure of plant and equipment may be carried out with the help of a variety of techniques including HAZOP (HAZard and OPerability) studies, FTA (Fault Tree Analysis) or FMEA (Failure, Mode and Effect Analysis). These studies are normally carried out by engineers specialising in the safety and reliability of processes.

7.4.3 Benefits of risk assessment over hazard assessment

In some situations, hazard assessment on its own can give a risk assessor or the regulatory authorities a good indication of the need, for example, to increase the integrity of containment, without proceeding to the more complex and costly risk assessment stage. In other cases it may not be conclusive and the risk assessor will need to proceed further. These further stages are time consuming and require expertise in the safety and reliability aspects of process technology, knowledge of the dispersion of pollutant plumes in surface and groundwaters, and ecotoxicological data on the impact of pollutants on aquatic biota, man and other potential receptors.

The additional cost of carrying out a risk assessment should be balanced against the potential benefits, which include:

- development of a clearer view of the events that could lead to loss of containment
- the ability to rank risks with more certainty
- following on from the above, the opportunity to allocate resources to containment works more effectively.

7.4.4 Examples of existing methodologies

Three of the more familiar existing hazard and risk assessment methodologies are outlined in Appendix A2. They represent a mix of hazard-based and risk-based approaches and vary in the extent to which they attempt to quantify the issues. The approaches are classified in the Appendix under the following headings.

1. Qualitative hazard assessment ('end-of-pipe' approach).

2. Semi-quantitative risk assessment.

3. Matrix scoring systems.

In addition, Appendix A2 summarises an alternative *prescriptive* approach. Although the German WGK system is cited as an example of this approach, it should be noted that this system is not necessarily recommended or acceptable to UK Regulators.

7.5 A FRAMEWORK FOR ENVIRONMENTAL HAZARD AND RISK ASSESSMENT

As has been shown in the previous section, risk assessment techniques are already well established and are used by risk assessors to predict the risk of an incident happening and the likely consequences of such an incident. This section presents a framework around which hazards and risks may be assessed systematically, considering the factors that need to be taken into account in designing a containment system - Class 1, 2 or 3 (as defined at Section 7.6) - which is appropriate to the circumstances. The process is summarised in the flowchart at Figure 7.2.

The purpose of carrying out the risk assessment is to establish the degree of hazard or risk so that the designer can make judgements on the type of containment system that would be appropriate in the light of the particular circumstances.

The starting point of the process is to establish hazards.

In assessing the hazard rating to the aquatic environment the assessor must consider:

the *source* - it is important to be clear that in the context of assessing hazard, *source* refers solely to the hazardous materials present on the site. Other site-related factors such as management, the presence or absence of alarms or fail-safe devices, and the effectiveness of existing primary containment and any secondary containment relate to the risk of release of pollutant;

the *pathway* - i.e. the means by which any pollutant can escape to the aquatic environment (including both surface and groundwater). *Pathways* may be internal (i.e. within the boundaries of the site) or external; in the latter case pathways can extend for several kilometres or more;

the *receptor* - i.e. the humans, animals, fish, plants and biota which would be affected (directly or indirectly) by the escape of the pollutants to the receiving water.

The concept of source, pathway and receptor is illustrated in Figure 7.3.

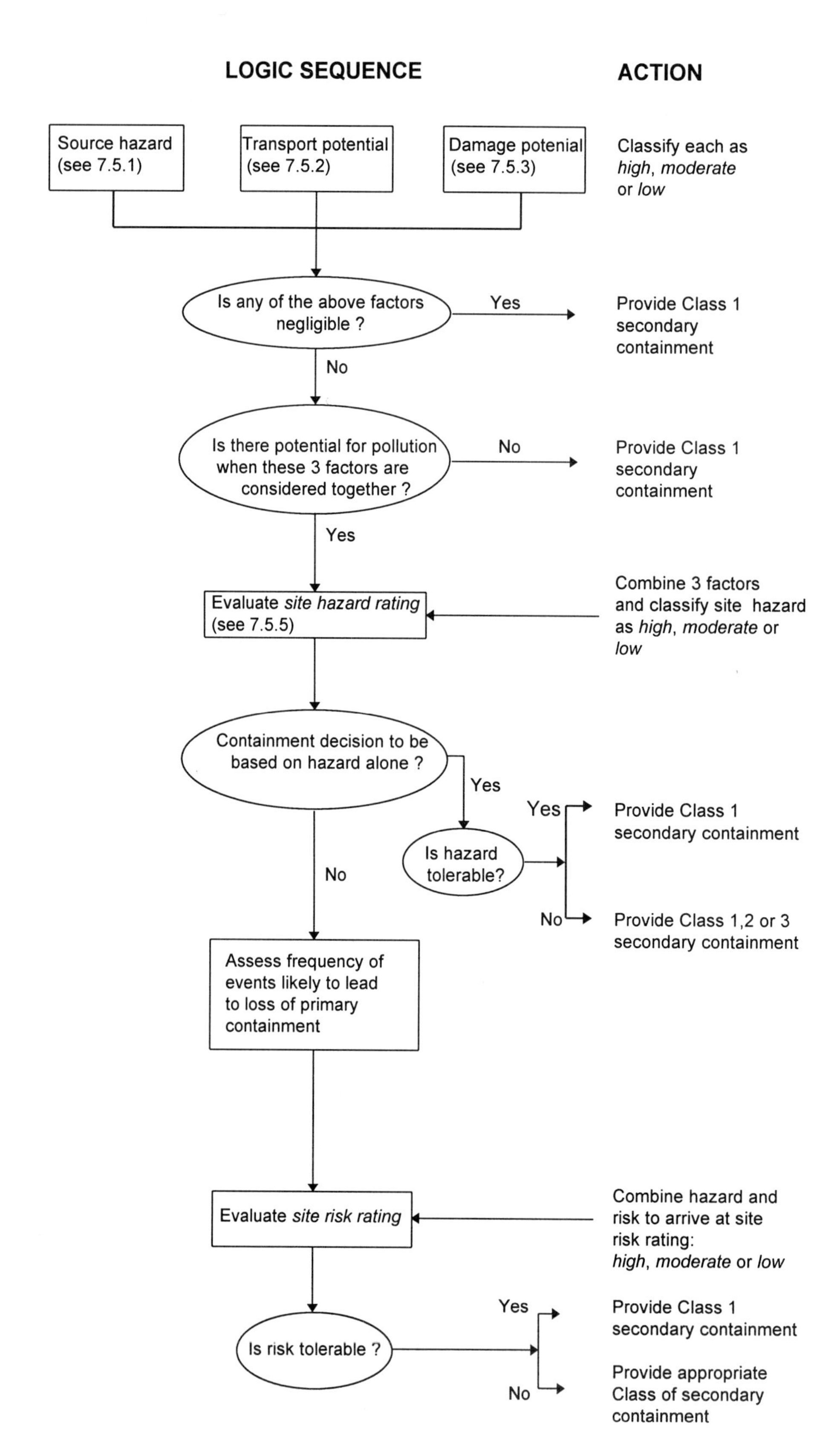

Note: this flowchart is intended to assist in the classification of secondary containment on the assumption that a decision to provide some level of containment has already been taken. For this reason, where the hazard or risk is assessed as negligible or tolerable the 'ACTION' is still to provide Class 1 secondary containment.

Figure 7.2 *Flow diagram of process relating containment classification to hazard and risk assessment*

With regard to the flowchart (Figure 7.2), there is currently a good deal of subjectivity in the interpretation of what is 'negligible' or 'tolerable' but these terms are likely to be defined by the Regulators in the future.

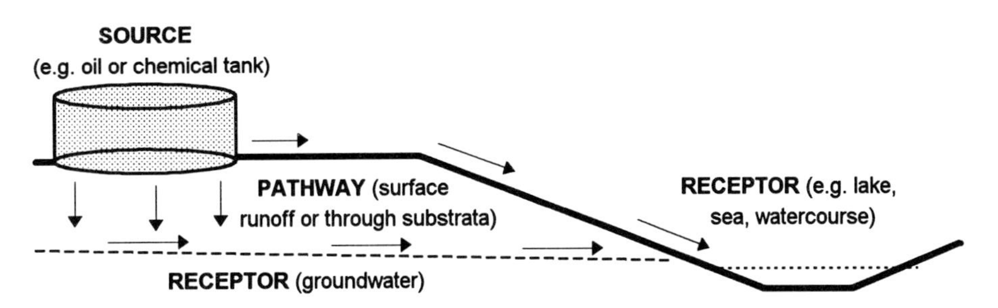

Figure 7.3 *Illustration of source, pathway and receptor*

7.5.1 Source

Nature and quantity of potential pollutants

The potential pollutants present on industrial sites will comprise a range of raw materials, products, fuels and wastes. The quantity and nature of these materials must be assessed in relation to their polluting potential, the extent to which their presence may trigger or exacerbate an incident (e.g. highly inflammable substances), and any physical or chemical properties that may call for special containment measures (e.g. corrosive materials that may damage concrete). In relation to pollution potential, the assessor must take into account a wide range of characteristics of the materials including:

- physical properties (e.g. density and viscosity)
- chemical and biochemical properties (e.g. BOD and pH)
- ecotoxicological properties
- bioaccumulation, biomagnification or persistence potential.

Properties of a wide range of substances may be found in the references given in Section 4.2.5. Further information on hazardous substances is given in (Edwards, 1992) and in Appendix A3 which includes lists of particularly hazardous materials on the EC black list and UK red list.

The quantity of hazardous material present on a site is clearly an important factor to be taken into account when considering secondary containment capacity. There may also be threshold quantities, however, (Appendix A2 gives some examples) below which the escape to a watercourse may not have a significant environmental impact.

Figure 7.4 illustrates the relationship between material quantities and toxicities, and the hazard they present. Materials present in quantities that represent a low hazard may not require secondary containment. Conversely, where material quantities and toxicities combine to create a high hazard rating, high integrity secondary containment may be required.

It is proposed that the source hazard for the particular site is classified as *high*, *moderate* or *low* according to the nature and quantity of the material present. Examples of substances which, if present on site, would suggest a *high* source hazard are listed in Box 7.1 (see also Appendix A3).

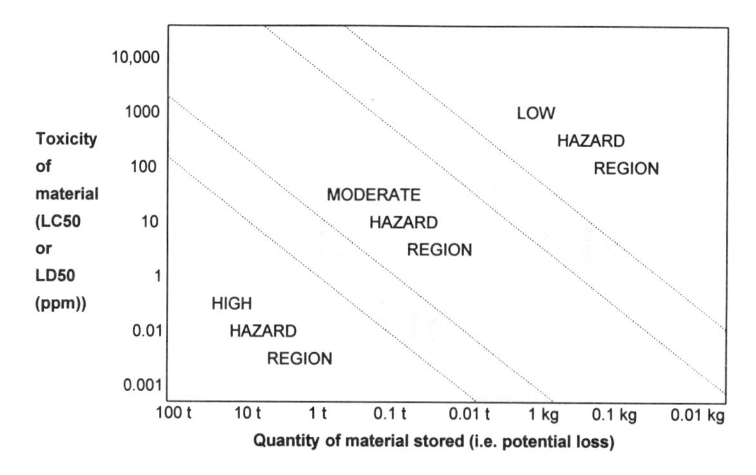

Note: LC50 is the concentration of an agent in water which kills 50% of the exposed aquatic organisms within a given time period, usually 96 hours. Similarly, LD50 is the concentration which kills 50% of an exposed population of test animals. There are many other possible measures of toxicity.

Figure 7.4 *Relationship between material quantities, toxicity and hazard*

Box 7.1 *Substances likely to lead to a high source hazard rating*

- Substances controlled by the EC Dangerous Substances Directive and Groundwater Directive (Edwards, 1992).
- Substances listed by the Oslo and Paris Commissions for control, or on the North Sea Conference list of priority hazardous substances (Edwards, 1992).
- All pesticides and herbicides.
- Other substances highly toxic to aquatic flora and fauna.
- Substances affecting the appearance of water (e.g. oils, fuels, detergents and dyes).
- Organic compounds with a high oxygen demand.
- Fire fighting water contaminated by any of the above materials.

It is important that the polluting effects of all the possible cocktails of materials that may arise during an incident are taken into account. This is a particularly important consideration in warehouse situations where many different materials may be stored together.

Effects of fires and fire fighting water

On many industrial sites, one of the most significant hazards is fire. The potential effects of a fire can alter the assessment of source hazard in a number of ways, including physical or chemical modification of the materials on site and damage to other primary containers which could result in further materials being released.

Fire may also affect any secondary containment on-site and cause changes to pathways, although these are not strictly issues to do with *source*.

By far the most important effect of fire, in the context of this report, is the introduction of potentially very large volumes of water used to fight fires and to damp down (collectively termed fire fighting water in this report) and, to a lesser extent, foams. Fire fighting water is likely to contain pollutants and it is therefore just as important to control its release to the environment as it is with any other hazardous substance. It follows from this, therefore, that sites where there is a greater potential for fire (perhaps as a result of the processes or materials involved, or the type of buildings, or absence of fixed fire fighting installations, etc.) will tend to attract a higher source hazard designation.

7.5.2 Pathways

Introduction

The *pathway* is defined as the physical route by which released pollutants may travel from the *source* to the aquatic environment, i.e. the receiving water, or to any other relevant target such as an effluent treatment works.

It is necessary to take into account all of the *pathways*, both within and outside the site boundary, by which any hazardous material may reach the receiving water or a treatment plant. Important considerations include:

- the distance between the *source* and the various potential receiving waters
- site layout, and the position and effectiveness of drains and other internal and external pathways
- geographical, geological and hydrogeological features that could either impede or facilitate escape of pollutants from the site
- weather conditions
- the direct effects of fire and the introduction of fire fighting water or foam
- the presence of treatment plants (on or off site)
- modification of pollutants during passage through the *pathway*.

The time it would take for the polluting material to reach the receiving water is an important factor. The potential of the material to damage the environment is higher if it reaches the receiving water quickly since:

- there will be less opportunity to contain the pollutant (either on site or off site) and prevent escape to the wider environment
- mitigation of the effects of the primary pollutant by such factors as evaporation or dilution will be reduced
- there will be less time to warn other organisations and individuals likely to be affected, e.g. the Environment Agency, downstream water users and sewage treatment plant operators.

By taking into account all of the foregoing factors, it is possible to derive what is termed in this report as a *transport potential* for the *pathway*. In the same way as for designation of *source* hazard, it is proposed that *transport potential* is designated as *high*, *moderate* or *low* according to the particular characteristics of the *pathways*. *Transport potential* relates primarily to (a) the time taken for the pollutant to travel from the point of escape from the primary (or, if present, the secondary) containment on the site to the receiving water, and (b) any mitigation of polluting potential by dilution or other means. The various factors are considered in more detail below.

Proximity of receiving water

It is important first to identify all of the possible receiving waters (surface and groundwater) and their locations in relation to the source in order to be able to assess the relevant *pathways*.

Surface waters include the sea and estuaries, rivers, lakes, streams and ditches. Some of these may be dry for long periods and it is important that they are not overlooked. Sewers, culverts and drains all have the potential to convey pollutants rapidly away from a site and to release them into the aquatic environment many kilometres from the site boundary.

Permeable substrata can also convey pollutants large distances where they can affect underground water resources. Information on the location and nature of groundwaters can be obtained from the Environment Agency. Where the geology is complex, the Agency should be consulted for advice on what constitutes the receiving water.

Site layout and drainage

The layout of the plant, buildings, roadways, hardstandings and other features, and the surface finish and permeability of the surfaces over which the pollutant may flow in the event of an escape, are all relevant factors in deriving *transport potential*.

The following factors will tend to increase a site's *transport potential*:

- hardstandings around hazardous installations sloping towards a surface receptor
- hazardous installations surrounded by flat or slightly sloping permeable ground permitting infiltration to groundwaters
- on-site effluent drainage systems which provide *pathways* to trade effluent outfalls, to public sewers, or to on- or off-site public treatment works
- the presence of below ground features such as services, ducts, pipelines, filled ground, tunnels, tanks or sumps
- the existence of other man-made *pathways* such as old mine workings, storm drains and gullies, culverted watercourses and land drains located close to the *source* or potential *pathway*.

Rainwater soakaways are a common feature on many sites. The proximity of soakaways to sources of potential pollutants, their location with respect to physical pathways, the possibility of contamination in the event of an incident, and even the use of soakaways in principle, needs to be carefully reviewed.

Geography, geology and hydrogeology

The topography of the site and the permeability of the ground will have an effect on the transport of pollutants to surface waters and infiltration to groundwaters. On large sites there may be a considerable variation in landform, soil type and geology across the site which will influence runoff and infiltration.

With the exception of some small sites where the ground conditions are well known, it is recommended that geotechnical and hydrogeotechnical surveys are carried out on sites where hazardous materials are to be present.

Climatic conditions

Climatic conditions, including precipitation and temperature, can affect ground conditions and permeability, vegetation and evapotranspiration, each of which can affect the *pathway*. Frozen or saturated ground will increase the tendency for rapid run-off from areas where, at other times, runoff may be very much slower or absent altogether. Surface cracks and fissures in dry conditions will increase infiltration and may provide direct pathways to permeable substrata and groundwater.

The consequences of the failure of a containment system during a period of heavy rain need to be taken into consideration. One possible beneficial effect may be that the rain will dilute the pollutant before it reaches the receiving water. Rainfall may also significantly modify the pathway taken by the pollutant. For example, in dry weather a pollutant may seep slowly into a trade effluent drain, whereas in wet weather it may be washed rapidly into a storm water drain. Some drainage systems, for example public sewers, are designed with storm water overflows to relieve the load on the treatment works under storm conditions.

Fire fighting water

Fire fighting water has the potential to dilute very significantly any released primary pollutant. In the same way as heavy rain, however, it may also affect the *pathway*. If, as is likely, the flow of fire fighting water is greater than the capacity of the site drains, the contaminated fire fighting water will find other pathways from the site. Forecasting the pathway of the polluting material, taking into account the effects of heavy rain or fire fighting water, or even both together, is a key factor in assessing *pathways*.

Another important consideration is the effect that fire may have on the flow properties of the pollutant, particularly in the site drainage system. Fire and heat may cause increases in viscosity or surface crusting so that flow through the system is slowed down or even stopped completely. Drains may also become blocked through debris flushed into them by the fire fighting water.

Treatment plants

Pathways may lead to, and include, effluent treatment works on the site, or sewage treatment works off site. The unplanned entry of highly polluting effluent into a treatment plant at a level which exceeds the treatment or containment capacity of the works, may cause major damage which effectively puts the plant out of action. The damage may be long-term. The resulting discharge from the damaged works may result in more serious pollution than would have resulted from a direct discharge of the primary pollutant.

Mitigating effects

When a pollutant flows out from a site towards receiving waters it may be subject to a number of factors which reduce its environmental impact potential, either by modifying its properties or its volume.

In assessing possible mitigating effects, the factors which should be taken into account include:

- likelihood of dilution in the drains
- possible dilution and treatment at on-site or off-site treatment plant (note that such plant could be severely affected by the pollutant and to that extent can equally be regarded as a *receptor*
- chemical reactions (e.g. materials may be highly reactive with water)
- evaporation (e.g. volatile solvents)
- absorption (some materials may be absorbed by soil or other solids)
- settling (some materials may settle in drains, interceptors or lagoons)
- existing retention capacity on the site.

The effect of holding capacity on- or off site is a key factor. The site may be provided already with secondary containment in the form of, for example, bunds, interceptors or lagoons. These facilities may be near to the source of the release or some distance away. The extent to which these containment systems retain some of the primary pollutant, or increase the transfer time to the aquatic environment, or permit other mitigating effects to take place, should be considered in the overall assessment of the *pathway*.

Factors affecting transport potential

Examples of *pathways* where the *transport potential* is high are given in Box 7.2.

Box 7.2 *Pathway characteristics suggesting a high transport potential*

- Short runoff time between *source* and surface water.
- Short migration time between *source* and groundwater.
- Direct drainage links between *source* and receiving water or treatment plant.
- Absence of holding capacity in drains and sewers.
- Highly permeable strata between *source* and groundwater.
- Absence of treatment facilities.
- Little to mitigate the effects of the released pollutant.

7.5.3 Receptors

Introduction

In order to assess the impact of a chemical release on the surrounding environment, it is first necessary to identify the receptors (or targets) in the vicinity of the site. Further guidance on the nature of environmental receptors is given in Box 7.3. Useful information on the presence and nature of environmental receptors in the vicinity may be held by local authorities, countryside and heritage commissions, and the Regulators.

Box 7.3 *Receptors that may be present in receiving waters*

Nature of water	Nature of effect	Receptor/Target
Surface	Direct	Fish/biota/aquatic plants Water quality Protected areas (e.g. SSSI or nature reserve) Aesthetic issues (colour, smell, appearance) Structures (boats, jetties, harbours)
	Indirect	Man (via drinking water) Man (via contact water sports) Livestock (via watering) Crops (via irrigation) Other users and uses of abstracted water (e.g. as cooling water)
Groundwater	Indirect	Man, livestock and crops (as above) Rivers/estuaries/sea (via recharge) Other users and uses of abstracted groundwater (e.g. food processing)

The presence and nature of the environmental receptor can be thought of as the fixed point in any hazard or risk assessment. Although the site operator is able to modify *sources* and, to some extent perhaps, on-site *pathways*, altering the receptor is more difficult. Short of diverting rivers or moving abstraction points, there is usually little that can be done to change the *receptors*.

Water quality standards are likely to continue to improve and the Environment Agency's aim is that eventually all *Grade D, E and F* surface waters should be upgradable to at least *Grade C*. It is important, therefore, to take into consideration any target improvements in water quality in the foreseeable future when assessing site containment systems.

As with *source* and *pathway*, it is proposed here to designate *receptors* according to their *damage potential*, giving three options: *high*, *moderate* and *low* damage potential. This section considers the various characteristics of *receptors* in relation to their damage potential.

Environmental sensitivity

It is recommended that potential *receptors* are discussed with the regulatory bodies and agreement is reached on which of them are sensitive to damage. It should be borne in mind that if there are several potential pollutants at a site (or a large number as could be the case with contaminated fire fighting water) then each of these could affect *receptors* in different ways.

Treatment works

On-site effluent treatment plant and off-site sewage treatment works may be considered as *receptors*. The normal functions of the plant could be impaired for a long period by the entry of materials incompatible with normal treatment processes. For example, all biological treatment plants depend on the activity of bacteria to break down complex organic compounds. The uncontrolled discharge of substances such as pesticides, hypochlorite, metals, organochlorine solvents and acids and alkalis is likely to kill the bacteria and halt biodegradation processes. It could take a treatment plant many weeks to return to its previous level of biological activity. In this period, the plant could not carry out its normal function and sewage and trade effluents would pass to receiving waters with little or no treatment.

The effects of pollutants on effluent treatment plants need therefore to be considered in the hazard assessment. Tolerable levels and loads of pollutants should be established in collaboration with the treatment plant operators as part of the development of the emergency plan for the site. In some situations it is possible that a treatment plant will be the critical *receptor*.

Dilution and mixing

Receptors may be located many kilometres downstream of the point at which the pollutant enters the receiving water, whether this is groundwater or surface water. The dilution which takes place in the receiving water provides mitigation of the impact of the pollutant on the *receptor*. When a substance is discharged to water it disperses through the aquatic environment before it reaches its target. The dispersion model used can be very simple, as outlined in Appendix A2, but a more rigorous analysis such as that used by the Environment Agency in its PRAIRIE model (AEA Technology, 1994) is possible. This would need to take account of:

- duration and mode of release
- flow of the receiving water, and dilution
- background levels of the pollutant
- tidal influences
- mixing characteristics, stratification

- density and solubility (is the substance miscible with water, will it float or sink?)
- climatic conditions.

Other factors

Other factors may reduce or increase the severity of the environmental impact. Mitigating factors include biodegradation, evaporation, photolysis, hydrolysis and absorption. Aggravating factors include bioaccumulation (e.g. in fish) and biomagnification (e.g. along a food chain). The rate at which these effects mitigate or aggravate environmental impacts depends on a variety of interacting circumstances, but in some cases it is quantifiable.

Uncertainties

There are considerable gaps in the knowledge when it comes to quantifying the effects of pollutants on *receptors*. In particular, toxicity effects on man and ecotoxicity effects on ecosystems are only readily available for those substances commonly used in industrial and manufacturing processes.

The potential pollutant may not be a single chemical, but may be, for example, a complex mixture of hydrocarbons as in fuels or oils. In such cases it will be necessary to consider whether to assess all individual chemicals separately or to treat the mixture as a single substance using whole product data.

Nature and classification of receiving waters

Various systems are in existence for classifying surface waters, of which those developed by the Environment Agency and SEPA are most relevant in the context of this report. These two schemes, which are described in Appendices A4 and A5, classify rivers, canals and estuaries as follows:

- rivers and canals are divided into six main classes, ranging from good to bad water quality, depending on a variety of chemical and biochemical criteria

- estuaries are divided into four classes, again ranging from good to bad quality, depending on biological and aesthetic factors, and on dissolved oxygen content.

Groundwaters have not been classified in a similar fashion, but the Environment Agency (NRA, 1991) defines major aquifers, minor aquifers and non-aquifers, and identifies *pathways* which affect the vulnerability of the groundwater resource (refer also to Section 7.5.2). Some aquifers may be naturally non-potable (e.g. brackish) and may not, therefore, require the same degree of protection as potable sources.

Factors affecting damage potential

Some examples of situations where the environmental impact on *receptors* is potentially high are shown in Box 7.4.

Box 7.4 *Examples of receiving waters and receptors which may be classified as having high damage potential*

- Rivers above potable supply intakes.
- Aquifers used for public supply.
- High quality waters with high grade game or coarse fisheries.
- Rivers where water is abstracted for agricultural or horticultural purposes.
- Waters with aquatic ecosystems of particular value.
- Waters used extensively for recreational purposes.
- Treatment works whose function could be adversely affected or whose capacity overwhelmed by a release of pollutant (example of *receptor* rather than receiving water).

7.5.4 Incident potential

So far, the *source* and *receptor* have been considered in isolation from each other. At this stage of the assessment it is necessary to consider whether there is the potential for pollution incidents of varying severity to occur given the characteristics and quantity of the *source* pollutant (as modified by mitigation effects during transport along the *pathway*) and the nature of the target *receptor*. For example, the loss of a few kilograms of a polluting substance stored near to a large, fast flowing, river may be insufficient to cause even minor environmental damage to even the most sensitive *receptor*. On the other hand, the same release to a smaller river might cause severe damage.

In assessing, therefore, whether the source hazard falls in the *low*, *moderate*, or *high* category, it is necessary to consider it in relation to each of the potential *receptors*.

The site operator will be concerned primarily only with the pollutants that could be released from the site in question. However, the Regulators may need to consider the potential effects of a number of incidents occurring at different sites but affecting, potentially, the same *receptor*. Although the potential effect of the individual incidents may be insignificant, when considered in combination they may be significant. In such circumstances, each *source* may need to be designated as a higher hazard than otherwise would be the case.

Where information is available, the *receptor* damage potential should be quantified.

Quantitative information is available on the effects of a wide range of substances on man (via drinking water), on fish and biota and, to a lesser extent, on plants. Where data are not available, the potential effects of the substances can often be estimated adequately from known data for similar or related classes of substances. For effects on man the LD_{50} is usually used (the dose or concentration which kills half the exposed population of test animals). Similarly for fish, the LC_{50} is usually used. Where other factors such as aesthetic considerations are involved, quantification may not be possible.

Box 7.5 shows some of the ways in which incidents in the environment can be categorised. Some of these effects are qualitative and based on broad assumptions and generalisations, while others lend themselves to quantification. An example of how environmental impacts may be quantified is given in Appendix A2. It should be noted that the incidents described in Box 7.5 are similar to but not identical with the definitions used by the Environment Agency. Further information on definitions of major incidents in the environment can be found in (DoE, 1991a).

Box 7.5 *Examples of Incident categories*

Impact	Major Incident	Minor Incident
Kill of fish or other aquatic life	Large number	Very few or none
Damage to freshwater or estuarine habitats	Long term effects on chemical or biological quality or to overall habitat	Short term effects with rapid recovery of aquatic ecosystem
Disruption of drinking water supply	Closures affecting large number of people	Event of brief duration with few people affected
Aesthetic effects	Extensive complaints from large number of people	Few complaints
SSSI or a national or marine nature reserve	Permanent or long term damage	Short term effects with rapid recovery
Public reaction	National media coverage	Local comment only
Harm to human health (via intake of water, fish or shellfish)	Likely hazard	Very unlikely
Human health (via contact water sports)	Likely hazard	Very unlikely
Impairment of effluent treatment plant	Severe, long term	Slight, transient
Damage to groundwater	Contamination preventing use for human or agricultural purposes, or significant adverse effects on surface waters to which it discharges	Short term effects with rapid recovery

Clearly both scientific rationale and subjectivity are likely to be necessary in the designation of incident categories. Views on what constitutes an incident may be very different as seen through the eyes of the discharger, the Regulators or the general public.

7.5.5 Overall site hazard rating

The contribution of each of the above factors - *source*, *pathway* and *receptor* - to the overall potential of the site to cause environmental harm can be categorised as *high*, *moderate* or *low* according to the particular circumstances. Some indication of the criteria which would lead to a high rating have been given in Sections 7.5.1-7.5.4. Clearly a contrary set of criteria would lead to a low rating, while an intermediate set of criteria would lead to a moderate rating. The assessments are necessarily to a large extent subjective. It is strongly recommended that risk assessors consult widely with the plant operators, the regulatory bodies and the containment designers throughout the process.

The three factors must be combined to obtain an overall *site hazard rating*, for which this report again recommends three categories: *high*, *moderate* or *low*. There are a number of ways in which the individual factors can be combined, particularly if a different weighting is given to each factor, as may be appropriate in some circumstances; but assessing the combined effects has to be a judgement based on knowledge, experience and the degree of confidence in the information available. Where there is uncertainty about the correct categorisation of any of the individual factors, it may be appropriate to move the overall site hazard category to the next higher group.

If any one of the three factors suggests negligible hazard, or if there is no potential for a pollution incident when the factors are considered together, then the site represents no hazard so far as environmental impact is concerned, although there may be other hazards with respect to other criteria such as health and safety of operators.

It must be stressed that it may be necessary to model a number of different scenarios. For example, there may be one *pathway* to groundwater and another to surface water, each of which needs to be considered separately. Similarly it may be necessary to consider several receiving waters and *receptors*, since it may not be clear initially which of these is the most environmentally sensitive.

Typically, if the three factors are given equal weighting, they may be combined in the way illustrated in Box 7.6. to give an overall site environmental hazard rating.

In consultation with the Regulators, the assessment may be ended at this stage if it is concluded that decisions regarding containment can be made on the basis of the overall hazard rating of the site.

Alternatively, the assessment may now move on to consideration of risk by considering the frequency of all events which could lead to loss of containment and quantifying their impact on the aquatic environment. The costs of proceeding to risk assessment may be considerable, but there are significant potential benefits as described in Section 7.4.3. An approach to assessing risk is set out in Sections 7.5.6 and 7.5.7.

Box 7.6 *Suggested combinations of hazard ratings to give overall site hazard rating*

Environmental hazard ratings

H = High rating
M = Moderate rating
L = Low rating

SOURCE	***PATHWAY***	***RECEPTOR***
(hazard rating)	(transport potential)	(damage potential)
may be	may be	may be
H or M or L	H or M or L	H or M or L

Possible combinations of ratings:		**Suggested consequent overall site hazard rating:**
HHH or HHM or HMM	→	HIGH
HHL or MMM or HML	→	MODERATE
MML or HLL or MLL or LLL	→	LOW

7.5.6 Events and circumstances leading to loss of containment

Sections 7.5.1 to 7.5.5 are concerned specifically with hazard assessment. In order to assess the risks it is necessary to consider systematically the events which may lead to the release of polluting material from the primary containment. There are two stages to this: (a) identification of all the events which are capable of causing loss of containment, and (b) assessment of the likely frequency of occurrence of each event, the amount of material which would be released as a consequence of it, and the duration of the release.

The potential failures and the reasons for failure include:

- operational failures - caused by operators, machine or plant
- structural failure - defects in design, materials, components, detailing or construction
- abuse - inappropriate change of use or other misuse
- impact - vehicular, etc.
- vandalism, terrorism, *force majeure*, etc.
- flood, fire or explosion
- geological factors - subsidence, etc.
- poor site management, supervision or monitoring
- inadequate inspection, auditing or testing
- lack of alarms and fail-safe devices.

7.5.7 Site risk rating

By analysing the events and circumstances which may affect a site it is possible to arrive at an assessment of the likely frequency of loss of containment and release of hazardous material. The combination of *site hazard rating*, expressed in environmental impact terms, with the frequency of loss of containment, provides an assessment of the risk presented by the site in its particular environmental setting. It is suggested that the expected frequency of loss of containment at a particular site is represented simply as *high*, *moderate* or *low*. Some of the ways in which the ratings for hazard and risk may be combined are shown in Box 7.7 for a situation where, as is recommended in this report, hazard and the frequency are given equal weighting.

Box 7.7 *Overall site risk rating as defined by combining ratings of site hazard and frequency of containment failure*

Site hazard rating may be *high* (H), *moderate* (M), or *low* (L) (see Box 7.6)

Frequency of loss of containment may also be *high* (*H*), *moderate* (*M*) or *low* (*L*)

Possible combinations of hazard and frequency:		**Suggested consequent overall site risk rating:**
H*H* or H*M* or M*H*	→	HIGH
M*M* or H*L* or L*H*	→	MODERATE
L*L* or M*L* or L*M*	→	LOW

As with the classification of hazards in section 7.5.4, there are many uncertainties and gaps in current information. Combining ratings for hazard and frequency of loss of containment as described above calls for skill, experience and judgement if sensible and useful conclusions are to be drawn. There are likely to be some distinctly high risk or low risk situations which are relatively easy to define and classify but the majority of situations will necessarily require some subjective judgements to be made. It is strongly recommended that risk assessors consult widely with the plant operators, the regulatory bodies and the containment designers throughout the process.

The hazard and frequency rating combinations H*L* and L*H* (Box 7.7) are perhaps the most difficult to classify in terms of overall site risk, although the suggestion here is that they are given a *moderate* rating. There are likely to be many situations where the hazard rating is high but where the probability of an event causing loss of containment is low. Views on how this situation should be treated may differ between the risk assessor and the Regulators and it is therefore recommended that there should be full consultation from the start. (An analogy may be drawn here with airflight; the hazard is extremely high but, thankfully, the probability of crashing is very low. The overall risk therefore may be classified as low or moderate. The crucial point is that high hazard does not necessarily mean high risk - otherwise we would not fly so willingly !)

Updating hazard and risk assessments

It is important to stress that hazard and risk assessments need to be updated regularly. The *source hazard rating* is likely to change frequently as the nature and quantity of materials in use at the site change. Similarly, the *transport potential* is likely to vary as the site is developed, new plant and drainage systems are constructed, or old buildings are demolished. The *transport potential* may also change if the location of dangerous materials on the site is changed, or as external factors change, such as off-site drains or treatment works. The damage potential to a *receptor* is also likely to change if the discharge position to a surface water is changed. The *receptors* themselves will be different if a discharge is switched from a surface water to a public sewer or the water is used for a different purpose.

7.6 CONTAINMENT SYSTEM CLASSIFICATION

7.6.1 Hazard and risk assessment and design classification

This report proposes classifying containment systems into three categories, each representing a different level of integrity to match the different requirements of *high*, *moderate* and *low* site risk situations.

A *Class 1* containment system would be purpose-designed to comply with normal British Standards and Codes of Practice to meet the requirements of the lowest site category. At the other end of the scale, a *Class 3* containment system would include specially designed civil and/or structural works, incorporating amongst other things fail-safe systems and higher than usual safety factors necessary to meet the more stringent needs of the highest site risk category.

Although there is no direct quantifiable link between the site hazard or site risk and the design of the containment system, it is suggested that the following simple relationship is appropriate in most circumstances:

- *low* site hazard or risk - containment type *Class 1*, i.e. normal degree of integrity
- *moderate* site hazard or risk - containment type *Class 2*, i.e. intermediate degree of integrity
- *high* site hazard or risk - containment type *Class 3*, i.e. highest degree of integrity.

The difference in performance between the three classes of containment can be expressed in terms of:

1. System safeguards (e.g. whether or not fail-safe alarms form part of the system).

2. System and component redundancies (e.g. whether there are back-up collection and storage facilities in the event of the failure of a bund, or whether drainage pipes are sleeved).

3. Structural integrity and quality of construction.

These differences are defined in more detail in section 7.6.2 and, where the differences are specific to a particular form of construction, later sections of the report.

The differences between the three classes are concerned both with system (or concept) design and with component design.

7.6.2 Engineering definitions of containment classifications

The proposed three classes of containment are defined in the following sections. The detailed design guidance presented later in the report is based on these classifications.

Class 1 Containment System - low hazard or risk

The quality of design and construction will be based on:

- normal British Standards and Codes of Practice
- normal British Standard Materials Specifications
- Building Regulations (where applicable)
- Eurocodes and CENs (where applicable)
- normal supervision
- drainage to BS 8031
- design life consistent with life of primary installation
- Safety Factors as normal British Standards.

Class 2 Containment System - moderate hazard or risk

The quality of design and construction will be as *Class 1* but with the addition of:

- additional redundancies built into the concept design (see Section 8)
- full soil report with test results
- full site survey, above and below ground, with all hazards identified
- full drain survey and test results
- any enhanced British Standards e.g. water retaining structures, BS 5750 and BS 8007
- higher Safety Factors
- construction: rigorous control, full supervision and certification of operatives
- subsidence control
- materials testing - including Agrément certification where appropriate
- tender and contract control.

Class 3 Containment System - high hazard or risk

The quality of design and construction will be as *Class 2*, but with the addition of:

- additional redundancies built into the concept design (see Section 8)
- duplication of systems and key components to provide a fail-safe overall system
- alarms and monitoring equipment.

8 Containment system options

8.1 INTRODUCTION

This section of the report is concerned with general aspects of the selection and design of secondary containment **systems**. Sections 10 through to 15 deal with the **detailed design** of systems and their components.

The importance of selecting a suitable system for containment cannot be over-emphasised. Provided a system is selected that is appropriate to the site and the hazards or risks that are present, and that the components are properly designed and constructed, the result should provide effective environmental protection. If, on the other hand, the system is inappropriate for the site, then no matter how much care is taken over the detailed design of the components, the result is likely to be unsatisfactory either in terms of environmental performance, or cost, or both.

This section summarises system design options and considers, in turn, the key factors, including hazard and/or risk rating (see Section 7), that should be taken into account by the designer. In addition, by reference to schematic diagrams of system layouts, it provides an 'index' to where detailed design guidance on specific components can be found in Sections 10 and following.

8.2 SYSTEM TYPES

Depending on the way in which they provide protection, all secondary containment systems may be categorised broadly as either:

1. *local* (i.e. bunds),
2. *remote*, or
3. *combined* (combined *local* and *remote*).

It is important for the designer to understand the differences between the three methods of providing secondary containment, the situations in which they may be suitable and the protection that they can afford, before moving on to detailed system and component design. The three categories are described in Sections 8.2.2 to 8.2.4.

The performance of primary containment tanks may be improved by, for example, providing them with double-skinned walls. Although such tanks may be considered to constitute a particular form of secondary containment (in this example the outer skin acting as a bund to the inner tank) they are still regarded in the context of this report as primary containment. As such, their detailed design is outside the scope of the report. For completeness, however, the added protection that enhanced primary containment systems can provide is considered briefly in the following section.

8.2.1 Enhanced primary containment systems

There are a number of variations on basic primary tank design, each of which can provide additional environmental protection. They are described in this report as *enhanced primary containment systems* (EPCSs) and may be further subdivided by type as follows:

1. double-skinned tanks (sometimes referred to as double- or twin-walled tanks),
2. tanks marketed as *self-bunding*, and
3. tanks marketed as *integrally-bunded.*

Examples of *self-bunding* and *integrall- bunded* tanks are illustrated in Figure 8.1, but it should be pointed out that there is currently little uniformity in terminology applying to these products. Tanks of the types illustrated are normally prefabricated in the factory and delivered to site as an item. Their size is therefore constrained by transportation considerations. Larger double-skinned tanks may be constructed on site.

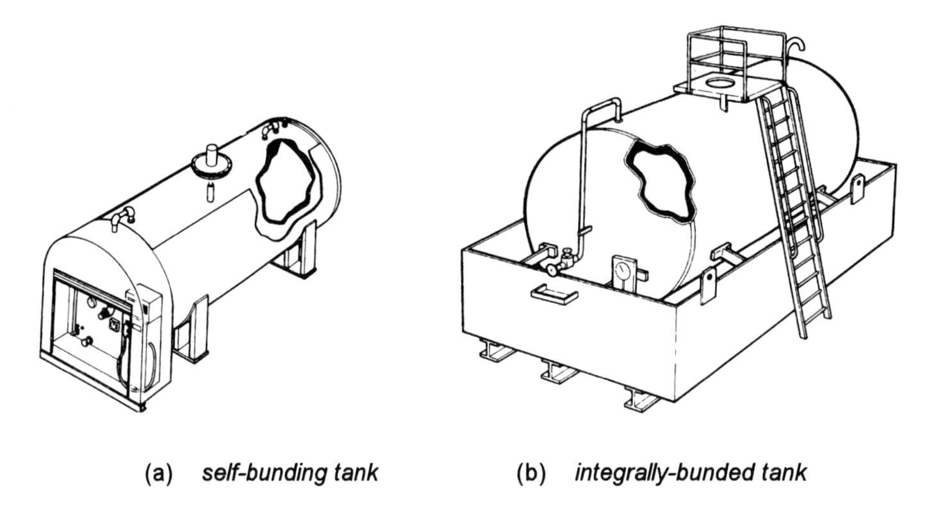

(a) *self-bunding tank* (b) *integrally-bunded tank*

Figure 8.1 *Enhanced primary containment systems*

An advantage of prefabricated EPCSs over site constructed secondary containment protection measures, such as bunds, is that they are built under factory conditions. The quality of construction is therefore potentially higher and their performance should be more reliable and predictable.

On the other hand, however, EPCSs do not and cannot on their own provide independent secondary containment since they are designed and fabricated, and act, as a composite item. It follows that there is a possibility that any defects in design, materials or fabrication in one skin of such a tank may be present in the other skin also. Similarly, it is probable that actions or forces on either the inner skin (e.g. excess pressure) or the outer skin (e.g. fire, vehicle impact) will affect the other skin, so that both skins may fail simultaneously.

While there is no doubt that EPCSs can provide improved environmental protection in many circumstances, the interdependence of their elements means that their effectiveness in the event of an 'incident', which is the subject of this report, will depend on the nature of that incident.

As this report is concerned with the provision of secondary containment and EPCSs are by considered in this report to be primary containment, their ability to provide environmental protection is not considered further here.

8.2.2 Local containment (bunds)

Bunds provide a second container around the primary container (which may be, for example, a tank or an area in which bulk chemicals are stored), designed to prevent the spread of any material that may escape from the primary containment. They contain the material at source, hence the term *local* containment.

Storage tanks or other areas used for storing or handling hazardous materials may be bunded individually or in groups. Although the layout of facilities usually dictates that bunds are built outside, they may also be built inside buildings. Some buildings, for example warehouses used for storing chemicals, may be specially built or modified so that the structure itself provides an effective bund.

An entire site may be bunded. Where a site is particularly sensitive, this may be carried out to provide a further level of protection in addition to that provided by bunds around individual tanks or storage areas within the site. Alternatively, a bund around the whole site may be the only means of protection. A bund should normally have no provision for gravity drainage except in *combined* containment systems where there would be provision for drainage to a purpose-built secondary containment tank (see Section 8.2.4). Typical bund arrangements are illustrated in Figure 8.2.

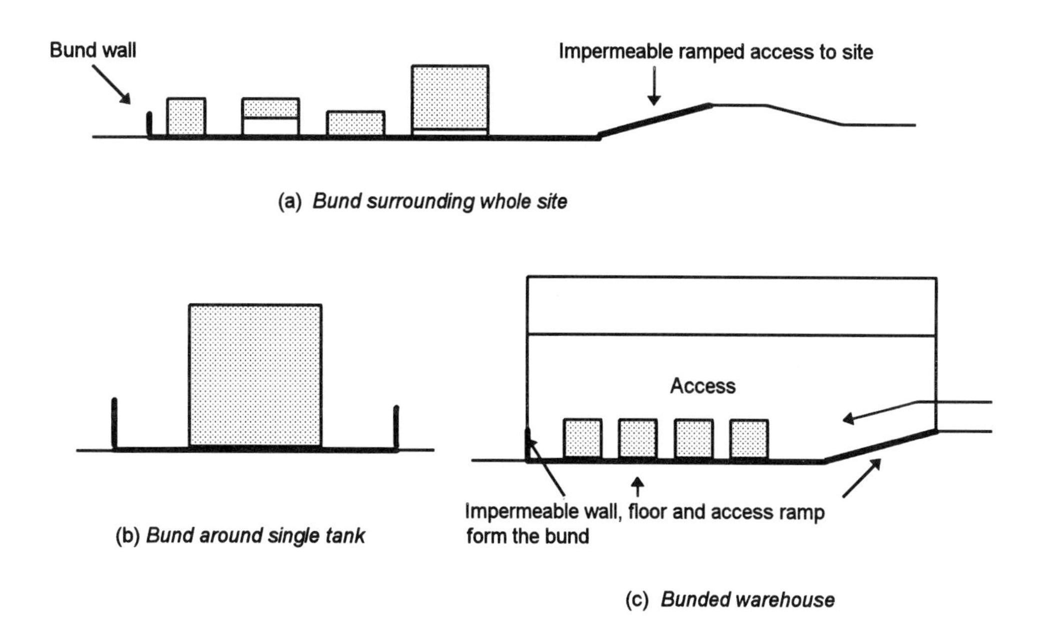

Figure 8.2 *Schematic sections showing typical local containment arrangements*

Another, less common, method of bunding is to surround a facility or even a whole site with diaphragm or sheet piling cut-off walls which extend down to an impermeable stratum. The impermeable stratum in effect takes the place of the impermeable floor of a conventional bund and, in the event of an incident, pollution is restricted to the ground bounded by the cut-off walls. A major disadvantage to this approach is that it may result in a large quantity of contaminated soil for which, ultimately, treatment or disposal arrangements would need to be made. Also, it is not possible to see what is going on underground and the integrity of the containment cannot easily be monitored. It is recommended, therefore, that cut-off walls are used only to provide additional protection (by implication, only on sensitive sites) where other types of secondary containment, described in this section, are already in place.

While bunds are normally associated with storage facilities, particularly tanks, they can also be used in linear form to protect against leaks from pipework. Such bunds may be built above or below ground level to suit the layout of the pipework. Since they will usually have only a small capacity, limited by space and cost constraints, and the volume of potential leakage from pipework can be large, it is good practice to arrange pipework bunds so that they drain or overflow into larger bunds built around storage areas or to suitable remote secondary containment (see Section 8.2.3).

Bunds built closely around primary containment may be the only safe option where highly toxic, flammable or volatile materials are stored and where their release into drains or over a wider area would create a hazard.

The degree of protection provided by a bund depends principally on:

- the extent to which all hazardous areas (i.e. those areas, tanks, pipelines etc. within which environmentally hazardous materials are stored or handled) are included within the bunded area
- the capacity of the bund
- the quality of design and construction of the bund
- maintenance.

8.2.3 Remote containment systems

With a remote containment system any material that escapes from the primary container is intercepted and transferred to a secondary containment facility which may be sited close to the primary or, more usually, some considerable distance away. This is illustrated in Figure 8.3.

In contrast to a bunded facility, the arrangements for intercepting any escaped material close to the source provide only limited containment capacity (generally less than the primary storage capacity), and the effectiveness of remote systems relies therefore on the rapid transfer of the material to the secondary container. Transfer may be through a drainage system (above or below ground) or over the surface of appropriately graded paved areas or formations. Drainage systems, pavings and formations must of course be impermeable and designed to cope with the rates of flow that would be associated with a sudden loss of primary containment.

Remote systems are suited best for sloping sites where the primary containment is located at a higher level than the secondary containment facility, so that escaped material can transfer under gravity. Where the levels and positions of primary and secondary containers are such that pumping would be necessary, remote systems are recommended only if the site hazard or risk rating (see Section 7) is low. A single remote secondary containment facility can be designed to serve a number of primary containment areas.

A remote system may comprise several stages of secondary containment, with one stage overflowing (or pumping) into the next as the loading demands (see, for example, the Case Study at Section 6.3). The first stage may be a relatively small capacity sump or small reception tank located near to the primary containment area and capable of containing small scale spills. The final stage may be a large capacity tank or lagoon capable of containing large scale spills and fire fighting water resulting from a major incident. In the case of smaller incidents the escaped material may be contained wholly within the smaller containment facilities upstream.

The key factors which determine the level of protection provided by remote systems are:

- the amount of local containment built into the system (i.e. the capacity within the catchment area to hold hazardous materials pending transfer through drains, etc. to the remote facility, see also Section 8.2.4 - combined systems)

- the ability of the system to transfer hazardous materials to the remote containment quickly, reliably and without leakage
- the impermeability of the catchment area
- the capacity and impermeability of the remote secondary containment.

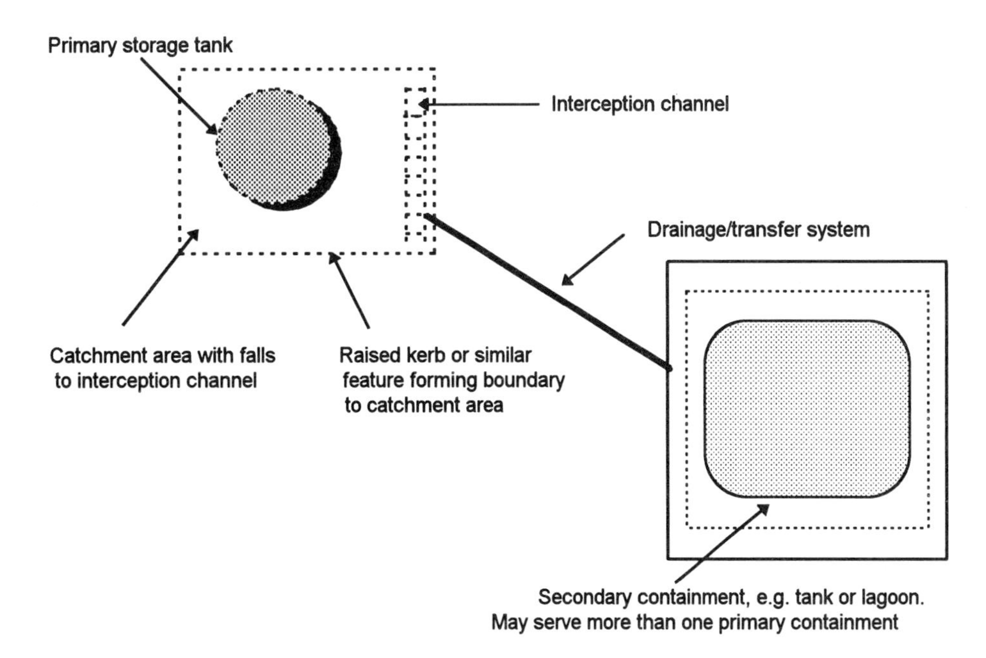

Figure 8.3 *Schematic plan of remote containment system*

8.2.4 Combined containment systems

Environmental protection at some large facilities combines features of both the *local* and *remote* systems described above. Such systems are defined in this report as *combined* systems. A typical schematic arrangement is shown at Figure 8.4. Combined containment systems are designed with the ability to contain some of the escaped material close to the source, as in local containment, but with facility also for transferring liquid (including escaped product, rainwater and fire fighting water) by gravity or by pumping to a secondary containment facility at a remote location.

A combined containment system may provide only limited local containment, in which case it becomes, in effect, a remote system as just described. At the other end of the scale, a combined system may include full secondary containment at source (see Section 9), in which case the additional facility to transfer contained material to remote secondary containment can provide an extra degree of environmental protection. The remote secondary containment may be regarded in this situation as tertiary containment.

The release of material from the bund to the remote containment may be controlled manually (for example, by operatives opening a valve when the bund appears to be in danger of overflowing) or automatically (for example with a valve operated by a float switch). Alternatively, an overflow weir may be incorporated into the bund design so that no intervention is required to relieve overfilling. Where the capacity of the bund is less than the recommended capacity for a local system, it is essential that the transfer control mechanism operates reliably and preferably incorporates backup or fail-safe devices. Failure of the control mechanism, or inadequate capacity of the transfer system, could cause a bund with a limited capacity to overflow.

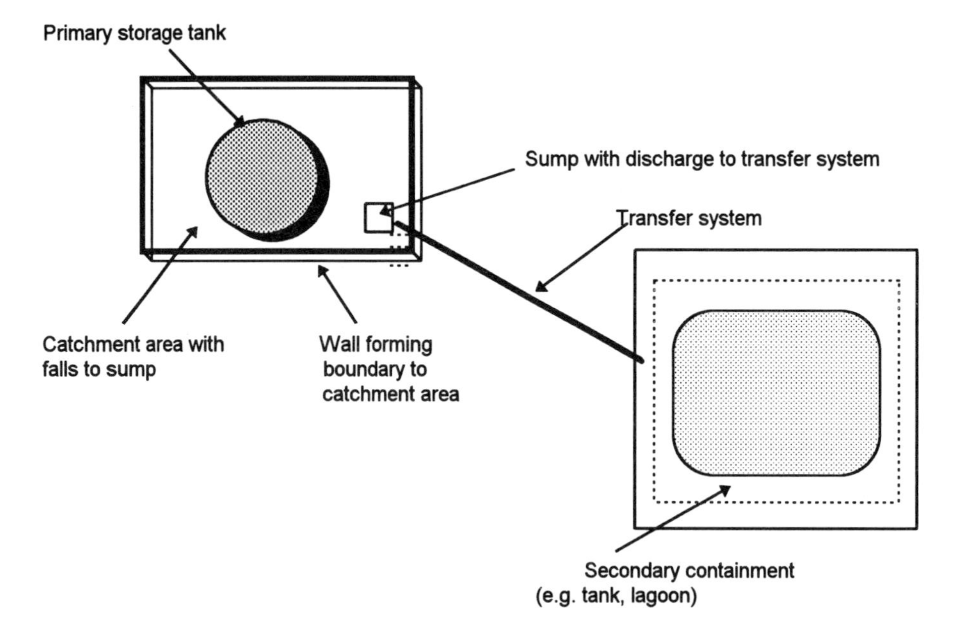

Figure 8.4 *Schematic plan of combined containment system*

A transfer system that operates under gravity, without the need for pumping, is preferable in so far as the risks associated with pump failure are avoided. On the other hand, gravity systems normally require transfer pipework to run at, or below, ground level and this introduces a separate set of risks associated with the difficulty of monitoring and maintaining inaccessible components. It may be preferable in some circumstances, therefore, to put in a pumped transfer system incorporating above-ground transfer pipework which can be easily monitored and, if necessary, bunded.

An example of the way in which the three types of containment system may be incorporated within a single site is shown schematically in Figure 8.5. The references in Figure 8.5 direct the reader to later sections of the report which provide detailed design guidance on the particular containment system or system component.

8.3 SYSTEM SELECTION

System options have been outlined in Section 8.2. The assessment of which type of system, or combination of systems, would be most effective and provide best value for money in a particular situation involves consideration of a wide range of factors including:

- the nature of the primary system, for example whether single tank, tank farm, process plant, warehouse, pipeline, loading point or drum store
- the potential for sharing containment facilities across different process areas of a site where a range of different materials may be present
- the potential for using or adapting existing containment facilities, for example interceptors, lagoons and bunds, and on-site or external treatment plant facilities which may have spare storage or treatment capacity
- the type of drainage system, including the method of disposal of trade effluents, sewage and storm water, and how the drains interconnect
- site topography
- physical constraints, particularly the space available for containment works
- future development plans for the site, either physical changes in layout, plant or buildings, or in the processes to be carried out

- quantity of material to be contained, in particular whether fire fighting water is included
- nature of the material present on the site and any 'cocktails' that may result from an incident
- cost constraints
- the site hazard or risk rating (as defined in Section 7).

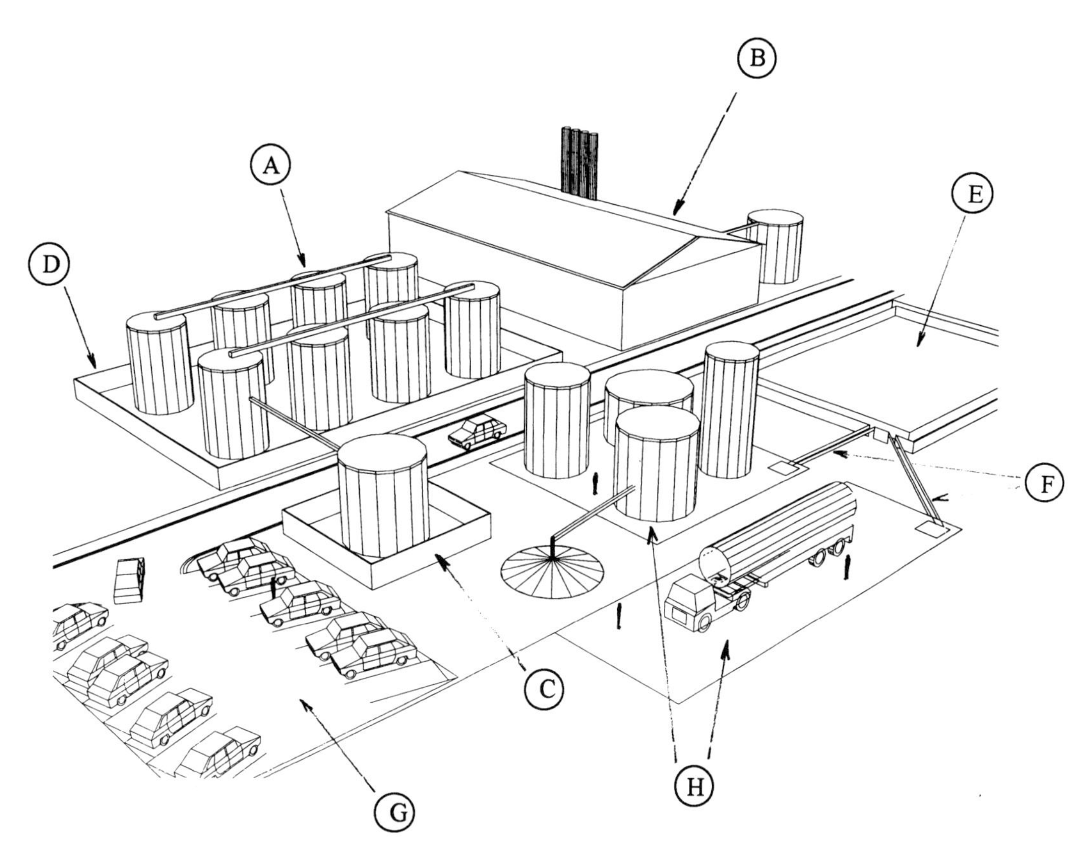

Reference	Description	Report Sections
A	Primary containment (bund)	-
B	Primary containment (warehouse)	-
C	Local secondary containment (bunded single tank)	9 and 10
D	Local secondary containment (bunded multiple tanks)	9 and 10
E	Remote secondary containment tank (one part of remote containment system)	11 and 12
F	Transfer systems (one part of remote containment system)	13
G	Sacrificial area (one part of remote containment system)	14
H	Catchment areas (one part of remote containment system)	15

Figure 8.5 *Containment systems' overview*

System design is an iterative process, beginning with a consideration of which systems would be capable of satisfying the main functional requirements and then choosing from amongst those systems, or adapting them, to provide the best practical solution given the other constraints, including cost. For secondary containment, the primary functional requirement is for it to contain hazardous materials safely and reliably.

In Section 7 a methodology has been developed to enable designers to rank the environmental hazard or risk presented by a site as low, moderate or high, and it was proposed that this should be reflected in the quality (in terms of degree of environmental protection) of the containment system adopted and the design of the components of the system. Three classes of containment were defined as summarised in Table 8.1.

Differentiation between the three classes of system **components** is achieved in the main by prescribing different material quality standards, design standards and levels of contract supervision. This is dealt with briefly in Section 7.7 and in more detail in Sections 10 to 14. Section 8.3.1 presents recommendations for relating site hazards or risks to the three types of secondary containment **system** which were identified and described in Section 8.2

Table 8.1 *Containment classification*

Environmental hazard or risk rating	Containment system classification	Component classification
High	3	3
Moderate	2 or 3	2 or 3
Low	1, 2 or 3	1, 2 or 3

8.3.1 System classification

If properly designed and constructed, each of the systems identified in Section 8.2 is capable of providing effective secondary containment. However, the degree of environmental protection provided by each system depends upon its reliability to respond to an incident in the way intended. The level of reliability of a system depends, in turn, on a number of factors including:

- complexity - the more there is to go wrong, the greater the risk
- whether intervention - manual or automatic - is necessary for the system to work
- ease of maintenance
- ease with which containment condition or integrity may be monitored and any defects or failures dealt with
- quality of site management.

The first four of these relate substantially to system design. Table 8.2 offers a broad characterisation of *bunded, remote* and *combined* systems in relation to each of these factors, from which an overall *system reliability rating* is proposed. It is important to stress that this reliability rating is a **relative** measure of the likelihood that the containment system will perform as it was designed to do, not only on the day it was commissioned but throughout the whole of its design life. The suggested *system classification* is based on this reliability rating and, in addition, on the ability of the system to provide some measure of protection in the event that the release of material exceeds (either in terms of volume or rate) that assumed in the design.

For example, a bund with a capacity based on the recommendations in Section 9 would have a 'high' reliability rating since it could be relied upon to contain safely the design capacity. However, in the event of a prolonged fire it is conceivable that there would be sufficient fire fighting water to cause the bund to overflow. In terms of environmental protection, therefore, it is given only a Class 2 rating. A *combined* system, on the other hand, could include a bund with the same 'high' reliability rating but with the additional facility to contain any overflow from the bund. It is suggested that such an additional degree of environmental protection warrants a Class 3 designation for combined systems which incorporate full capacity bunds.

Table 8.2 *Proposals for classifying containment systems*

Type of system	Summary of system characteristics				Suggested relative reliability	Suggested system classification
	Complexity	Intervention	Maintenance	Monitoring		
Local	• simplest system • fully passive • does not rely on operation of valves, transfer systems etc	• no intervention required in response to an incident • intervention necessary only to ensure bund kept free from accumulated rainwater and to empty following incident.	• relatively easy • all parts accessible • no valves, pumps, transfer systems, etc. to maintain.	• relatively easy • major defects obvious leakage from primary containment into bund easy to detect.	**High**	**Class 2:** except if no possibility of prolonged fire and consequent large volumes of fire fighting water causing design capacity of bund to be exceeded., in which case **Class 3**
Remote	• system has more components, i.e. local catchment area, transfer system and remote containment • additional components with pumped transfer system.	• none required with gravity transfer system • non-automatic pumped transfer systems require intervention at start of incident • intervention necessary to ensure remote containment free from accumulated rainwater, and to empty following incident.	• relatively difficult • systems more 'extensive', likely to include inaccessible transfer system • on pumped systems, valves, pumps, etc. require maintenance.	• less easy since three potential leakage areas to monitor: catchment area, transfer system and remote containment • transfer system particularly difficult.	**Low:** except where transfer system gravity operated, or pumped above-ground system, in which case **medium.**	**Class 1:** except where: (a) transfer system gravity operated, or (b) transfer system pumped above-ground system , **and** (c) no possibility of sufficiently rapid release of stored material, polluted rainwater, etc. to cause local catchment to overflow, in which case **Class 2**
Combined	• same as *remote* system	• same as *remote* system but need also to ensure any bunded area free from accumulated rainwater.	• same as *remote.*	• same as *remote.*	**Low** or **medium**, with same caveats as *remote* systems.	**Class 1** or **2** as *remote* system: except where capacity of bund equals that required for a *local* system, in which case **Class 3.** (In latter case, *remote* containment acts as back- up to bund and may be regarded as tertiary system)

Alternatively, if the nature of the site or facility was such that a prolonged fire could not happen, the bund and the combined system may both provide the same level of environmental protection and be designated Class 3. It is evident, therefore, that system classification is dependent not only on the intrinsic characteristics of the system but also on the circumstances in which it is used.

It must be stressed that Table 8.2 necessarily simplifies what is a complex design process and there may be site factors or particular characteristics of the operation that would lead the designer to different conclusions regarding relative reliability and, probably more so, classification.

The designer must balance the advantages and disadvantages of particular features and characteristics of a system and this requires a full and thorough understanding of the site and operations likely to be carried out on it. For example, with a remote or combined system there may be distinct advantages in terms of both maintenance and monitoring in having a piped transfer system which is raised above ground level.

However, raised pipework may mean that the system requires pumping (introducing the risk of system failure through pump failure) and the pipework and its supports may be vulnerable to accidental vehicle impact damage. The designer must balance these and other factors to arrive at the most effective solution.

A suggested approach to system design for secondary containment is outlined in the flowchart in Figure 8.6.

8.4 CONSULTATION

The need for full and early consultation with the relevant environment agency, the Local Planning Authority, the HSE and the Fire Service, at all stages of scheme planning and design, cannot be overemphasised.

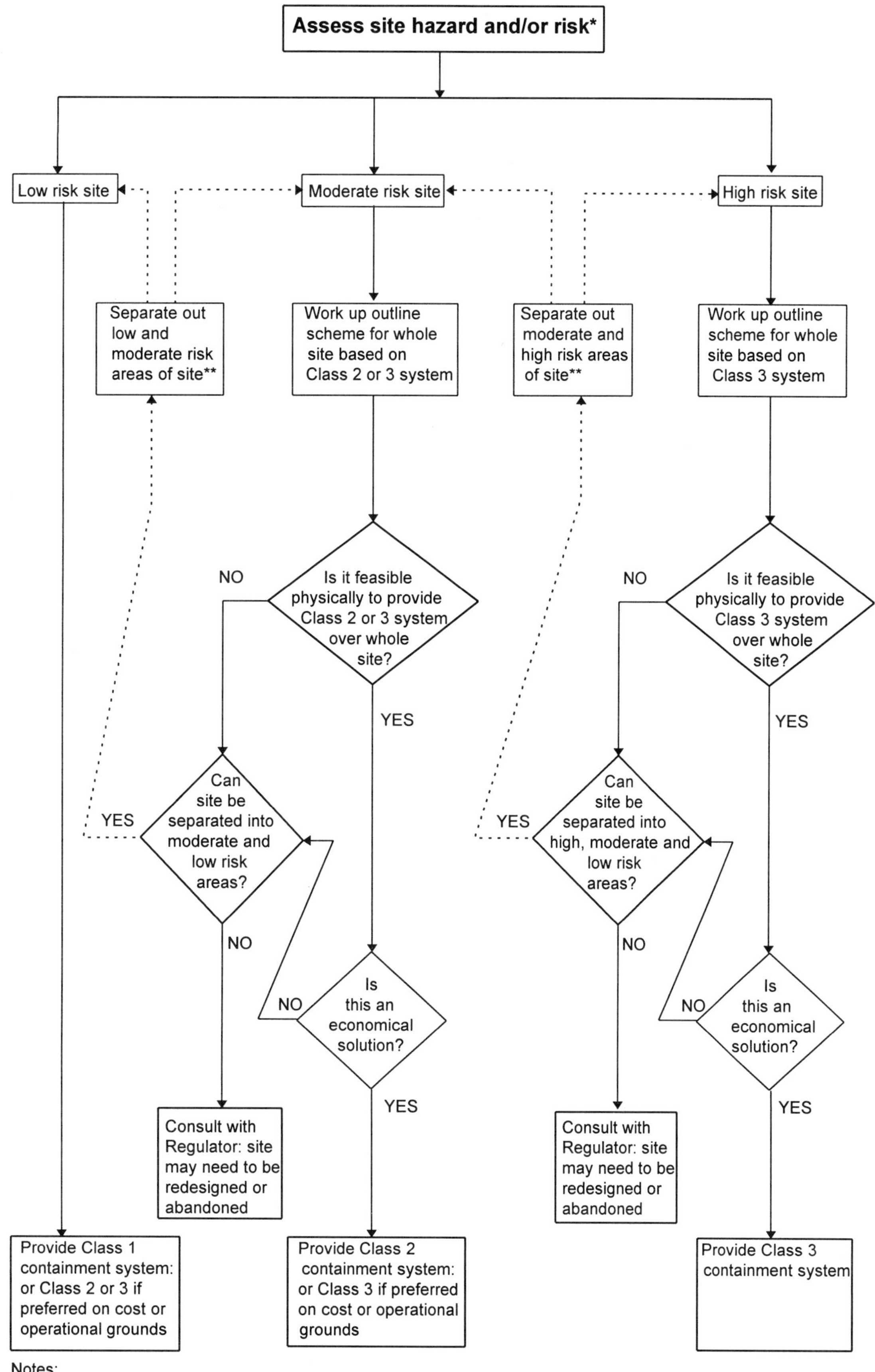

Notes:

* for sake of brevity, the expression 'hazard and/or risk' is abbreviated to 'risk' throughout the flowchart

** beyond these points in the iteration, for 'site' and 'whole site' read 'area' and 'whole area' respectively

Figure 8.6 *Containment system classification*

9 Containment system capacity

9.1 INTRODUCTION

Determining the correct capacity for a containment system is one of the most important parts of the design process. If the capacity of a system is too large, money which might otherwise have been used in other ways to improve environmental protection is wasted. If a system is too small, the considerable cost of installing it may also be wasted if it is incapable of providing effective protection in the event of an incident.

This section reviews and comments on current industry practice for assessing the required capacity of secondary containment systems. It is shown that the widely adopted 110% capacity rule can result in containment systems, particularly bunds, having very little spare capacity or freeboard to cope with fire fighting water and other fire fighting agents, rainwater and any dynamic effects.

An alternative, more rigorous, method for assessing containment capacity is presented. It is applicable to all types and sizes of containment system and its use is particularly encouraged on sites with a moderate or high hazard or risk rating.

The alternative method considers separately the containment provision likely to be necessary for fire fighting water and other fire fighting agents, rainwater, wave action and other dynamic effects, as well as the contents of the primary containment. In many situations, the provision of sufficient allowance for fire fighting water is the most significant factor in fixing containment capacity, so this difficult and, in many respects, indeterminate issue is dealt with in some detail.

Since the assessment of capacity involves so many imponderables, for example how much fire fighting water would be likely to be used in the event of a major fire, it is important to consult widely with the regulatory authorities, particularly the Fire Service, to ensure that a reasonable balance is achieved.

9.2 CURRENT INDUSTRY PRACTICE

The basis for much existing industry practice is the '110% rule', which recommends that for bulk liquid storage secondary containment has a capacity of 110% of the primary containment. Industry and Regulators are very familiar with the approach which has been in use for many years. The 110% rule is usually used in the context of calculating *local* containment capacity i.e. bunds, but it is also used as the basis for many remote and combined systems also.

9.2.1 110% rule

Where a single bulk liquid tank is bunded, the recommended minimum bund capacity is 110% of the capacity of the tank.

Where two or more tanks are installed within the same bund, the recommended capacity of the bund is the greater of:

(a) 110% of the capacity of the largest tank within the bund, or

(b) 25% of the total capacity of all of the tanks within the bund, except where tanks are hydraulically linked in which case they should be treated as if they were a single tank.

Table 9.1 shows the sizes of bunds that would be required according to these recommendations to protect a range of typical individual primary tanks. The column headed 'freeboard' is included to show how much freeboard would be available to contain rainwater, fire fighting agents or dynamic effects (see Section 9.5 and following) in the event that the contents of a full primary tank escaped into the bund. In these examples, the freeboard is calculated on the assumption that the bund is rectangular on plan and that the plan area is the smallest possible consistent with (a) not exceeding the HSE's 1.5 m maximum height recommendation (which is based on safe access and egress considerations) and (b) leaving a minimum of 750 mm clearance between the bund and primary tank (see Section 10).

Table 9.2 shows the sizes of bunds that would be required according to these recommendations for multi-tank installations.

Table 9.1 *Typical bund dimensions dictated by the 110% rule (single tanks)*

Primary tank dimensions			Minimum plan dimensions			Bund height	Freeboard
Capacity (m3)	Length (m)	Diam. (m)	Length (m)	Width (m)	Area (m2)	(m)	(mm)
(a) Cylindrical primary tank: axis horizontal							
2	2	1	3.5	2.5	9	0.2	18
16	5	2	6.5	3.5	23	0.8	69
71	10	3	11.5	4.5	52	1.5	137
589	30	5			432	1.5	137
(a) Cylindrical primary tank: axis vertical							
6	2	2	3.5	3.5	12	0.6	51
16	5	2	3.5	3.5	12	1.4	128
126	10	4			92	1.5	136
14139	20	30			10376	1.5	136

Table 9.2 *Typical bund dimensions for multi-tank installations*

Primary tank capacity (m^3)	Primary tank height (m)	Primary tank diameter (m)	Number of tanks in bund	Typical plan areas of bund[(1)] (m^2)	Bund height[(2)] (m)
			1	12	0.6
6	2	2	2	22	0.4
			4	39	0.2
			1	30	1.9
50	4	4	2	56	1.0
			4	105	0.6
			1	92	1.5
126	10	4	2	92	1.5
			4	105	1.4
			1	15553	1.5
21209	30	30	2	15553	1.5
			4	15553	1.5

Notes:

(1) primary tanks circular on plan.
Minimum clearance of 750 mm between primary tanks
Minimum clearance of 750 mm between tanks and bund wall

(2) bund height (up to 1.5 m) calculated on a bund capacity which is the greater of:
(a) 110% of the largest primary tank in the group, or
(b) 25% of the total capacity of the primary tanks in the group

9.2.2 Comment on 110% rule

Although the existing recommendations on bund capacity are not underpinned by any clearly defined rationale, they do have the benefit of being well established and therefore well known and understood by the industry and Regulators.

The existing 110% recommendation for single tanks and hydraulically linked multi-tank installations implies a safety margin of 10%. The recommendation for other multi-tank installations - the 25% rule - is based on the assumption that it is unlikely that more than one tank will fail at any one time. This may be reasonable in circumstances where the contents escape from a primary tank as a result of, for example, tank corrosion or operator error, which is likely to affect only one tank at any one time. It is difficult to justify, however, when considering the potential effects of vandalism, or a major incident such as fire, which could affect all of the tanks within a bunded area.

The 10% safety margin is interpreted by industry and Regulators to cover a range of factors including:

- prevention of overtopping in the event of a surge of liquid following the sudden failure of a primary tank
- prevention of overtopping which may be caused by wind-induced wave action during the time that the bund is full following failure of a primary tank
- an allowance for fire fighting agents, including a foam blanket on the surface or fire fighting water
- protection against overfilling
- an allowance for rain that might collect in the bund and reduce its net capacity, or for rain that might fall coincident with, or immediately following, the failure of the primary containment.

It should be noted that in relation to protection against overfilling (overflowing), the 110% capacity rule results in different safety margins depending on the relative heights of the bund wall and the primary tank. This can be illustrated by considering two situations (see Figure 9.1). In the first, it is assumed that the top of the bund wall and the top of the primary tank are at the same level. In this situation, if the bund capacity, including the contained primary tank, is 110% of the primary tank capacity, the bund will overflow if the primary tank is overfilled by more than 10%. In the second situation it is assumed that the primary tank is raised within the bund so that its top is higher than the top of the bund wall. Although the capacity of the bund would be the same as before (i.e. 110% of the primary tank) less of it would be taken up by the primary tank. Consequently, the bund would not overflow unless the primary tank was overfilled by more than 110% less the volume taken up by the tank supports and that part of the tank below the top of the bund wall.

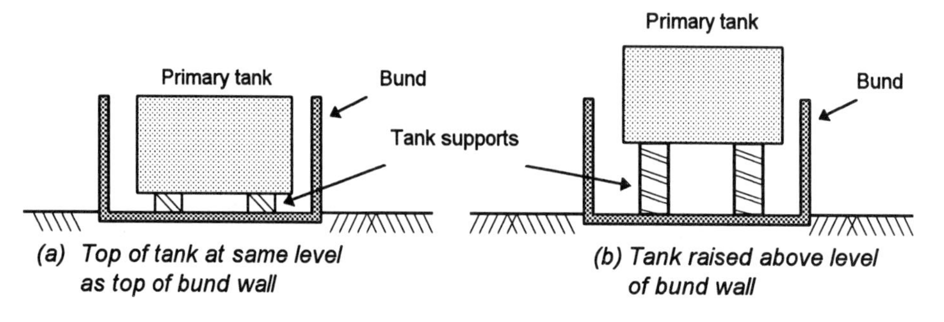

Figure 9.1 *Interpretation of the 110% capacity rule*

In practice, therefore, particularly for smaller tank installations, the existing 110% capacity recommendation does not give a uniform safety margin against overfilling.

Table 9.1 shows that the 110% rule results in a variable freeboard depending on the size and configuration of the primary tank and the plan area of the bund. If the plan area is increased, for a given primary tank capacity the required height of bund will decrease and there will be a commensurate reduction in freeboard. It is therefore recommended that if the capacity of a bund is to be determined using these rules, the designer should consider increasing the freeboard to not less than 100 mm, which is the freeboard recommended by the Fire Service for the purposes of retaining fire fighting foam. The alternative approach to assessing bund capacity, presented in Section 9.3 and following, considers specifically the need for adequate freeboard.

Table 9.2 shows that where primary tanks are not very tall, the 110% rule can result in very shallow bunds where more than one tank is enclosed within a bund. In view of this, and the possibility that more than one tank could be affected by a single incident, the existing rules cannot be recommended for multi-tank installations unless the total capacity of all of the contained primary tanks is used to calculate the bund capacity.

9.2.3 Environment Agency recommendations

The Environment Agency has published recommendations on secondary containment capacities in PPG2 (NRA, 1994b) which deals specifically with oil storage tanks, and in PPG18 (NRA, 1994i) which deals more generally with chemical storage and fire fighting runoff.

PPG2 recommends the 110% rule as described in Section 9.2.1. PPG18 also recommends the 110% rule for what it describes as 'primary bunding', i.e. bunding provided to contain the material stored in the primary containment. It recommends also the provision of 'secondary bunding', over and above the primary bunding, to contain fire-fighting runoff. No guidance is given on the volume of runoff water other than to recommend discussion with the relevant parties, including the site owners, the HSE, the Environment Agency and the Fire Service.

PPG18 lists other secondary containment systems that may be used instead of, or in combination with, bunds but does not give any recommendations on capacity.

9.2.4 HSE recommendations

HSE Safety Guidance Notes HS(G)50 (HSE, 1990a), HS(G)51 (HSE, 1990b) and Specialist Inspector Report Number 39 (HSE, 1993a) recommend calculating bund capacity using the 110% rule. HS(G)52 (HSE, 1977) recommends 100% bunding capacity of the largest tank within a bund for tank farms or tank installations with a total capacity in excess of 10 000 m^3.

HSE have published guidance (1993b) to help operators of CIMAH sites ensure they have adequate arrangements in place to deal with water used to fight fires on their sites.

The HSE recommends also that bunds should not exceed 1.5 m in height in order not to hinder unduly the natural ventilation of the bund area, hamper fire fighting or impede means of escape.

9.2.5 MAFF/ BASIS recommendations

The *Code of Practice for Suppliers of Pesticides* (MAFF, 1990) adopts the 110% rule. However, in what the Code describes as 'environmentally sensitive situations' it recommends an additional 75% capacity allowance on top of the 110%. The Code's recommendations are included in BASIS (1992), the registration scheme for the pesticide industry.

9.3 ALTERNATIVE METHOD FOR ASSESSING CONTAINMENT CAPACITY

This section sets out an alternative approach for assessing the required capacity for secondary containment and, particularly for bunds, freeboard. It is a more rigorous approach and may be more appropriate, especially for sensitive sites.

The method is based on the principle that secondary containment should be capable of containing:

- the total volume of substance that could be released during an incident
- the maximum rainfall that would be likely to accumulate in the secondary containment either before or after an incident
- fire fighting agents (water and/or foam), including cooling water
- where bunds are used they should have sufficient freeboard to minimise the risk of substance escaping as a result of dynamic factors such as surge and wave action.

Each of these components is considered in turn below.

9.4 VOLUME OF SUBSTANCE

The net capacity of the containment (in the case of bunds, allowing for tank supports and other intrusions) should be not less than the maximum storage capacity of the primary containment. Where a secondary containment facility serves more than one primary containment area, it should be assumed that an incident would affect all of the primary containment within those areas. In the case of bulk liquid primary storage, capacity should be taken as the brimful capacity (i.e. the level at which the containment would overflow) rather than the nominal capacity.

This recommendation is considerably more onerous than the 25% rule (see Section 9.2.2). The latter results in adequate secondary containment only for such incidents that would affect the primary containment selectively or independently; for example, operator error or structural corrosion. There are, however, other types of incidents, for example fire, major impact, sabotage or vandalism, which could affect all of the primary containment at the same time.

If it is decided to assess the capacity of the secondary containment on only a fraction of the primary capacity, it is essential that the designer identifies all of the incident scenarios that would not adequately be covered and assesses the likely consequences should such an incident occur. It is important that this issue is discussed fully with the relevant regulators, the Fire Service and the plant operators, which may lead to a lower containment capacity should other measures be in place.

9.5 RAINFALL

The following recommendation is based on the assumption that secondary containment is regularly inspected and that any rainwater that has collected is removed daily. If this is not the case, capacity should be increased to allow for the forecast rainwater accumulation (net of evaporation) over the period chosen for routine removal of the rainwater. Bunds and remote containment are considered separately below.

9.5.1 Allowance for rainfall in bunds

The capacity of an uncovered bund should allow for rain falling on the bunded area immediately preceding an incident (i.e. before it could be removed as part of routine operations) and immediately after an incident (i.e. before substance which had escaped from the primary could be removed from the bund).

The amount of rainfall should be calculated using *The Wallingford procedure* (HR Wallingrord, 1986) (see Figure 9.2) for the specific location and assuming a 10 year return period and the following rainfall durations:

(a) immediately preceding an incident - 24-hour duration
(b) immediately after an incident - 8-day duration

From Figure 9.2 it may be seen that, depending on geographical location, the rainfall depth for (a) is between 29 mm and 106 mm and for (b) between 54 mm and 288 mm. Appropriate additional allowances must be made where bunds may be affected by rainwater runoff from adjacent buildings. Where rainfall, or dealing with accumulated rainwater, is likely to present a significant problem, consideration should be given to providing a roof over the bund.

9.5.2 Allowance for rainfall in remote and combined systems

The capacity of the storage component (i.e. tank, lagoon, sacrificial area etc.) of a remote or combined containment system should make allowance, as described above for bunds, for rain falling directly on to it. In addition, appropriate allowance must be made for rain falling on areas of the site which drain into the secondary storage. The arrangement of the site drainage system and catchment areas will affect how much rainwater enters the secondary containment and this is considered further in Section 13. The *Wallingford procedure* is again recommended for calculating rainfall amounts.

9.6 FIRE FIGHTING AGENTS

It is difficult to give general recommendations for capacity allowance for fire fighting agents since so much depends on the nature of the site and the manner of response to an incident. However, since many incidents are likely to involve fire, and almost all 'worst case' scenarios involve fire, making adequate provision for retention of fire fighting and cooling water is of critical importance.

In making recommendations, a distinction has to be made between bunds and remote or combined containment in terms of what is practicable.

9.6.1 Allowance for fire fighting agents in designing bund capacity

It would normally be impracticable to design a bund with sufficient capacity to contain the quantities of fire fighting and cooling water that would be used in a major fire. In

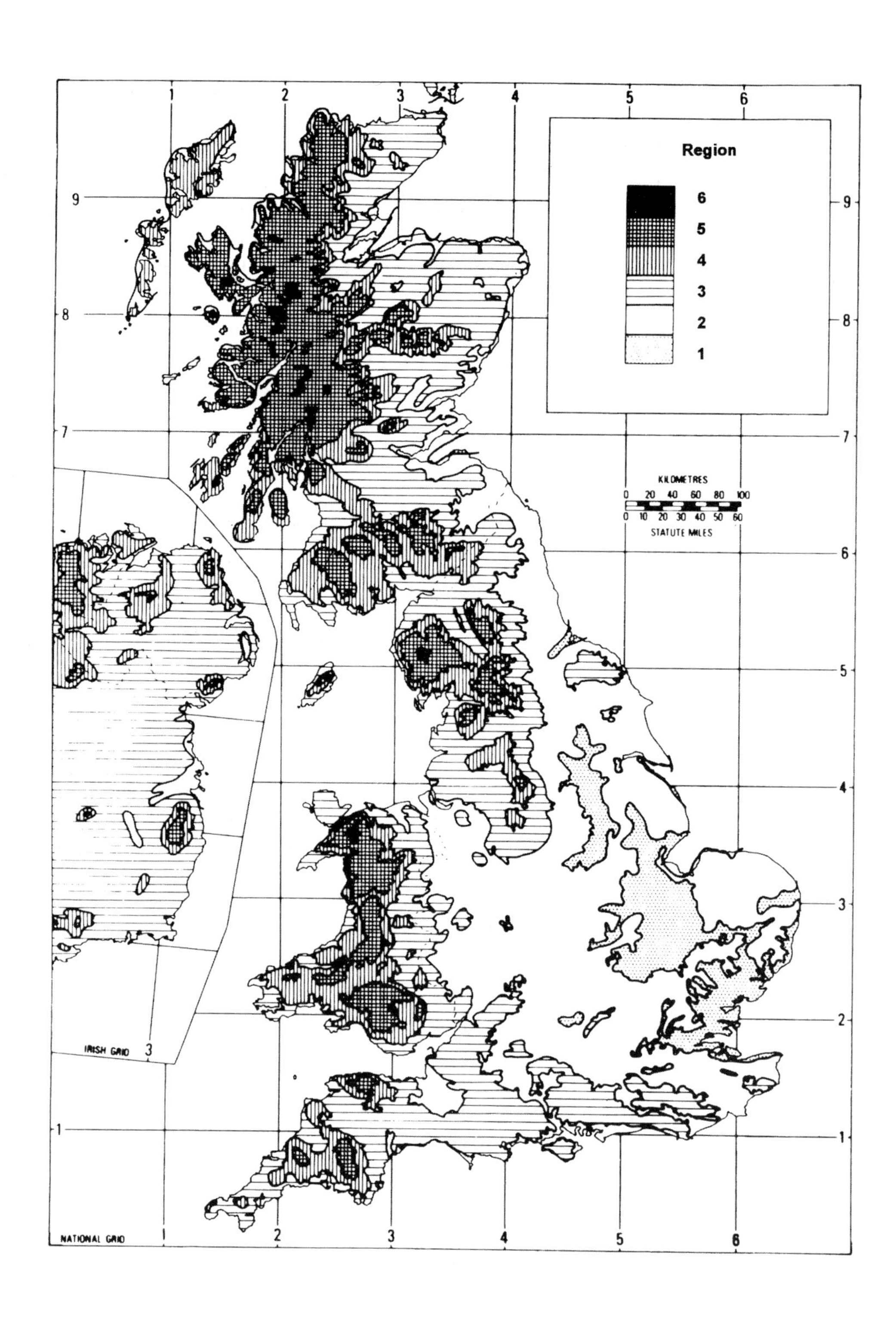

Region	Standard Annual Average Rainfall (mm)	Rainfall depth (mm) 10-year return period	
		24-hour duration	8-day duration
1	<600	29	54
2	600-800	32	65
3	800-12 00	41	95
4	1 200 - 1600	52	120
5	1 600 - 3 200	88	231
6	.3 200	106	288

Figure 9.2 *Average rainfall depths* (map reproduced courtesy of HR Wallingford, 1986)

many cases this would require the bund to be several times larger than dictated simply by the capacity of the primary containment, and on most sites this would translate into very much higher bunds. This would cause construction, operational and safety problems, not least the ability to fight fires, and would not therefore be a feasible option. For bunds, therefore, no separate allowance is recommended other than to provide sufficient freeboard (above the height of bund required to contain released substances and rainwater) to retain a blanket of fire fighting foam, and additional capacity or other provision for cooling water as discussed below.

Fire fighting foam: the amount of freeboard required for fire fighting foam should be agreed with the Fire Service but in no case should it be less than 100 mm.

Cooling water: in practice, in the event of a major fire, cooling water would be sprayed on to any threatened primary container, either through hoses or a fixed sprinkler or deluge system, or both. As a consequence the bund could be partly or totally filled with cooling water leaving insufficient or no residual capacity to cope with any release of substance if the primary containment failed. In this situation the Fire Service should be consulted and one of the following options agreed:

- procedures for recycling cooling water so that there would be no build-up in the bund: this would have bund design implications, such as the provision of suitable pick-up sumps
- the maximum amount of cooling water that would be likely to be used in the worst case scenario, and increasing the capacity of the bund accordingly. Where cooling water is supplied through fixed installations the maximum possible application will equal the capacity of the supply reservoir. Such installations are normally designed for a water delivery rate of not less than 10 litres/m^2 per minute. Where cooling water is applied by hose, perhaps to supplement the fixed installation, assessment of the volume must necessarily rely on assumptions about application rate and duration. As discussed above, in many situations it may be impracticable to provide the additional capacity.

Alternatively, where neither of these options is feasible, it will be necessary to incorporate the bund as part of a *combined* system.

It follows from the above discussion that, in most cases, unless a bund is part of a combined system it cannot be regarded as a satisfactory form of secondary containment where a major fire is one of the possible incident scenarios.

9.6.2 Allowance for fire fighting agents in the design of remote and combined systems

Remote and combined systems must have sufficient capacity to retain such fire fighting and cooling water as could reasonably be expected to be used in a major fire. It is essential to consult fully with the Fire Service to arrive at a reasonable estimate, taking into account:

- the size and layout of the plant
- the nature of the materials present and the processes carried out
- the fire detection and response systems (e.g. fixed sprinkler installations, on-site fire fighting capability) on (or proposed for) the site
- the Fire Service's contingency strategy for dealing with an incident
- the Fire Service's own delivery capability.

In the light of this information the designer must then decide, in consultation with the regulators and the plant operators, the capacity of containment required.

On sites which attract a high or medium hazard rating (see Section 7) it is suggested that sufficient capacity should be provided to contain all of the fire fighting and cooling water that could reasonably result from a worst case scenario fire, as occurred for example in the Allied Colloids incident (see Section 6). Where this would not be possible because of physical constraints (on existing sites) or where it would be difficult operationally or financially, it would be necessary to review with the Fire Service ways of reducing the anticipated quantities of fire fighting and cooling water to a level that could be contained. This may mean, for example, installing additional fixed fire sprinklers, or compartmentalising the plant or site, or gaining the agreement of the Fire Service, the regulators and the operators to a 'controlled burn' response strategy. However, even with the best laid plans the real life response to a major incident can be chaotic, and protection of life (and, to some extent property) will be the priority. In terms of assured environmental protection, there really is no substitute for a large secondary containment capacity.

For sites with a low hazard or risk rating it will be harder to justify the costs of full containment for fire fighting and cooling water and the designer should seek to strike a reasonable balance between protection and cost, in consultation with the Fire Service, the Regulators and the site operators.

The amount of fire fighting and cooling water released at a site during an incident is limited by the capacity of the fixed water delivery installations on the site (e.g. sprinklers, deluge systems, ring mains supplied by on-site storage tanks) together with the delivery capability of the fire brigade using tendered or pumped-in water. The maximum quantity that can be delivered by fixed installations equates to the capacity of the delivery storage tanks or reservoirs, with due allowance for replenishment rate through, for example, water company mains. In contrast, the delivery capability through fire brigade hoses may be effectively unlimited so that the quantity of fire fighting and cooling water delivered would depend solely on the demands of the incident.

Table 9.3 summarises water delivery rates from fire hydrants and tenders. In order to calculate the quantity of fire fighting or cooling water that might be used in an incident the designer must make assumptions concerning the number of appliances and duration of delivery.

Table 9.3 *Typical water delivery rates for hydrants, hoses and fire tenders*

Delivery appliance	Flow rate (litres/min)	Delivery capacity (litres)	Head (m)
Fire hydrant	> 550 (delivery rate at hydrant)	dependent on supply storage	> 3
Hose (25 mm)	24	dependent on supply storage	> 4
Fire tender (average) 24 mm nozzle	dependent on number of hoses	1200 to 2000	up to 50
Fire tender (large) 24 mm nozzle	dependent on number of hoses	up to 4500	up to 50

The following sections summarise three of the most widely used approaches to calculating amounts of fire fighting and cooling water:

- BS 5306: Part 2 (BSI, 1990a)
- the German VCI method (VCI, 1988)
- the European CEA method (CEA, 1993).

It is important to stress that they are all based on a forecast of the amount of extinguishing water that would be required assuming that:

- stores are designed and constructed in accordance with the requirements of the respective standard (e.g. compartmentalisation, storage density and fire detection and protection)
- that any fire will be contained within the defined fire compartment, i.e. that it is effectively controlled.

Section 9.6.6 compares typical results obtained using each of the approaches.

In contrast, Section 9.6.7 presents a summary of guidance produced by ICI (ICI, 1986) which is concerned with the amount of water required to fight major fires affecting whole plants.

Lastly, Section 9.6.8 provides a summary of the containment capacities that have been installed recently at some major sites.

9.6.3 BS 5306: Part 2

BS 5306 is concerned with the specification and design of fixed sprinkler systems in buildings. It categorises buildings, or discrete parts of buildings, as *light hazard (LH)*, *ordinary hazard* (*OH*) or *high hazard (HH)* according to the fire hazards associated with the materials present or the processes carried out (see Table 9.4). The potential of the substances to pollute the environment is not taken into account.

For each category of building BS 5306 specifies:

- an *assumed area of maximum operation (AAMO)*, which is the maximum area over which it is assumed, for the purpose of designing the sprinkler system, that the sprinklers will operate in the event of a fire
- the minimum water discharge density, i.e. the rate at which the sprinklers must be capable of discharging
- the minimum capacity of the water storage tank required to feed the sprinklers or, alternatively, the required supply characteristics of a continuously fed supply, for example a town main.

Although BS 5306 does not specify the minimum durations for sprinkler operation, the recommendations for minimum capacity and minimum water discharge density imply delivery durations in the ranges 47-58 min, 88-222 min and >115 min for *light*, *ordinary* and *high* hazard categories respectively. In most cases the designer of the secondary containment will need to know only the maximum delivery quantity. Table 9.4 summarises the main recommendations regarding water discharge rates and storage capacity.

BS 5306 does not cover fire fighting water that may be used other than in fixed sprinkler installations. In order to calculate the total retention requirement for fire fighting and cooling water the designer must therefore make assumptions about the quantity likely to be delivered by hose in circumstances where the sprinkler system has not been able to extinguish the fire.

Table 9.4 *Summary of BS 5306 water supply and storage requirements for sprinkler systems* (Source: BS 5306)

Hazard category[1]	Minimum water discharge density (mm/min)	Assumed Area of Maximum Operation (AAMO) (m^2)	Minimum feed storage capacity[2] (m^3)
Light	2.25	84	9 to 11
Ordinary (Group I)	5	72	55 to 80
Ordinary (Group II)	5	144	105 to 140
Ordinary (Group III)	5	216	135 to 185
Ordinary (Group III special)	5	360	160 to 185
High (process area)	7.5 to 12.5	260	> 225
High (storage area)	7.5 to 30	260 to 300	> 225

Notes:

(1) Hazard categories in relation to industrial premises are defined as follows:

Light Hazard: no industrial premises in this category. They are all categorised as *Ordinary* or *High hazard*.

Ordinary Hazard: industrial occupancies involving the handling, processing and storage of mainly ordinary combustible materials, which are unlikely to develop intensely burning fires in the initial stages. Examples of *Ordinary Group I* to *III* categories are as follows:

Group I: cement works

Group II: chemical works (ordinary)

Group III: soap/candle factory, garage, paper mill

Group III Special: match factory, oil/ cotton mill

High Hazard: industrial occupancies having abnormal fire loads, sub-classified as:

Process high hazard: likely to develop into rapidly and intensely burning fires

High-piled storage hazards

Potable spirit storage hazards

Oil and flammable liquid hazards.

(2) Smaller capacities permitted where pressure tanks used: refer to the source.

9.6.4 VCI guidelines

The VCI guidelines present rules for calculating the quantities of water required for fixed sprinkler systems for chemical warehouses and the containment of runoff water in the event of a fire. It groups hazardous substances into nine separate storage classes (designated LGK1 to 9) and prescribes the maximum quantity in each class that can be held in a warehouse, or in each compartment of a properly compartmentalised (in terms of spread of fire) warehouse. Provision for fire fighting water retention is specified for each such warehouse, or compartment, relating the capacity required to the pollution potential of the materials stored in it (four categories, designated WGK 0, 1, 2 and 3) and the following aspects of fire protection provision:

- level of fire detection and alarms (two categories, designated B1 and B2)
- response time of the fire brigade (three categories, designated F1, F2 and F3)
- category of fire extinguishing plant (two categories, designated LA1 and LA2)
- the rate of water supply available (three categories, designated L1, L2 and L3).

Eleven different combinations (*stages*) of these four fire protection factors are identified. For each *stage* a recommendation is given on the containment capacity for fire fighting water, ranging from 2.50 m^3 per tonne of storage capacity for *stage 1* (where the fire protection provision is minimal) down to 0.05 m^3 per tonne for *stage 11* (where there is full fire protection provision). Table 9.5 shows the containment recommendation for each *stage*.

The VCI guidelines go on to specify the minimum *stages* (as defined in Table 9.5) of fire detection and fire fighting according to the storage capacity of the warehouse (or each separate storage compartment), as governed by the LGK designation, and the WGK category of the stored substances. This is summarised in Table 9.6.

Table 9.5 *Relationship between fire fighting water quantity and fire detection/fighting measures* (Source: VCI)

Combinations of fire detection and fire fighting measures associated with warehouse(1)	Recommended capacity(2) of containment for fire fighting water for each category(3) of substance stored (m^3 per tonne of storage capacity)			
	WGK 0	WGK 1(4)	WGK 2	WGK 3
Stage 1: L1 + F1	0	2.50	3.75	5.00
Stage 2: L1 + B1 + F1	0	1.50	2.25	4.50
Stage 3: L2 + B1 + F2	0	1.00	1.50	2.00
Stage 4: L3 + B1 + F3	0	0.75	1.125	2.25
Stage 5: L2 + B2 + F2	0	0.50	0.75	1.00
Stage 6: L2 + B2 + LA1 + F2	0	0.25	0.375	0.50
Stage 7: L3 + B2 + F3	0	0.15	0.225	0.30
Stage 8: L3 + B2 + LA1 + F3	0	0.10	0.15	0.20
Stage 9: L1 + LA2 + F2	0	0.10	0.15	0.20
Stage 10: L2 + LA2 + F2	0	0.05	0.075	0.10
Stage 11: L3 + LA2 + F3	0	0.05	0.075	0.10

Notes:

(1) Fire extinguishing water supply categorised as L1 (>800 l/min), L2 (>1600 l/min) and L3 (>3200 l/min)
Response time of fire brigade categorised as F1 (>15 mins), F2 (5 to 15 mins) and F3 (<5 mins)
Fire detection and alarms categorised as B1 (hourly control) and B2 (automatic control)
Fire extinguishing plants are categorised as LA1 (semi-stationary) and LA2 (automatic)

(2) Capacity based on assumption that density of storage in warehouse averages 1 tonne of substance per m(2)

(3) WGK class relates to potential of substance to pollute water: WGK 0 (very low), WGK 1 (slight), WGK 2 (moderate) and WGK 3 (high)

(4) No containment for fire fighting water required where warehouse (or compartment) storage capacity is less than 200 tonnes.

Table 9.6 *Minimum fire precautions in relation to storage capacity and substances' pollution potential* (Source: VCI)

Storage capacity of each storage compartment (tonnes)	Minimum fire precautions (*stages*) required for each category of substance stored			
	WGK 0	WGK 1	WGK 2	WGK 3
1 - 5	Stage 1	Stage 1	Stage 1	Stage 1
5 - 20	Stage 1	Stage 1	Stage 2	Stage 2
20 - 200	Stage 1	Stage 2	Stage 5	Stage 5
200 - 800	Stage 1	Stage 5	Stage 6	Stage 6
800 - 1600	Stage 2	Stage 6	Stage 7	Stage 8
1600 - 2400	Stage 5	Stage 7	Stage 8	Stage 10
> 2400	Stage 5	Stage 8	Stage 10	not admissible

The guidelines specify broadly the water supply requirements to allow the delivery quantities shown in Table 9.5 to be safely achieved. These are summarised in Table 9.7.

Table 9.7 *Fire fighting water supply and storage requirements recommended by VCI* (Source VCI)

Warehouse type	Minimum supply rate(1) (litres/min)
Small	800
Warehouse complex	3200

Note:

(1) If these supply rates cannot be achieved through the mains, a reservoir must be provided for both warehouse types with capacity to deliver at least 3 200 litres per minute for 120 minutes or, where highly flammable materials are stored, 6 000 litres per minute for 90 minutes and 3 200 litres per minute for 30 minutes thereafter.

9.6.5 CEA draft guidelines

The European Insurance Commission (CEA) has prepared guidelines (currently in draft form) which are similar in principle to the VCI method. The main difference between the two approaches is that the CEA relates the required fire fighting water retention capacity to the fire-load of the stored chemical substance rather than to its potential to pollute water.

The CEA guidelines apply to stores which contain hazardous substances in quantities exceeding those shown in Table 9.8 and cover:

- block and rack stores in single and multi-storey buildings
- high rack stores and outdoor stores for solid, liquid and gaseous substances in mobile vessels and packages.

Tank installations, silos and bulk goods stores are not covered.

Table 9.8 *Storage capacities covered by the CEA approach*

Nature of stored substances	CEA code designation	Minimum quantity
Substances posing a serious threat to air	Z1	100 kg
Substances posing a serious threat to water	P1	1 tonne
Substances posing a threat to water	P2	10 tonne
Substances posing a slight threat to water	P3	100 tonne
Substances posing a threat to air	Z2	100 tonne

The guidelines differentiate between stores fitted with automatic extinguisher systems and those without. The recommendations covering each of these situations are summarised below.

Stores with fixed extinguishing systems

Recommended retention capacities for stores with substances stacked up to 12 m high are shown in Table 9.9, based on:

- the size of store or fire compartment area
- the degree of fire protection (five categories, designated K1 to K5)
- the fire load of the stored substances (six categories, designated F1 to F6).

For stores where substances are stacked more than 12 m high, additional structural and fire protection measures are specified and in this case the fire fighting water retention recommendations are as shown in Table 9.10.

For external storage the retention volumes given in Table 9.9 apply, subject to a number of stipulations on the maximum size of storage block (according to the nature of the substance stored) and protection measures.

Stores without fixed extinguishing systems

For stores with no fixed extinguishing system the CEA recommends the following formula for calculating the capacity of fire fighting water retention:

Table 9.9 *Retention volumes for block and rack stores with stack heights up to 12 m* (Source: CEA)

Fire compartment area (m^2)[3]	Retention volume (m^3)[4]													
	K1[1]			K2			K3			K4/K5 Stack height ≤ 6 m			K4/K5 Stack 6 to 12 m	
	F1/ F2[2]	F3/ F4	F5/ F6	F1/ F2	F3/ F4	F5/ F6	F1/ F2	F3/ F4	F5/ F6	F1/ F2	F3/ F4	F5/ F6	F1 - F4	F5/ F6
50	50	25	10	50	25	10	35	25	10	25	15	6	15	6
100	100	50	20	100	50	20	75	50	20	45	30	12	35	14
150		90	35	180	90	35	120	80	30	70	45	18	60	24
200		140	55	280	140	55	165	110	45	90	60	24	90	35
250			80	400	200	80	210	140	55	110	75	30	130	50
300			110	540	270	115	270	180	70	150	100	40	200	80
400			160		400	160	375	250	100	180	120	50	230	90
500			200		500	200	450	300	120	210	140	55	240	100
600			240		600	240	450	300	120	240	160	65	250	100
900					900	360		300	120	300	200	80	300	120
1200						480		300	120	400	250	100	300	120
1600						650			120		300	120	300	120
1800						720			120		300	120	300	120
2400						960			120		300	120	300	120
3600									120		300	120	300	120
4800												120		120
7200												120		120

Notes:

1. Fire protection categories K1 - K5 defined as follows:
   K1: Structural - small fire compartments
   K2: Surveillance - fire compartment formation
   - automatic fire **detection** system with automatic alarm transmission to public fire brigade

   K3: Surveillance with company fire brigade on permanent stand-by
   - fire compartment formation
   - automatic fire **detection** system with automatic alarm transmission to factory/company fire brigade

   K4: Extinguishing system
   - fire compartment formation
   - automatic **extinguishing** system with automatic alarm transmission to public fire brigade

   K5: Extinguishing system with company fire brigade on permanent stand-by
   - fire compartment formation
   - automatic **extinguishing** system with automatic alarm transmission to factory/company fire brigade
2. Fire load categories F1 - F6 defined as follows:
   F1: solids - readily flammable and extremely rapid burning; liquids - flash point < 21°C; gases - combustible
   F2: solids - flammable and rapid burning; liquids - flash point in range 21°C - 55°C; gases - combustible
   F3: solids - readily combustible; liquids - flash point in range 55°C - 100°C; gases - combustible
   F4: solids - moderately combustible; liquids - flash point > 100°C; gases - combustible
   F5: solids - combustible with difficulty; liquids - combustible with difficulty; gases - combustible with difficulty
   F6: solids - non-combustible; liquids - non-combustible; gases - non-combustible
3. Assumes storage density in warehouse ≤ 1.2 tonnes/m^2
4. Retention volumes take no account of rainwater or contaminated process water

Table 9.10 *Extinguishing water retention volumes for high rack stores (i.e. greater than 12 m high)* (Source: CEA)

Maximum stack height (m) (approx. pallet layers)	Catchment volume (m^3) F1/2 and F3/4	Catchment volume (m^3) F5/6
18 (10)	350	175
24 (13)	450	225
32 (17)	550	275
40 (22)	650	325

$V = b\,(1 + Z)\,qf\,t_F\,A_B$

where V = volume of the extinguishing water retention (litres)
b = dimensionless 'fire factor' (see Table 9.11)
Z = dimensionless 'supplement for outflowing material' (see Table 9.12)
qf = water output of fire brigade (litres/min per m^2) (see Table 9.13)
t_F = operational time of fire brigade (min) (see Table 9.13)
A_B = fire compartment area (m^2)

Table 9.11 *CEA method: values for fire factor 'b'*

Fire load category	Value of factor *b* for protection category		
	K1	K2	K3
F1/2	2.0	2.0	1.5
F3/4	1.0	1.0	1.0
F5/6	0.4	0.4	0.4

Table 9.12 *CEA method: values for supplement for outflowing material 'Z'*

Fire compartment area A_B (m^2) or storage capacity (tonnes)	Value of supplement Z (protection categories K1, K2 and K3)
up to 100	0.67
100 - 150	0.33
150 - 200	0.20
greater than 200	0.10

Table 9.13 *CEA method: values for fire brigade factors 'qf' and 't_F'*

Fire compartment area A_B (m^2)	Specific water output qf (litres/min per m^2)	Maximum extinguishing time t_F for protection category (min)	
		K1/K2	K3
< 100	10	30	30
200	10	60	45
300	9	90	60
400	7.5	120	75
500	6	150	90
> 600[(1)]	5	180	90

Note:

(1) For fire protection category K3 it should be assumed that the extent of the fire is restricted to not more than 500 m^2

9.6.6 Comparison of BS 5306, VCI and CEA approaches

The three approaches each assume that a fire will be contained within a defined fire compartmented area of an industrial site and extinguished using the fixed sprinkler installations on site. In addition, the CEA method makes recommendations on the amount of extinguishing water that would be required in the absence of a fixed sprinkler installation, i.e. delivered by a fire brigade.

Table 9.14 attempts to compare the results arising from the four approaches for a typical situation. Direct comparison is difficult, however, since each approach relates the amount of fire fighting water to a different set of variables. For example, the BS 5306 and the CEA approaches both relate the quantity of extinguishing water and, therefore, the amount of water to be contained, to the fire hazard of the substances stored. The VCI approach, on the other hand, takes no account of substance fire hazard but instead links the required retention quantity to the polluting potential of the substance. As a result, the differences highlighted in Table 9.14 are likely to change depending on the assumptions made about the site and the substances stored, particularly their rated fire hazards and pollution hazards.

It is evident from Table 9.14 that the CEA method forecasts similar quantities of extinguishing water whether or not an automatic sprinkler system is installed.

It is important to stress that these approaches to calculating extinguishing water all assume that:

- stores are designed and constructed in accordance with the requirements of the respective standard (e.g. compartmentalisation, storage density and fire detection and protection)
- that any fire will be contained within the defined fire compartment.

Table 9.14 *Comparison of BS 5306, VCI and CEA approaches for a typical situation*

Method	Size of store (m^2 or tonnes)	Fire hazard of stored substance	Ecotoxicity of stored substance	Level of fire detection and/or protection	Volume of extinguishing water (m^3)
BS 5306	250	Ordinary(1) (Group III)	not applicable	not applicable	135 -185
VCI	250	not applicable	WGK 2(3)	Stage 6(2),(4)	94
CEA (with fixed extinguisher)	250	F4(1)	not applicable	K2(2)	200
CEA (without fixed extinguisher)	250	F4	not applicable	K2	206

Notes:

(1) BS 5306 and the CEA define substance fire rating differently: *F4* is nearest equivalent to *Ordinary Group III*, and vice versa.

(2) VCI and CEA define fire protection differently: *Stage 6* is nearest equivalent to *K2*, and vice versa.

(3) See Tables 9.5 and 9.6.

(4) See Table 9.5.

9.6.7 ICI's guidelines for fires involving whole chemical plants

ICI have produced guidance (ICI, 1986) for internal use, on the demand flow rate and duration for fires at chemical plants. The forecast of the total amount of fire fighting water that might be used in the event of a fire affecting the whole of a chemical plant (as distinct from just a discrete area or fire compartment as assumed in the preceding approaches) is shown in Table 9.15.

Table 9.15 *Forecast of fire fighting water needed to tackle major chemical plant fires* (Source: ICI)

Plant hazard rating(1)	Fire fighting water demand
High severity	Total demand 1 620 - 3 240 m^3/hr for 4 hrs
Medium severity	Total demand 1 080 -1 620 m^3/hr for 4 hrs
Low severity	Total demand 540 - 1 080 m^3/hr for 4 hrs

Notes:

(1) High severity includes plants with:

- over 500 tonnes of flammable liquid above its flashpoint
- over 50 tonnes LPG above its boiling point and over 50 bars
- over 100 tonnes combustible solid with ready flame propagation
- other factors which increase severity.

Low severity includes plants with:

- less than 5 tonnes flammable liquid above or below flashpoint
- less than 100 kg flammable gas under 1 bar or a flash liquid
- less than 5 tonnes readily combustible solid
- other factors which decrease severity.

Medium severity covers plants which fall between high and low severity ratings.

9.6.8 Capacity of fire fighting water retention basins: examples of recent practice

In the course of researching this report, a number of major industrial sites have been visited. Where fire fighting water retention basins have recently been installed, their capacity has been found generally to be in the range 2 000 to 7 000 m^3 (see Section 6 - Case Studies - for more detail). This is very much larger than the capacity arrived at using the BS, VCI and CEA methods since in these examples the site operators have attempted to take account of a major fire affecting the whole plant rather than just isolated areas. On the other hand, the capacity provided at these sites appears in many cases to be significantly less than the ICI guidelines suggest for a *high severity* plant hazard rating.

9.6.9 Conclusions on forecasting fire fighting water

It is clear from the foregoing that forecasting the likely volume of fire fighting water is difficult. Methods such as the BS, VCI and CEA may be suitable for smaller sites where there is effectively only one (or very few 'fire compartments') or in those circumstances where, for a number of reasons, it may be agreed with the Regulators that it is necessary only to provide protection against an incident of limited extent.

Where it is agreed with the Regulators that it is necessary to take account of more major incidents (e.g. where a fixed sprinkler installation fails to contain a fire locally), the results obtained using the BS, VCI and CEA methods must be interpreted appropriately to reflect the assumptions on which the forecast incident scenario is based.

ICI's general recommendations for incidents affecting whole plants appear to give a conservative answer, considerably exceeding current practice on many sites where the amount of containment provided has been assessed as carefully as possible in the light of the particular circumstances.

The conclusion is that each site, plant and situation must be considered in relation to the particular hazard it presents, drawing as appropriate on the guidance given in the approaches discussed above. It is essential that proposals are discussed at an early stage with the Regulators and Fire Service, since a scheme based on an assessment of capacity which is subsequently found unacceptable would be very costly to remedy.

9.7 FREEBOARD IN BUNDS TO ALLOW FOR DYNAMIC EFFECTS

The sudden failure of a bulk liquid container could result in a surge or wave of liquid which in some circumstances might overtop a bund wall. It is difficult to give general recommendations to cover such dynamic effects, since so much depends on the shape and size of the bund and the position of the primary storage tank(s) in relation to the bund walls. A detailed explanation of the topic is presented in Wilkinson (1991) which, on the basis of small-scale model tests, suggests that the total fraction spilled, Q, could be written as a function of the dimensionless variables:

$$Q = Q\,(h/H,\ r/H,\ R/H,\ \theta)$$

where h = bund height
r = bund radius
H = height of fluid in the tank
R = tank radius
θ = bund inclination

It is also suggested that sloping the bund towards the primary tank can considerably reduce spillage.

In the absence of a detailed analysis taking into account the particular characteristics of an installation, it is recommended that a freeboard of 250 mm is provided to protect against dynamic effects. For bunds with earth embankments it is recommended that the freeboard is increased to 750 mm to reduce possible erosion due to wave action.

When calculating the overall height of a bund wall to achieve this freeboard it is not necessary to take into account the allowance for rainfall immediately following an incident (see Section 9.5.1), or the 100 mm freeboard allowed for fire fighting foam (see Section 9.6.1), since the major dynamic effect, surge, is only transitory.

9.8 SUMMARY OF ALTERNATIVE METHOD

The foregoing recommendations on containment capacity requirements are summarised in Table 9.16.

Table 9.16 *Summary of retention capacity recommendations*

Factor to take into account	Bund capacity/freeboard recommendations	Remote and combined system capacity recommendations
Primary storage capacity (i.e. possible storage inventory)	Capacity at least 100% of primary capacity. Include capacity of all primary tanks in multi-tank installations. (Section 9.4)	Capacity at least 100% of primary capacity. Include capacity of all primary tanks in multi-tank installations. (Section 9.4)
Rainfall	For uncovered bunds only, provide sufficient freeboard for: (a) immediately preceding an incident - 10-year return, 24-hour rainfall, and (b) immediately after an incident - 10-year return, 8-day rainfall. Freeboard component (b) does not act cumulatively with allowance for dynamic effects. (Section 9.5.1)	As for bunds plus an allowance for rain falling directly on to remote containment and areas of the site draining into it. (Section 9.5.2)
Fire fighting and cooling water	No allowance specifically for fire fighting water. Allowance for cooling water, or procedures for recirculating cooling water, to be agreed with the Fire Service. Quantity of cooling water normally based on a delivery rate of 10 litres/m^2 per min. (Section 9.6.1)	Allowance for extinguishing and cooling water delivered through fixed and non-fixed installations based on BS 5306, VCI, CEA and ICI methodologies, with appropriate adjustments in the light of the particular circumstances. Consultation with the Regulators and the Fire Service essential. (Sections 9.6.2 - 9.6.9)
Foam	Allow freeboard of not less than 100 mm. (Section 9.6.1)	Allow freeboard of not less than 100 mm. (Section 9.6.1)
Dynamic effects	Carry out detailed analysis. In absence of detailed analysis provide freeboard of not less than 250 mm (750 mm for earth walled bunds). (Section 9.7)	Provide 750 mm for earth walled bunds. (Section 9.7)

The designer of the containment system must take into account the probability of a number of events occurring simultaneously. The worst case scenario for containment is represented by the design return period rainfall (e.g. the rainfall that is likely to occur, say, once in 50 years) coinciding with the sudden and total loss of primary containment and a fire involving applied fire fighting water. At low risk sites or sites where it can be demonstrated that the probability of a simultaneous occurrence of events is sufficiently low, it may be possible to apply less stringent capacity requirements. Such relaxations must be subject to the designer's and site operator's discretion and the agreement of the various regulatory bodies in the light of the particular circumstances.

10 Bunds

10.1 INTRODUCTION

This section sets out a performance specification for bunds, detailed recommendations and guidance on the design and construction of bunds built *in-situ*, and more general advice on prefabricated bunds. The design recommendations differentiate between *Classes 1*, *2* and *3* containment (as defined in Section 7) by specifying (a) a range of structural design standards and (b) physical arrangements to permit effective monitoring specific to each Class.

The purpose of a bund is to contain any substance that may escape from the primary containment until remedial and safety measures can be put in place. Properly designed and constructed bunds can provide an effective safeguard against pollution. In many situations they also contribute to health and safety, and reduce the possibility of product wastage.

Bunds provide a passive defence against escape of pollutant in that they do not require operator intervention in the event of failure of the primary containment. For this reason they are usually seen as the main line of defence against pollution, frequently in combination with *remote* secondary containment systems (see Section 8) or other active control measures such as high-fill-level alarms.

Bunds provide visible protection and that may be both an advantage and a drawback. Visibility is an advantage in so far as a bund's superficial condition and fitness for purpose (or alternatively, defects) will be readily apparent. Also, the physical presence of a bund is a constant reminder to operatives that hazardous materials are present. On the other hand, visibility is a disadvantage in that it can lead to a false sense of security, resulting in the view that other protective measures, including safety and environmental management procedures, need not be taken seriously. It is important, therefore, to get over the message that bunds do not, and cannot, provide the whole answer to environmental protection.

Partly because bunds are a passive measure, and partly because they are not directly related to plant operations and contributing to productivity (they may actually be seen by some to be hindering efficient operations), their monitoring and maintenance is frequently neglected.

An incident causing leakage of a hazardous liquid into a bunded area may endanger the life of persons accessing the area during normal duties. The HSE have therefore provided recommendations (1977, 1990a) (see also Section 4.2) on the safe design of bunds, covering the height and construction of certain bund walls. In some situations, the HSE recommendations may also influence the extent of a bunded area.

In order to be effective, a bund must be leakproof at all times. Bund walls and floors must therefore be properly designed and built to ensure that they are impermeable. Openings in bund walls and piercing bunds for pipework should be avoided wherever possible (an 'essential' recommendation for *Class 3* containment (see Section 10.4.5)).

10.2 DEFINITION OF BUND

There is no universally recognised definition of the term 'bund'. In the context of this report, a bund is defined as:

a facility (including walls and a base) built around an area where potentially polluting materials are handled, processed or stored, for the purposes of containing any unintended escape of material from that area until such time as remedial action can be taken.

A bund provides *local* secondary containment in the sense that the source of the potential pollutant is within the bund, as distinct from *remote* secondary containment where any escaped material is transferred to a separate facility.

Bunds may be used in a number of different situations, providing local secondary containment at, for example, small single tank installations (Figure 10.1), and extensive tank farms and large chemical processing plants (Figure 10.2). In both of these examples the bund wall is shallow in relation to the height of the primary container and there is a large clearance between the two. An alternative form of bund, commonly termed a *collar bund*, is shown at Figure 10.3. Full height collar bunds (Figure 10.3(a)) are built close to the primary vessel, with a height equal to the maximum storage depth of liquid in the primary vessel. Three-quarter collar bunds are not so high but cover a larger area.

Bunds may also be provided to protect against the escape of materials from pipelines, both above and below ground (Figure 10.4). Although they are more usually associated with outdoor situations, bunds may also be used to protect against the failure of primary containment within buildings (Figure 10.5). The recommendations presented in this section are applicable to each of these situations.

Figure 10.1 *Typical example of bund around individual tank* (Reproduced courtesy of Elf Atochem)

Bunds may be designed and built in a number of different ways and using a variety of materials. In all cases, it is recommended that they are built in accordance with the performance specification set out in Section 10.3. Guidance on meeting this performance specification is presented in the sections following 10.3.

Figure 10.2 *Typical example of bunds around a tank farm* (Reproduced courtesy of GATX)

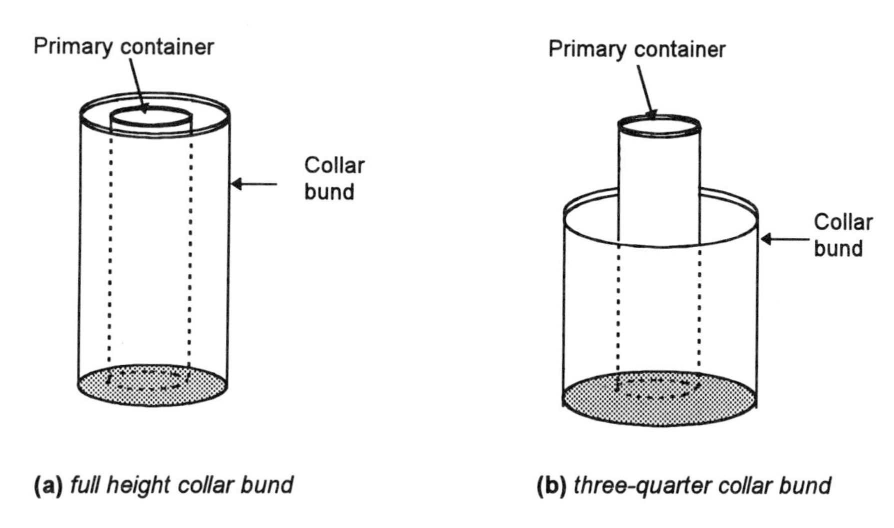

(a) *full height collar bund*

(b) *three-quarter collar bund*

Figure 10.3 *Collar bunds*

Figure 10.4 *Example of bund around pipeline* (Reproduced courtesy of Ciba)

Figure 10.5 *Typical example of a bund inside a building* (© courtesy of Hickson Timber Products Ltd.)

(Note: in some sectors of industry the word *catchpit* is used in place of *bund*. The term implies generally a small facility. Since this report deals with all sizes of facility from small to very large, the term *bund* is used throughout.)

10.3 PERFORMANCE SPECIFICATION

It is recommended that all bunds as defined in Section 10.2 should be designed and constructed to comply with the performance criteria which are summarised in Box 10.1 and considered in more detail below.

The designer of a bund should assess and take into account in the bund design:

- all possible modes of escape of pollutant from the primary containment
- all possible modes of failure of the bund
- all possible incident scenarios.

10.3.1 Height of bund walls

A bund wall should be high enough to retain the contents of the primary storage, with appropriate allowances for rainwater and firefighting water as detailed in Section 9. In addition, Section 9 recommends 'fixed' freeboard allowances for fire fighting foam (not less than 100 mm) and surge (in the absence of a detailed analysis, not less than 250 mm) which are independent of bund capacity. While these considerations dictate the minimum wall height, health and safety considerations must be taken into account in deciding the maximum height.

Box 10.1 *Performance criteria for bunds*

Aspect of performance	Recommended criteria
• General arrangement	Size and layout should take account of all possible modes of failure of primary containment. There should be no provision for gravity discharge of the bund unless part of a *combined* system (see Sections 10.3.2, 10.3.4)
• Capacity	At least 100% of primary storage capacity + allowance for rainfall (pre- and post-incident) + allowance for cooling water + freeboards for fire fighting foam and dynamic effects (see Sections 10.3.7, 9.3)
• Retention period	Capable of retaining maximum design volume of prescribed material for not less than 8 days (see Section 10.3.8)
• Impermeability	For earth structures: not less than the equivalent of 1m depth of soil with a permeability coefficient of 10^{-9} m/ sec All other forms of construction: 'watertight' as defined by compliance with the British Standards or other recognised standards appropriate to the form and/or materials of construction and containment class (see Section 10.3.9)
• Strength	Capable of withstanding the static and dynamic loads associated with: - release of liquid from primary storage tanks - release of water from hoses during fire fighting operations - wind (50-year design life) Bund floor to be capable of withstanding loads from activities within bunded area and the effects of differential settlement (see Section 10.3.10)
• Durability	Capable of resisting the effects of weather, aggressive ground conditions and abrasion (in each case assuming a durability life of 50 yrs unless otherwise specified), fire and, depending on the primary storage inventory, corrosive materials (for the duration of the specified retention period) (see Section 10.3.11)
• Structural independence	Bund walls to be structurally independent from the primary containment. Wherever possible, raised bunds to be supported independently from primary containment (see Section 10.3.12)
• Accessibility	Walls and, where practicable, floors to be sufficiently accessible to permit inspection and for maintenance to be carried out Where access to parts of the floor is not practicable (e.g. large tanks sited directly on the bund floor) provision should be made to detect any leakage through the base of primary containment (see Sections 10.3.4, 10.3.13)

The following two sections deal in more detail with the issue of 'surge' and the health and safety limitations.

Surge

The following findings emerged from a series of experimental and theoretical studies reported by Wilkinson (1991) which were carried out to determine the quantity of liquid that would overflow a bund in the event of a sudden and catastrophic failure of a primary storage tank :

- where bund and primary tank capacity were equal, the dominating factor was the ratio of the height of the bund h to the initial height of liquid in the tank H. With $h/H = 1$ there was no overflow; with $h/H = 0.5$ about 25% of the liquid overflowed
- for given values of h/H the overflow decreased as the bund was sited further away from the primary tank
- amount of overflow was strongly influenced by the angle of inclination of the bund wall. The overflow where the bund was inclined at 30° away from the primary tank was up to double the overflow over a vertical bund of the same height
- a 'throwback' lip reduced overflow
- small 'tertiary bunds' inside the bunded area reduced overflow.

These findings are summarised in Box 10.2. It should be stressed that these very significant overflows are associated with the very rapid release of liquid from the primary containment.

In practice, constructing bunds which slope inwards or which incorporate some form of throwback lip is likely to add considerably to the cost. A more practicable option in many situations will be to build dividing walls, or baffles, inside the main bunded area, between the primary tank and the main bund, and this ought to be considered for sites which have a high hazard or risk rating. Such 'bunds' need not be bunds as defined at the beginning of this section, and they need not comply with the performance criteria. Their purpose is to act as a baffle, slowing down any surge of escaping liquid. Strength and stability is a more important requirement than impermeability.

It should be noted that subdividing or compartmenting a large bund, a feature of bunds surrounding tank farms, is unlikely to provide any protection against surge since to be effective the baffles must be between the primary tank and the outer bund. Subdividing bunds would mitigate surge only in respect of those primary tanks towards the centre of such a facility; there would be no additional protection in respect of those tanks nearest the outer bund wall. Subdividing large bunds does offer other advantages, however, as discussed in Section 10.3.2.

Health and safety

It is recommended generally by the HSE that bund walls should not exceed 1.5 m in height. The purpose of the height limitation is:

- to avoid hindering fire fighting operations
- to ensure relatively easy egress from a bunded area in the event of an emergency
- to encourage natural ventilation of the bunded area.

There will be circumstances where, for operational or conflicting safety reasons, it will be necessary for a bund wall to exceed 1.5 m high and in such cases the HSE should be consulted. It will be essential to provide adequate means of escape for personnel and to make deep bunds secure against unauthorised access.

Box 10.2 *Effect of bund wall geometry on overtopping*

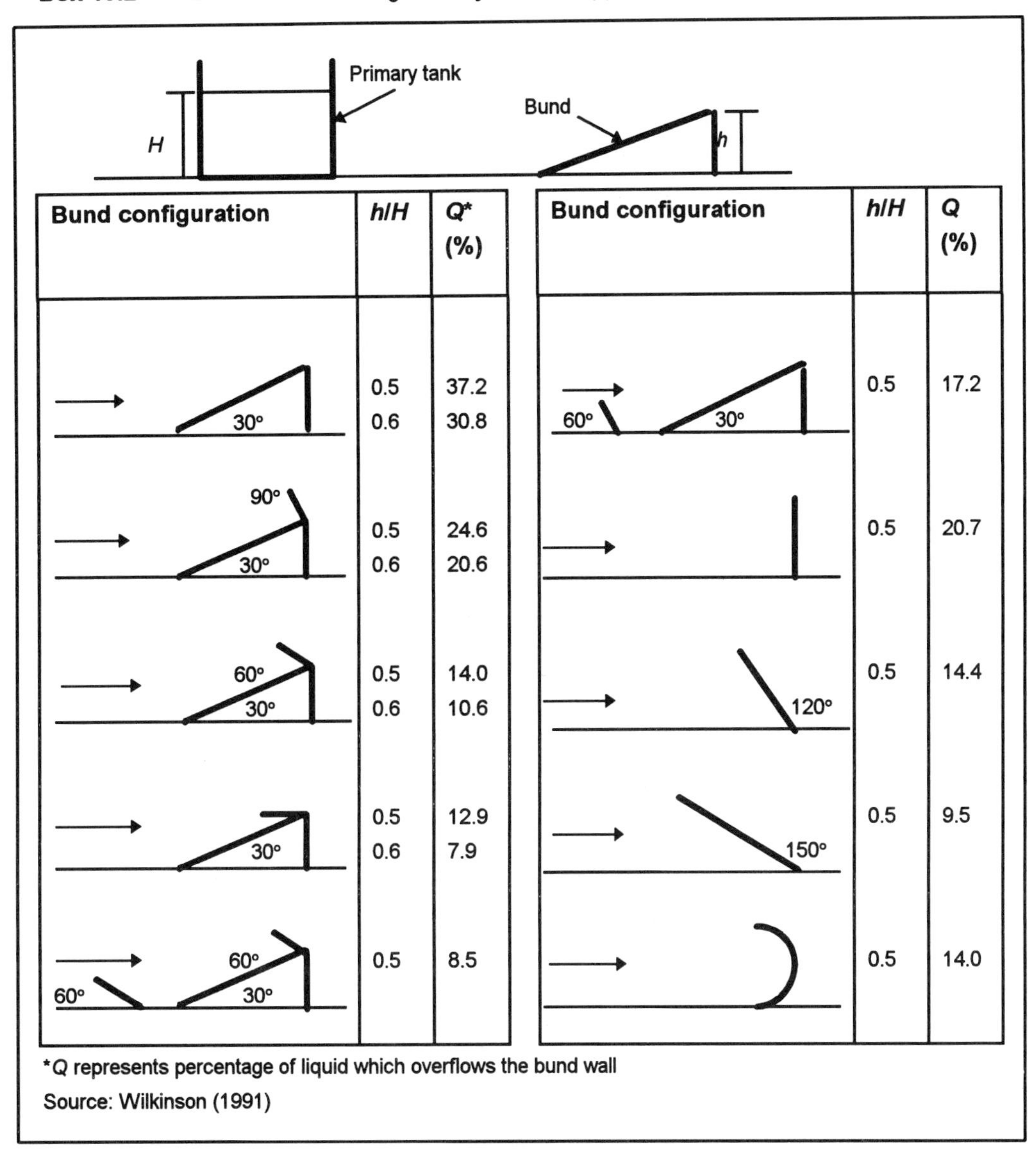

Bund configuration	h/H	Q* (%)
30°	0.5 0.6	37.2 30.8
90°, 30°	0.5 0.6	24.6 20.6
60°, 30°	0.5 0.6	14.0 10.6
30°	0.5 0.6	12.9 7.9
60°, 60°, 30°	0.5	8.5

Bund configuration	h/H	Q (%)
60°, 30°	0.5	17.2
	0.5	20.7
120°	0.5	14.4
150°	0.5	9.5
	0.5	14.0

*Q represents percentage of liquid which overflows the bund wall

Source: Wilkinson (1991)

Clearly the HSE height recommendation cannot be applied to collar bunds where the issues to do with fire fighting, access by personnel and ventilation are very different and must be addressed specifically for each installation.

10.3.2 Bund shape and compartmentation

The shape of a bund should be kept as simple as possible, taking operability requirements into consideration. Complicated footprint shapes tend to lead to difficult structural detailing at corners with a consequent increased risk of failure. The footprint area will be determined by the assessed capacity requirement for the bund (see Section 8) together with any height limitations, for example the general 1.5 m height recommendation of the HSE. In some circumstances it may be better, easier and cheaper to provide a larger bunded area than the capacity requirement dictates, simply to avoid complicated detailing.

Compartmentation of large bunds (e.g. bunds around tank farms) means that escapes from the primary containment may be confined to a relatively small area, typically the immediate vicinity of a leaking tank. This has several advantages including:

- a reduction in the risk of escape of polluting material from the bund (because the area it occupies is smaller)
- easier recovery of the material from the bund, and less contamination of that material
- making the recirculation of fire fighting or cooling water easier (greater depth for pick-up for a given volume of water)
- a reduction in the risk of damage to neighbouring tanks and facilities.

A typical example of compartmentation is shown in Figure 10.6

Figure 10.6 *Compartmentation of a bund* (Reproduced courtesy of GATX)

It is unnecessary for the dividing walls in a bund to be impermeable to the same extent as the bund walls. However, they must be strong enough to withstand a full hydrostatic head of liquid and, on sites with a high hazard or risk rating, it is recommended that they are designed to withstand surge loads (see Box 10.5).

10.3.3 Proximity to primary storage

The greater the distance between a bund wall and the primary containment, the less is the risk of failure or bund overflow through:

- surge (see Section 10.3.1)
- a damaged bund wall falling on to, and damaging, primary containment (and *vice versa*)
- jetting (see below).

These advantages must be weighed against the additional cost of construction, possible additional operational problems, and the increase in the amount of rainwater that will collect in the bund.

Given the wide variability of sites, it is not possible to give hard and fast rules on the minimum distances. The guidance given here is therefore limited to sites with a high hazard or risk rating, where it is recommended that the bund wall is situated so that no structure within the bund is closer to the wall than a distance equal to its own height.

Jetting

The failure of a storage tank through, for example, a rupture or corrosion of the side wall, could result in the escape of a jet of liquid with sufficient force that it projects over the bund wall, even though the capacity of the bund and the height of the bund walls are in accordance with the recommendations in Section 8. This phenomenon is referred to as *jetting*.

Loss of liquid through jetting is rare and when it does occur the amount lost may be relatively small. In addition, jetting failure is likely to be highly visible and it would be possible to take emergency action (e.g. using temporary baffles) to deflect the jet of liquid into the bunded area. In view of this, and given the additional costs involved in building a bund to contain all potential jetting situations, it is recommended that design to contain jetting should be a requirement only on those sites with a moderate or high hazard or risk rating.

The potential for failure through jetting is minimised by:

- keeping primary storage tanks as low as possible
- increasing the height of the bund wall
- building the bund wall as far away from the tank as necessary.

The recommendation given in the preceding section with regard to the location of bunds on sites with a high rating, would remove the possibility of a bund failure through jetting.

It is recommended that bunds for sites with a medium or high hazard or risk rating take account of jetting. Box 10.3 provides a method for calculating the minimum height of a bund wall, or the minimum distance from a tank to a bund wall, to ensure that any discharge through jetting is contained within the bund. Figure 10.7 shows two typical examples of installations that would be vulnerable to jetting failure.

Figure 10.7 *Examples of installations vulnerable to jetting failure*

Box 10.3 *Method for calculating bund geometry to prevent jetting*

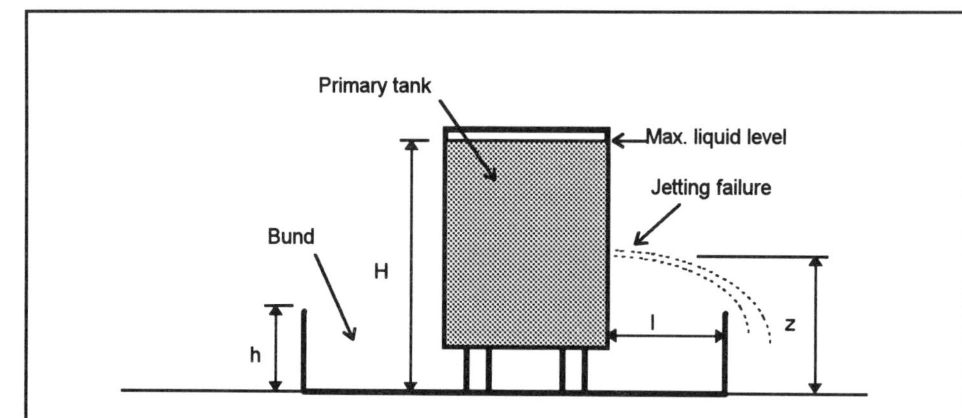

For a small diameter sharp edged discharge orifice, it can be demonstrated that:

$l^2 = 4\, C_v^2 (z - h)(H - z)$

where C_v = coefficient of velocity

In practice, $C_v \cong 0.99$. Assuming $C_v = 1$ leads to the conservative solution:

$l = [4\,(z - h)(H - z)]^{0.5}$

For a given value of h, it may be shown that l is a maximum when

$z = 0.5H + 0.5h$

which leads to the solution:

$\boldsymbol{l_{max} = H - h}$

10.3.4 Drainage and leakage detection

This section considers arrangements for drainage within bunds, detection of leakage from primary containment within bunds, and drainage from bunds.

Drainage within bunds

Drainage within a bund should be provided to:

- collect any liquid that enters the bund (e.g. rainwater), ready for disposal
- drain spilled material away from the immediate vicinity of the primary tank to another part of the bund where it will be less of a hazard and easier to deal with. This is particularly relevant to larger multi-tank sites where flammable or volatile liquids are stored.

On small bunds, the floor should be laid to a slight cross fall of approximately 1% to prevent any rainwater or leakage from the primary tank from ponding. A sump in the bund floor makes emptying easier, but makes construction more difficult. A slot sump (i.e. one running the whole length or width of a bund) is generally easier to empty and clean than a pocket sump.

In the case of extensive bunds it is not normally practicable to provide a fall across the whole site and, in any case, this would be undesirable since it could encourage the spread of escaped liquid across the whole site. Where extensive bunds are compartmented it may be possible to incorporate a sump within the compartment (the cover of such a sump may be seen in Figure 10.6) but care must be taken over its construction to ensure that it is not a potential source of leakage. Where oil may enter a bund, care should be taken to ensure that, through suitable design of the collection sump and the pump, emulsification during pumping is minimised.

Leakage detection

It is important to be able to detect if there is leakage from primary containment tanks so that remedial action may be taken. If a tank is supported clear of the floor of a bund, any leakage should be relatively easy to detect. However, if the base of the tank rests directly on the bund floor, any leakage is likely to go undetected until there is sufficient to seep out from beneath the tank and form a visible accumulation in the bunded area. Even then, the leakage could be mistaken in some circumstances for rainwater. Also, if there was a defect in the bund floor in the area beneath the primary tank, any leakage could escape to the environment without detection.

It is therefore recommended that in all cases where there is insufficient clearance between the base of a primary liquid storage tank and the bund floor, a means of detecting leakage should be installed. Where the site hazard or risk rating is assessed as *moderate*, leakage detection is *strongly* recommended. Where the hazard or risk rating is *high*, leakage detection is considered essential.

The design of systems for leakage detection from primary containment is outside the scope of this report. The American Petroleum Institute (1988) provides guidance on leakage detection systems for large petroleum tanks and a number of systems recommended by the API are illustrated in Appendix A7.

Drainage from bunds

Provision must be made to empty rainwater and other liquids from bunds using mobile or fixed pumps. It is recommended that these are switched manually. It is recommended that bunds should not be equipped with means for gravity discharge, even if lockable valves are provided, unless the bund is part of a properly designed *combined* system.

As it is probable that any rainwater in a bund will become contaminated it is recommended, therefore, that it should be routinely sampled and analysed so that it may be disposed of in an appropriate manner.

10.3.5 Pipework and associated equipment

Piercing the walls or floor of a bund, particularly for pipework, introduces a source of potential leakage, and should therefore be avoided (with the exception of overflow pipes - see below) unless there is no practical alternative. Routing pipework over the top of the bund wall, rather than through it, is regarded as essential for *Class 3* containment and is highly recommended for *Class 2*.

Figures 10.8 and 10.9 illustrate good and poor practice respectively.

Where bunds may be required to retain flammable liquids which are less dense than water, they should incorporate overflow arrangements which, in the event of the bund capacity being exceeded (e.g. by fire fighting water) will prevent burning liquid spilling over and thereby spreading the fire to other parts of the site. This is illustrated in Figure 10.10. In this situation it is recommended that the overflow pipework does pierce the bund but only in the freeboard zone which would normally be above the level of liquid.

Figure 10.8 *Pipework diverted over a bund wall* (Reproduced courtesy of GATX)

Figure 10.9 *Example of practice to be avoided - pipework piercing bund wall*

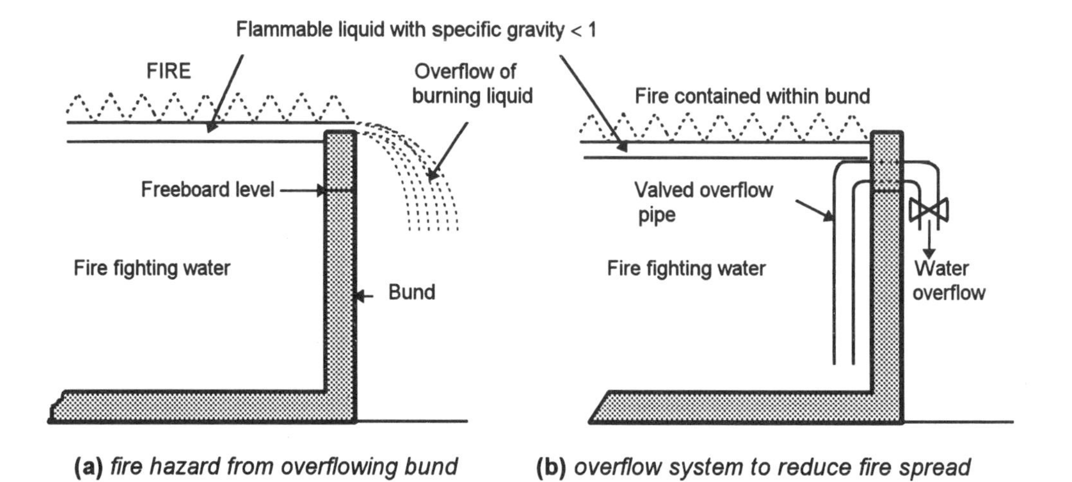

(a) *fire hazard from overflowing bund* **(b)** *overflow system to reduce fire spread*

Figure 10.10 *Bund overflow arrangements to inhibit spread of fire*

Pipework, pumps, valves and associated equipment are frequently the elements of a plant most vulnerable to failure and leakage. In many situations it may be sensible, therefore, to include this sort of equipment within the bunded area, having due regard for health and safety issues (e.g. the need for spark suppression on pumps in bunds which may be subject to build-up of flammable vapours) and operability. On sensitive sites a solution could be to bund individual tanks with a *Class 3* bund and, in addition, to bund the whole site with a *Class 1* bund to protect against the potentially lower volume leakage from pipework.

10.3.6 Summary of general arrangement recommendations

Table 10.1 summarises the foregoing general arrangement recommendations for bunds, differentiating where appropriate between *Classes* of containment. Figure 10.11 illustrates the recommended general arrangement for a bund for a typical small fuel tank situation.

Table 10.1 *Summary of general arrangement recommendations*

		Containment Class		
	Recommendation	***Class 1***	***Class 2***	***Class 3***
(a)	provide not less than 750 mm clearance between primary tank and bund walls for maintenance access	desirable	strongly recommended	essential
(b)	system to detect leakage from primary tank in situations where not practicable to provide clearance between base of tank and bund	desirable	strongly recommended	essential
(c)	no structure within bund to be closer than its own height to the bund wall	not necessary	desirable	strongly recommended
(d)	pumps*, valves, couplings, delivery nozzles and other items associated with the operation of a primary container to be located inside the bund	desirable	recommended	strongly recommended
(e)	pipework associated with primary tanks to be routed over the top of bund wall rather than through it	desirable	strongly recommended	essential
(f)	no provision for rainwater draw-off via a valved outlet in bund wall	essential	essential	essential
(g)	vents from primary tanks directed vertically downwards into bund	essential	essential	essential
(h)	take account of possible *jetting* failure	not necessary	desirable	essential
(i)	dimension bund walls to take account of surge or provide feature to mitigate it (e.g. baffles)	not necessary	desirable	essential
(j)	avoid complicated 'footprint' shapes	desirable	strongly recommended	strongly recommended

* health and safety implications must be taken into account where pumps operate in bunds where flammable vapour may collect.

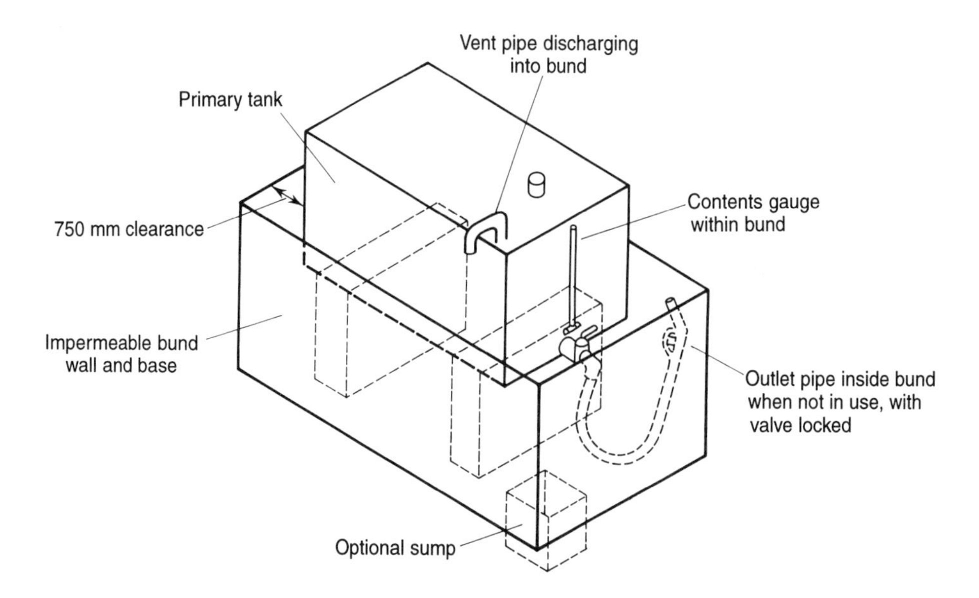

Figure 10.11 *Recommended general arrangement for bund around small fuel tank*

10.3.7 Capacity

Detailed guidance on calculating the required capacity of bunds is presented in Section 9 and summarised in Table 9.16. It should be noted that this recommends moving away from the traditional 110% rule for calculating bund capacity and adopting instead a more rigorous approach based on estimates of primary storage inventory, rainwater, fire fighting and cooling water, and dynamic effects. Further guidance on dealing with dynamic effects, principally surge, is given in Section 10.3.1.

In calculating the required capacity for a bund, it is recommended that no distinction is made between *Class 1*, *Class 2* and *Class 3*.

10.3.8 Retention period

Any liquid that has collected in a bund following an incident should be removed as soon as possible to minimise the possibility of subsequent leakage (from the bund) or damage to the bund caused by aggressive materials which were part of the primary storage inventory. Where an incident involves only an individual storage tank, or a discrete area of a plant, it should be possible to make provision for emptying the bund (and of course safe disposal of the contents) within a few days. In the case of major incidents, however, where there are likely to be competing demands on the clean-up operation, emptying and disposal of the contents of a bund may take far longer.

It is recommended, therefore, that bunds are designed so that they are capable of retaining the primary storage contents for a period of not less than eight days. This has particular implications for the selection of building materials or the choice of protective coatings where aggressive materials are to be contained.

10.3.9 Impermeability

It is impossible to make a bund completely impermeable or 'watertight' and the performance specification should ideally include recommended levels of impermeability. The impermeability of a bund is a function of:

- the intrinsic porosity of the material(s) used in its construction (e.g. concrete, earth, steel)
- the way in which the bund is designed and constructed or fabricated using those materials.

In practice, the latter is usually the most significant in the context of bund design, where the concern is short-term retention of liquids. Within limits, materials' porosity is normally only an important consideration in circumstances involving longer-term storage.

It would be impracticable (a) to build a bund to a precisely specified level of impermeability and (b) to measure whether that impermeability had been achieved. Consequently, with the exception of soil structures, this report does not include a performance criterion for impermeability but prescribes instead approaches to design and construction which will ensure an adequate level of impermeability.

For bunds which are prefabricated using steel sheet, or built *in-situ* using concrete or masonry, adequate impermeability will be achieved so long as they are designed and constructed in accordance with the British Standards and Codes of Practice and the other recommendations given in Sections 10.4, 10.5 and 10.7.

Where bunds are constructed using earth, the porosity of the material (soil) is an important factor since it can be sufficiently high to allow significant leakage through the works. While such leakage may present a direct pollution threat, even more important is the damage it can cause to the integrity of the earth structure through internal erosion, which could lead to a sudden collapse. For this reason the performance criterion for earth bunds includes a requirement for an impermeability equivalent to that provided by a layer of soil not less than 1 m thick with a permeability coefficient not exceeding 10^{-9} m per second. This is considered further in Section 11.2.

Impermeability testing

It is recommended that bunds up to 25 m^3 capacity and constructed from concrete or masonry, are tested for leakage as described in Box 10.4. Testing of prefabricated bunds is dealt with in Section 10.7.4. For larger bunds, testing by filling with water becomes increasingly impractical in terms of dealing with both the supply and disposal of the quantity of water involved. These difficulties should be discussed with the appropriate Regulators in the light of the particular circumstances.

Box 10.4 *Testing impermeability of small bunds (i.e. < 25 m^3 capacity) constructed from concrete or masonry*

1. On a dry, cool day the bund should be filled to brimful capacity (i.e. until the bund begins to overflow) with water containing a marker dye.

2. The bund should be covered (to reduce evaporation) and left for six hours after which there should be no drop in the water level.

3. If the bund is found not to be watertight then it should be drained and treated as described in Section 5.6 below.

4. The test should be repeated until the bund is found to be watertight.

British Standard 8007 (BSI 1987a) contains additional guidance on the testing of structures designed to contain aqueous liquids.

10.3.10 Strength

Bunds should be designed to withstand:

- the static and dynamic loads that would be exerted by the escape of liquid in the event of the failure of the primary containment
- the weight of the primary containment when filled with liquid, and any other forces arising from activities carried out within the bunded area, acting on the base of the bund. (Note: it is not permitted to support the primary tank on the bund walls, for example on joists spanning across the top of the walls, since the primary tank and the bund would then be structurally connected. If the tank was supported in this way, any impact on the bund may have an adverse effect on it, possibly causing
- it to fall)
- wind loading
- stresses induced by ground conditions, for example, differential settlement
- thermal and shrinkage stresses.

These loadings and actions are considered in more detail below.

Hydrostatic loads

The hydrostatic load should be calculated using the formula:

$$p = 9.81 \text{ x } sg \text{ x } h$$

where p = horizontal pressure on the wall of the bund (kN/m^2)
sg = specific gravity of the liquid that could enter the bund
h = depth below surface of liquid to point under consideration (m)

The specific gravity should be taken as not less than 1.0, even where the contained liquid has a lower specific gravity. This is to allow for the possibility that during an incident a bund will be filled with fire fighting or cooling water.

(Note: The designer must ensure that any primary storage vessels are adequately secured to prevent floatation in the event of a bund filling with water.)

Hydrodynamic loads

The sudden failure of a primary liquid storage tank can result in a wave or surge of liquid across the bunded area. At the same time, because its mass is far less than its liquid contents, the ruptured tank will be propelled in the opposite direction to the main release. There are therefore two loading components to be considered:

- the hydrodynamic force of the wave of liquid hitting the bund wall
- the impact, on the bund wall and possibly other primary tanks, of debris from the ruptured primary tank.

These loads are difficult to quantify. Wilkinson (1991) has reviewed work on model experiments that predict peak hydrodynamic pressures of up to six times the hydrostatic head in collar bunds (see Figure 10.12) and later theoretical work which predicts three times hydrostatic head. It is reported that experiments in which the annular width between the primary container and the bund was varied from 1 to 3.5 m had little effect on the dynamic pressures. The impact force of any debris from the damaged primary tank will clearly be heavily dependent on the form of construction of the primary container and the nature of the incident.

No published information has been found on the dynamic pressures on shallow bunds but it is probable that these will be considerably lower than for collar bunds since energy will be dissipated as the wave of liquid has further to travel. In addition, the wave may overtop a shallow bund, further reducing the dynamic pressure.

It is concluded that there is insufficient information at present to give recommendations on the loads on collar bunds. It follows from this that it is not possible to predict with sufficient confidence how a collar bund will react in an incident and this is reflected in Table 10.1 which relates the general arrangement of bunds to their classification.

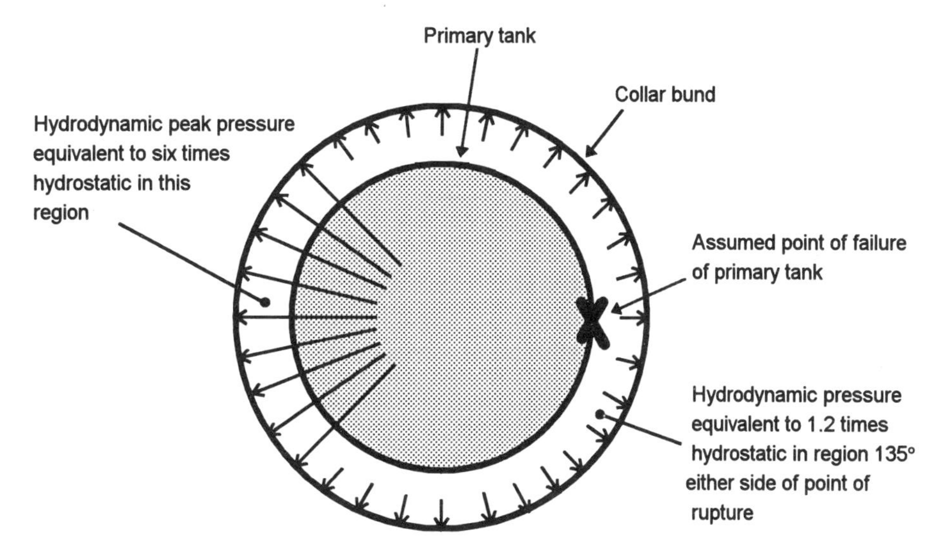

Figure 10.12 *Hydrodynamic loads on collar bunds* (after Wilkinson, 1991)

So far as shallow bunds are concerned, it is suggested that hydrodynamic forces may be taken into account by designing the walls for a hydrostatic head calculated in relation to the bund height inclusive of the recommended freeboard (250 mm) for dynamic effects. On a typical bund wall designed for 1.3 m depth of liquid, the inclusion of a 250 mm freeboard in the pressure calculations results in an increase in thrust of 36% and an increase of 69% in bending moment at the foot of the wall (Box 10.5).

Box 10.5 *Calculation of hydrodynamic loads on bund walls (excluding collar bunds)*

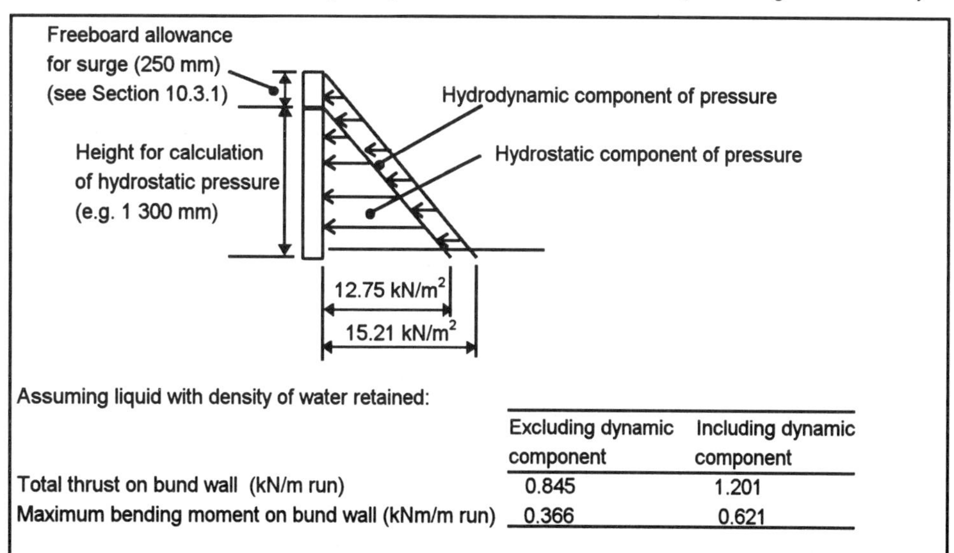

Assuming liquid with density of water retained:

	Excluding dynamic component	Including dynamic component
Total thrust on bund wall (kN/m run)	0.845	1.201
Maximum bending moment on bund wall (kNm/m run)	0.366	0.621

Hydrodynamic pressures may be reduced significantly by subdividing a large bunded area with a number of internal bunds to slow down any wave of liquid.

Water from fire brigade hoses

The force of water issuing from fire brigade hoses is unlikely to be a significant issue except in the case of bunds with earth walls. Methods for stabilising earth banks to resist these and other loads are considered in Section 11.

Wind loading

Wind is likely to be a significant component of loading only on collar bunds, where the recommendations of CP3 Chapter V, Part 2 (BSI, 1972a) should be followed. Collar bunds should be designed for a 50-year return period loading, making the assumption that the annular space between the primary tank and the bund is empty. Where the annular space is open to the atmosphere, particular attention should be paid to wind suction forces.

Stresses induced by differential settlement

Uneven ground bearing pressure can result in differential settlement which may cause cracking and leakage in bunds. It is important to assess the likelihood of differential settlement, particularly across large sites, and to ensure that the forces are properly taken into account in the design.

10.3.11 Durability

A bund must be capable of withstanding:

- weather: in most situations a bund wall will be exposed to the weather on both faces and must be designed accordingly (e.g. 'severe' exposure as defined in BS 5628 (BSI, 1992, 1985) for masonry). Where bunds are located on industrial sites where corrosive materials are handled, the effects of atmospheric corrosion should be taken into account
- aggressive materials present in the ground: the most frequent problem is with naturally occurring sulphates, which attack concrete (see Section 10.4). As a consequence of previous industrial use the ground may be contaminated with other materials which are harmful to some construction materials. Ground conditions should be assessed and appropriate precautions taken
- abrasion: floors of bunds may be subjected to traffic abrasion where materials are moved around within the bund. Heavily loaded fork lift trucks can be particularly damaging. Surface treatments and finishes should be designed accordingly. Where surfaces have become eroded, crack detection is much more difficult
- fire: a bund must be able to withstand the effects of a fire of the anticipated maximum duration and intensity, without collapsing or leaking
- material which escapes from the primary storage: a bund must be able to resist the effects of damaging materials which escape from the primary containment, without collapsing or leaking, for the specified retention period (normally eight days).

Appendix A6 lists the chemical, fire and weathering resisting properties of a range of protective coatings that are available for protecting concrete and masonry.

10.3.12 Structural independence

The purpose of a bund is to prevent the escape of material in the event of the failure of containment facilities within it. To ensure its effectiveness, a bund must be built so that it acts independently of the primary storage. If it does not, any action that results in failure of the primary containment may also cause failure of the bund. An example of this would be where a bund is attached to a primary tank with struts in order to improve its stability. A typical incident scenario could involve the bund suffering damage as a result of accidental impact by a passing vehicle and, consequently, the strut piercing the primary tank. The resulting simultaneous damage to both containers clearly creates a pollution threat. An alternative scenario would be for the primary tank to become overstressed from, say, a build-up of vapour pressure, and for the struts to transfer those stresses to the bund. Again, there would be a possibility of damage to both containers simultaneously.

Although it would be possible to reduce the risk of external impact damage to a bund by erecting bollards or other physical protection, it should be recognised that each such supplementary safeguard added to make a system sufficiently safe introduces an additional level of uncertainty and risk in terms of the performance of the system as a whole.

Similar arguments apply to support arrangements where primary containers and bunds are supported above the ground. Separate supports are essential to ensure that the failure of one set of supports does not cause the collapse of both the primary container and the bund.

Although, it is not possible to avoid altogether the risk that the failure of the primary container will affect the bund (e.g. it may simply fall on it), or *vice versa*, structural independence minimises that risk significantly.

10.3.13 Accessibility

Adequate accessibility is important for three reasons:

- to permit visual monitoring for leakage from the primary containment
- to allow inspection of the inside face of the bund for signs of deterioration
- to facilitate maintenance of the bund.

Where practicable, a minimum clearance of 750 mm should be provided to allow access to the inside face of a bund wall, and 600 mm for access to the floor. Larger storage tanks are usually built off a prepared base resting on the bund floor, so access to the bund floor is not possible. In such cases it is recommended that leakage detection measures are installed to give early warning of leakage from the base of the primary tank (see Section 10.3.4). This is particularly important because it is precisely that area of the bund which, since it cannot be inspected and maintained, is most vulnerable to failure. In the absence of leakage detection, chronic leakage through the base of the tank and through a cracked bund could go unnoticed.

10.4 REINFORCED CONCRETE BUNDS

This section gives general guidance on designing and building bunds using reinforced concrete and presents a number of detailed model designs and a specification.

Designing and constructing impermeable concrete structures requires the exercise of skill and care on the part of the designer and the contractor. A good design executed by a poor contractor is likely to be unsuccessful, as is a poor design, no matter how carefully implemented. This section considers the main factors that affect the quality of concrete in bunds.

There is no single correct way of designing impermeable concrete bunds. Designing impermeable, durable and buildable bunds requires careful consideration of the way concrete as a material behaves, and how concrete structures react to loads, temperature changes, drying shrinkage and differential movement of supports.

10.4.1 Design approach

Reinforced concrete bunds should be designed and built to comply with the requirements of BS 8007 (BSI, 1987b) or *Class 2* or *3* containment, but BS 8110 (BSI, 1985a) which is slightly less onerous may be used for *Class 1* containment. Conformance with these standards will ensure an adequately impermeable bund through:

- specification of concrete mixes that can be well compacted, resulting in low permeability
- specification of details to control structural cracking.

BS 8007 is concerned specifically with water containment. Where aggressive substances may be present, as in many bund situations, additional corrosion protection may need to be considered. Although adequate for *Class 1* containment, concrete bunds built to BS 8110 cannot be expected to achieve the same degree of impermeability as BS 8007 structures; the concrete itself may be more permeable and, more significantly, BS 8110 does not provide details to control structural cracking to prevent liquid leakage.

While the requirements of BS 8110 are well known to most designers, BS 8007 is likely to be less familiar. Some of the main provisions that are specific to BS 8007 are therefore summarised in Box 10.6.

10.4.2 Concrete mix specification

In order to achieve sufficiently impermeable and durable concrete it is essential to ensure that the component materials (cement, aggregate etc.) and mix proportions are specified appropriately.

British Standard mix specification

Concrete mix specifications are covered in detail in BS 5328 (BSI, 1990, 1991). BS 8007 permits only a grade C35A mix, with a cement content between 325 kg/m^3 and 400 kg/m^3 and a maximum water cement ratio of 0.55. For *Class 1* bunds constructed to BS 8110, the mix should be either C40 or C45 depending on the amount of reinforcement cover provided.

Cements and cement content

The types of cement available, and their properties, are listed in Table 10.2. Bunds will usually be constructed using Ordinary Portland Cement (OPC). The other cements are generally used to overcome specific problems associated with the site, the form of construction (e.g. massive sections) and the proposed method of working (e.g. a need to strike formwork quickly).

Box 10.6 *Summary of the main provisions of BS 8007*

Provision	Clause in BS 8007
• explicit assumptions about qualifications and experience of designers and contractors	Foreword
• maximum design surface crack width of 0.2 mm for reinforced concrete	2.2.3.3
• only those aggregates with low or medium coefficients of thermal expansion to be used. Shrinkable aggregates to be avoided	2.6.2.2
• cement content to be the minimum consistent with durability requirements	2.6.2.2
• cements with low rates of heat evolution to be used	2.6.2.2
• concrete to be prevented from drying out	2.6.2.2
• minimum reinforcement area and spacing are prescribed	2.6.2.3
• reinforcement to be as near to surface as is consistent with cover for durability (40 mm minimum)	
• designer to provide maintenance and operation instructions	2.7.2
• methods of providing movement and construction joints, including their locations, are to be prescribed by the designer (this is a fundamental requirement of BS 8007)	Section 5
• waterstops and joint sealing compounds to be used in expansion and contraction joints	Section 5
• concrete to be grade C35A	6.3
• cement content not to exceed 400 kg/m^3 of Ordinary Portland Cement. Other limits specified for partial cement additions	6.3
• water/cement ratio not to exceed 0.55	
• where wall or floors are founded on the ground a 75 mm (minimum) concrete screed of not less than C20 concrete to be placed over the ground	6.6
• completed structure to be tested for impermeability	9.1/9.2

The principal limitation of OPC is its low resistance to attack by sulphates. Where sulphates are likely to be present in the ground or groundwater, sulphate-resisting or supersulphated cements should be used.

Rapid-hardening cements are frequently used in precast works to enable early striking of moulds. They may also be used in cold weather working where their high rate of heat evolution reduces the risk of frost damage. Low-heat Portland cement is used mainly for concrete in massive sections where it is necessary to limit heat build-up in order to reduce thermal stresses.

Blended cements, in which OPC is partially replaced by pozzolanic materials such as ground granulated blastfurnace slag (ggbs), fly ash (pfa) or micro-silica (silica fume), produce generally slower hardening concrete which, it is claimed, has less tendency to early thermal cracking and greater sulphate resistance.

Table 10.2 *Properties of cements* (Source: BRE Digest 325. Reproduced by permission of the Controller of HMSO: Crown Copyright)

Cement type	BS No.	Rate of strength development	Rate of heat evolution	Resistance to sulphates
Main types of Portland cement				
Ordinary (OPC)	12	Medium	Medium	Low
Rapid-hardening (RHPC)	12	High	High	Low
Sulphate-resisting (SRPC)	4027	Low - medium	Low - medium	High - v. high
Other types of Portland cement				
Ultra-high early strength	-	High - very high	High - v. high	Low
Low heat (LHPC)	1370	Low	Low	Medium - high
White	12	Similar to OPC	Similar to OPC	Similar to OPC
Coloured	12	Similar to OPC	Similar to OPC	Similar to OPC
Cements containing blastfurnace slag				
Portland-blastfurnace (PBFC)	146	Low - medium	Low - medium	Low - medium
Low-heat Portland-blastfurnace (LHPBFC)	4246	Low	Low	Medium - high
Supersulphated (SSC)	4248	Medium	Low	High - v. high
Portland pfa	6588	Low - medium	Low - medium	Medium - high
Pozzolanic cement	6610	Low	Low - medium	Medium - high
High-alumina cement (HAC)	915	V. high but will decline	V. high	Low if 'converted'

The total amount of cement in a mix is an important factor in determining impermeability and durability. A higher cement content generally produces a denser, less permeable and more durable concrete. The recommendations for minimum cement content for OPC in BS 5328 range from 220 kg/m^3 for unreinforced concrete in 'mild' exposure conditions to 400 kg/m^3 for reinforced concrete subject to 'most severe' exposure.

High cement content mixes stiffen more quickly and, without admixtures, are more difficult to handle than leaner mixes. The concrete is also more likely to be affected by thermal cracking (owing to its higher heat of hydration) and shrinkage cracking during drying. This is clearly undesirable in structures intended to be impermeable and that is why BS 8007 specifies a maximum cement content of 400 kg/m^3 in addition to the minimum of 325 kg/m^3.

Aggregates

The size, proportion and composition of aggregates all affect concrete properties.

A 20 mm coarse aggregate will be suitable for the construction of most bunds. Larger aggregate may be suitable in some situations depending on the size of section, reinforcement detailing and cover.

Selection and testing of aggregates for concrete is covered in a number of British Standards (BSI, 1973), (BSI, 1975), (BSI, 1983b). Aggregates may consist of crushed or uncrushed naturally occurring rock or crushed concrete. Crushed aggregates would normally be used in higher-strength concrete (>50 N/mm^2). Generally, concretes containing uncrushed aggregates take less water and are easier to work.

In order to achieve dense and impermeable concrete, it is essential that aggregates are clean, well graded and dimensionally and chemically stable.

Water/cement ratio

Water/cement ratio is one of the most important factors affecting concrete durability and impermeability. The relationship between concrete permeability and water/cement ratio is illustrated in Figure 10.13 which shows the importance of minimising the amount of water in a mix.

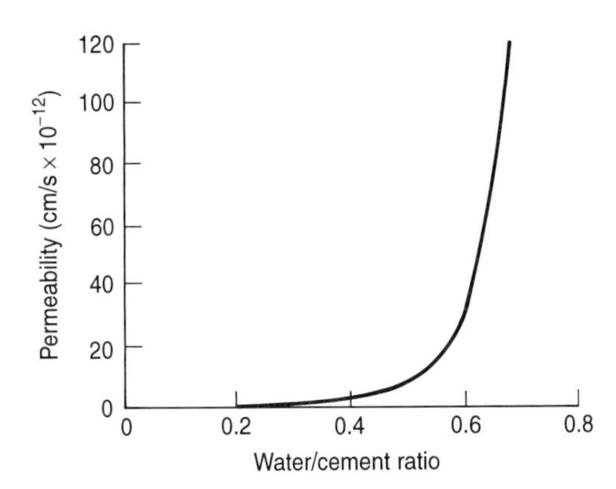

Figure 10.13 *Effect of water/cement ratio on concrete permeability*

Recommendations on maximum water/cement ratios are given in BS 8110 (0.45 to 0.65 according to mix specification) and BS 8007 (0.55). Water/cement ratios may be reduced significantly, without creating workability problems, by incorporating water reducing agents into the mix (see following section). For bund construction to either BS 8110 or BS 8007 a maximum water/cement ratio of 0.55 is recommended.

Admixtures

There is a wide range of admixtures designed to modify the properties of both wet and hardened concrete. The main types, which are listed and described in Box 10.7, are all covered by BS 5075 (BSI, 1982a) apart from integral waterproofers.

Box 10.7 *Admixtures for concrete*

Admixture	Effect and application
Accelerators	Accelerate setting and hardening times. Increase heat of hydration. Used to permit more rapid striking of formwork and also for cold-weather concreting. Accelerators based on calcium chloride must not be used in reinforced concrete work.
Retarders	Delay setting and reduce heat of hydration. Often used in conjunction with water-reducing admixtures. Used to increase the time available for placing and working concrete.
Water reducers (plasticisers)	Reduction in the amount of water needed for a given workability; increased workability for a given water content. Widely used to improve impermeability and durability of concrete.
Super-plasticisers	As water reducers, but higher dosages may be applied to produce free-flowing or very low water/cement ratio concretes.
Air entrainers	Entrain bubbles of air in the concrete during mixing. Permit reduction in water/cement ratio for a given workability. Resistant to the effects of freezing and thawing and to de-icing chemicals.
Integral waterproofers	Lower the porosity of concrete by reducing or blocking the capillary pores. Often used in combination with water-reducing admixtures. Improved durability and chemical resistance is achieved as aggressive agents are not able to permeate the concrete. Microsilica and stearate-based admixtures are included within this group.

Admixtures must be used strictly in accordance with the manufacturers' instructions and particular care must be exercised where it is proposed to use a combination of admixtures. Many admixtures produce more than one effect (e.g. accelerators also increase the heat of hydration) and it is important that all likely effects are taken into account.

10.4.3 Crack control

The achievement of impermeable concrete structures requires cracking in the finished product to be controlled and the elements of the structure to be properly joined. In many structures, deterioration of the concrete, and in the case of bunds consequent potential leakage paths, begins at cracks in the floor slabs and at joints between walls and floors. Cracks may be associated with:

- stresses due to applied loads
- thermal expansion or contraction
- shrinkage as the concrete dries, hardens and cures
- settlement of the concrete in its wet state
- poorly constructed daywork joints
- differential settlement.

Guidance on limiting cracking is given in BS 8110 and BS 8007.

The size and location of cracks induced by applied loads can be predicted reasonably accurately using the methods described in the codes. The size of these cracks can be restricted to within prescribed limits through the design process, or alternatively they can be sealed if there is a danger of leakage or ingress of aggressive substances.

Thermal and shrinkage cracking are more difficult to predict. In the absence of effective crack control measures, cracks of various sizes will occur at random. These are difficult to seal effectively and therefore present a potential leakage path. Aggressive substances entering such cracks can lead to more general and rapid deterioration of the concrete.

There are essentially two ways of controlling the cracks arising from contraction and shrinkage. The first approach is to provide a high percentage of reinforcement. Although more reinforcement does not prevent the occurrence of cracks, it has the effect of encouraging a large number of very fine cracks as against the relatively large number of wider cracks that would otherwise occur. With the correct amount of reinforcement, preferably in the form of closely spaced small diameter bars (rather than larger diameter bars at wider spacings), any cracks will be sufficiently fine not to present a problem. The design of slabs and walls by this method (termed 'fully restrained') should be undertaken only by an experienced engineer following the guidance given in BS 8007 and BS 8110.

The second method of crack control is to induce, or build in, controlled cracks (generally termed 'contraction joints') at predetermined positions in the slab. This method is described in more detail below.

Contraction joints

Contraction joints are provided to relieve the stresses that would otherwise cause uncontrolled cracking. In a reinforced concrete floor to a bund, contraction joints should normally be provided at between 7 and 15 m centres depending on the reinforcement detailing. Cracks may be induced by pressing a crack former into the top surface of the wet concrete, or by using a waterstop incorporating a crack inducer in the underside of the slab (see Box 10.8). A variety of proprietary crack inducers is available for creating such joints.

Box 10.8 *Application of waterstops in typical construction joints* (Reproduced courtesy of the British Cement Association)

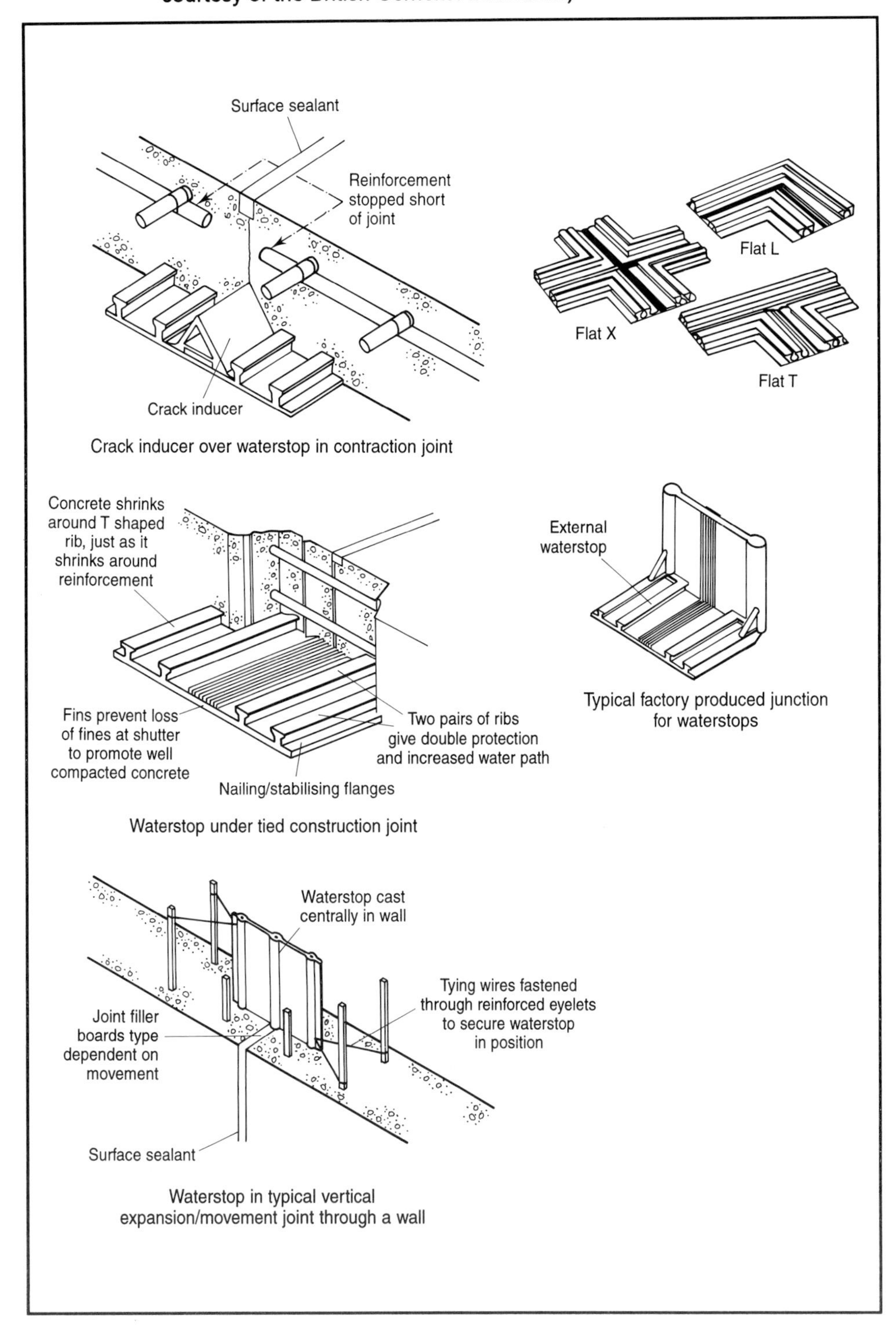

A cross-section through a typical contraction joint is shown at Figure 10.14. Provision of a pre-formed waterstop in the bottom face and a suitable sealant in the groove in the top face are both essential to prevent leakage.

The detail in Figure 10.14 shows the top sealant applied over a 'bond-breaker' to prevent it adhering to the bottom of the groove. Dowel bars should be incorporated in discontinuous contraction joints to prevent out-of-plane movement between the joined sections. One half of each dowel must be coated with a debonding agent so that there is no restraint against in-plane contraction.

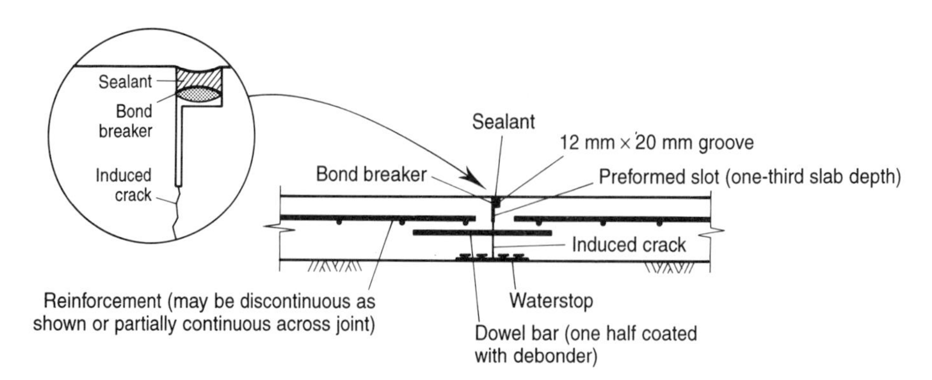

Figure 10.14 *Contraction joint detail*

Expansion joints

Expansion joints should be provided where bund bases butt up to other elements, such as column bases or bund walls, and in long runs of concrete exceeding about 90 m. Figure 10.15 shows a section through a typical expansion joint. As with contraction joints, sealants and waterstops are required to ensure impermeability. In contrast to contraction joints, expansion joints are required irrespective of the amount of reinforcement provided.

Dowel bars should be incorporated in expansion joints, as for contraction joints, but in this case the bars must be fitted with end caps containing a compressible filler.

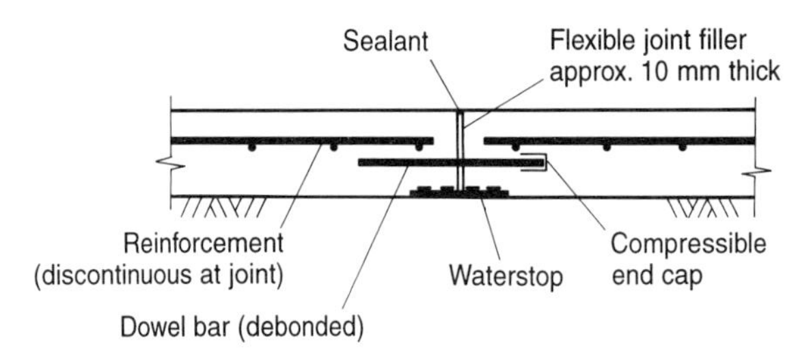

Figure 10.15 *Expansion joint detail*

Construction joints

Properly formed construction joints must be provided wherever concreting work is temporarily discontinued and between areas of concrete that are laid at different times. A section through a typical construction joint is shown at Figure 10.16. In a carefully made construction joint, in which laitance is removed from the face of the first cast before the second is made, a waterstop may not be necessary, but a sealed groove at the top of the joint is recommended. If it is not possible to provide continuity of reinforcement, the joint should be designed as a contraction joint as shown in Figure 10.14.

Slabs should be laid in as short a time as possible to minimise differential shrinkage and thermal strains between sections.

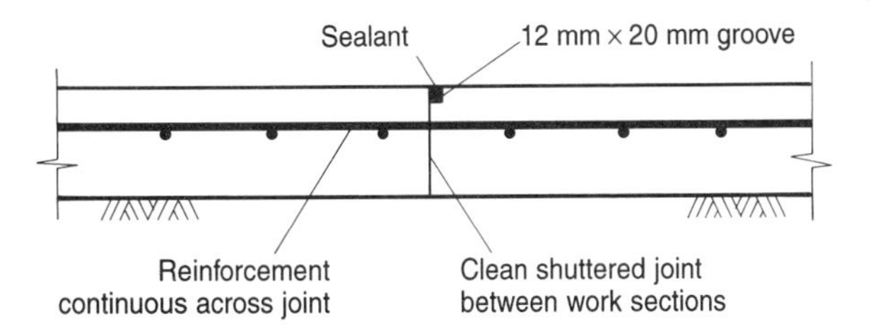

Figure 10.16 *Construction joint detail*

Joint sealants

Careful specification and application of joint sealants is necessary to ensure the satisfactory performance of a bund.

There is a wide variety of sealants available. While some are designed for general-purpose applications others have been developed for very specialised and specific purposes. Sealants differ in physical and chemical composition, methods of application, durability and cost. Guidance on sealant selection and application is given in BS 6213 (BSI, 1992b) and CIRIA Special Publication 80 (CIRIA, 1991).

In order to ensure satisfactory performance of sealants, the manufacturers' instructions should be followed closely. General points to note are as follows:

- concrete receiving sealants should be clean and surface-dry when primed
- recesses and butting faces that are to be sealed should be sound and regular
- the shape of the recess should normally allow for a sealant bead with a width twice its depth (but note that for some types of sealant the bead should be deeper than it is wide)
- a bond breaker is required to prevent adhesion of the sealant to the bottom of the recess
- the sealant should be finished slightly lower than surrounding surfaces in order to avoid mechanical damage
- two-part sealants must be mixed accurately
- surfaces usually require priming using the manufacturers' specified primer to ensure satisfactory adhesion. It is important to apply the sealant within the specified time period
- in bunds, sealants must be adequately supported to resist the hydrostatic pressures likely to occur during an incident.

10.4.4 Reinforcement cover

The high alkalinity of concrete protects embedded steel reinforcement against corrosion. There are three principal causes of possible breakdown of that protection:

- cracks in the concrete allowing water and oxygen, and possibly other corrosive agents, to reach the reinforcement
- the gradual loss of concrete alkalinity due to the permeation of atmospheric carbon dioxide ('carbonation')
- the presence of chlorides in the concrete. The main sources of chlorides are de-icing salts and sea water.

The degree of corrosion protection provided by the concrete is related to the depth of cover to the reinforcement, crack widths and the permeability of the concrete. For bunds designed to either BS 8007 or BS 8110 a cover of not less than 40 mm is recommended.

10.4.5 Surface treatments

Surface treatments may be applied to hardened concrete to provide physical protection or to improve its properties. There are two categories: surface coatings and penetrating sealants. The way they act, and the difference between the two categories, is summarised in Box 10.9.

Surface coatings provide a physical barrier that protects the concrete surface. Their effectiveness in protecting bunds depends on:

- the impermeability of the coating system
- the ability of the coating system to withstand the physical and chemical conditions to which it may be exposed
- the adhesion of the coating to the concrete substrate.

Some coating systems have a relatively short service life (as little as two years in some cases), after which re-coating is required. The service life of each subsequent coating is likely to decrease owing to the gradual deterioration of the concrete substrate.

Box 10.9 *Surface treatments*

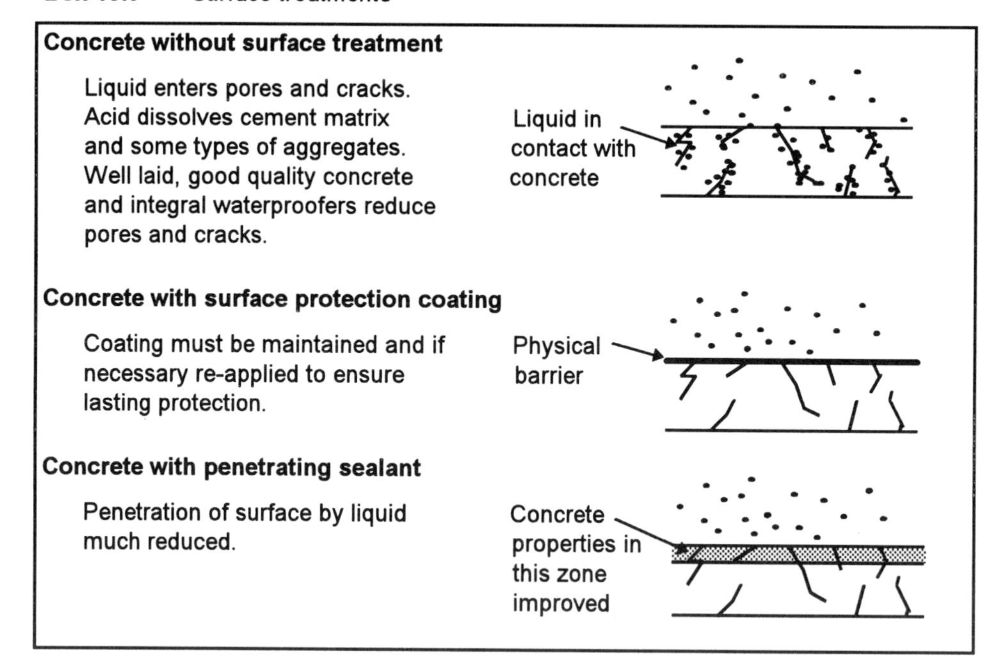

Penetrating sealants act by blocking the surface pores of the concrete to make it less permeable (see Box 10.9). Aggressive agents are prevented from entering the substrate and attack is therefore limited to the surface of the concrete and the rate of attack is substantially reduced. As with surface coatings, penetrating sealants have a limited life and re-application is needed from time to time. Some penetrating sealants are applicable only to new concrete and require the concrete to be artificially dried.

10.4.6 Concrete bund construction

The designer of the works should specify the concrete mix, the thickness of the sections, the reinforcement detailing, joint layout and design, and all other aspects affecting the performance of the finished bund. If anything has been omitted from the specification, the contractor should ensure that the necessary additional information is provided before starting construction. This section of the report draws attention to the principal factors that determine whether or not an acceptable quality of construction will be achieved.

Concrete production

Ready-mixed concrete should be obtained only from depots that are accredited under the Quality Scheme for Ready Mixed Concrete. Where mixing is carried out on site, all batching should be by weight rather than volume.

BS 8000: Part 2 (BSI, 1989,1990) provides useful guidance on mixing and transporting concrete, materials' handling and concreting in adverse weather conditions.

Shuttering and formwork

Shuttering for bund walls must be robust and adequately secured by means of through-ties, braces and props. Figure 10.17 shows a typical arrangement. The design and construction of shuttering is covered in BS 5975 (BSI, 1982b).

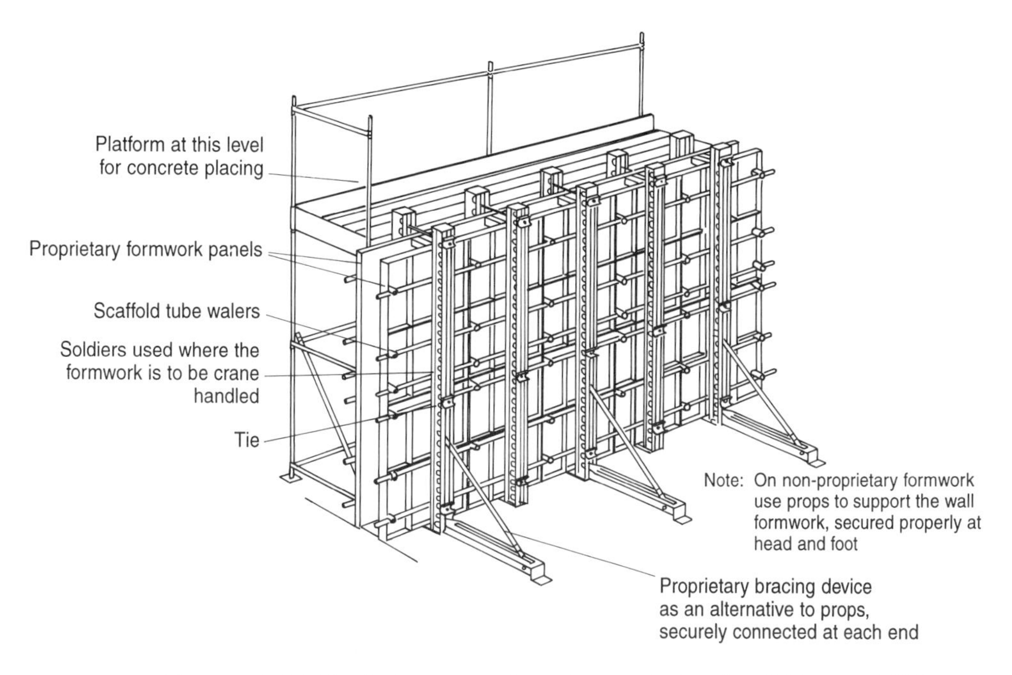

Figure 10.17 *Shuttering for reinforced concrete bund wall* (Source: HSE)

Where the size of a bund makes it practicable, the concrete base should be laid in one continuous pour, requiring formwork to be provided only around the perimeter. Larger bases may be laid using the alternate strip method requiring formwork running the length of each strip. Straight, clean and undamaged roadforms should be used for strip slab shuttering, the top edge providing a smooth and even surface for the vibrating beam and a sharp right-angled arris to the concrete. Roadforms must be pinned securely to the ground to withstand the vibration of machinery used to compact the concrete.

Reinforcement fixing

Reinforcement must be fixed accurately and securely to prevent it being displaced while the concrete is being placed and compacted. Displacement can result in serious structural weakening and/or durability problems. For example, if reinforcement in a 150 mm thick bund wall is fixed with 50 mm cover instead of, say, 40 mm specified by the designer, the section will be approximately 16% less strong.

Waterproof membrane

A waterproof membrane of 1000-gauge polyethylene or similar material should be laid beneath bund floor slabs in order to:

- prevent loss of cement and fines from the concrete mix
- protect the concrete from aggressive ground or groundwater conditions, particularly sulphates
- provide a smooth and regular slip surface to minimise resistance to thermal and shrinkage movements of the bund floor (but see below).

The membrane should be laid on sand blinding to prevent puncture by the underlying material. Adjacent sheets should be joined using either taped or double folded laps to make them waterproof.

Where a bund floor is designed as a fully restrained slab, provision of a slip plane should be avoided. Instead, the bund floor should be cast on top of a lean-mix concrete (Grade GEN2 or ST2) blinding laid on top of a prepared hardcore subgrade without a polyethylene membrane. As a result, the bund floor, blinding layer and subgrade act monolithically.

Waterstops

Waterstops decrease the risk of leakage by increasing the length of the potential leakage path. In order to be effective they must be positioned carefully and joined using the methods specified by the supplier (usually by heat welding using a special jig). The concrete must be thoroughly compacted around them.

Concrete placing and compaction

On exposed sites, one or more of the bund walls should be constructed in advance of the floor slab in order to minimise drying crosswinds.

Concrete should not be placed when the shade air temperature exceeds 30°C, or is less than 5°C, unless special precautions are taken. Low-temperature concreting may require the use of rapid-hardening cements or accelerators and/or insulating the works to prevent frost damage.

The concrete should be placed as quickly as possible. Double-handling should be avoided in order to minimise segregation. On no account should extra water be added to aid workability as this would adversely affect the permeability and durability of the hardened concrete (see Section 10.4.2).

It is essential that the concrete is thoroughly compacted to remove any air pockets and small air bubbles, otherwise the forecast design strength, impermeability and durability of the bund will not be achieved. It is important also to avoid over-compaction, particularly in bund floor construction, where bleeding of fines and water to the surface can lead to an unsatisfactory finish.

Walls should be cast full height in one pour in order to avoid horizontal joints which are a potential leakage path.

Concrete curing

Immediately after compaction, the concrete floor slab should be protected with a curing membrane to slow drying. Over-rapid drying results in a poor-quality surface with reduced strength and durability and can lead to plastic shrinkage cracking.

Impermeable sheet (usually polyethylene) or proprietary spray-on membranes may be used. Sheet membranes must be carefully sealed and securely anchored over the work to prevent the formation of wind tunnels. Where spray-on curing membranes are used, it is important to check that they will not interfere with any surface treatment it is intended to apply once the concrete has hardened.

Joint formation

Provision of construction, contraction and expansion joints as described in Section 10.4.3 is one of the most critical operations, since aggressive agents penetrating joints can cause rapid deterioration of the concrete.

Concrete finishing

In most situations, bund floor slabs should be trowel finished to close surface pores and cracks, making it more difficult for liquids to penetrate.

10.4.7 Maintenance and repair

This section describes the main problems that are likely to affect concrete bunds and the repair techniques that are available.

Defects

Defects may be categorised broadly as cracks, local surface deterioration, and general surface deterioration.

Cracks in a bund can allow corrosive agents to penetrate the concrete. Wide cracks are generally deeper and these may allow corrosive agents to attack the steel reinforcement, causing expansion forces which result in concrete spalling. Deterioration of the concrete then accelerates, leading to a serious loss of strength and durability, and increased permeability.

Cracks may be caused by thermal or shrinkage stresses, or by overloading. In bund floors, cracking may be the result of uneven or inadequate ground support.

Local surface deterioration may be caused by:

- variability in quality of the laid concrete (e.g. poor compaction in a particular area)
- local exposure to aggressive agents (e.g. small spills from a storage tank)
- mechanical abrasion.

Where acidic agents are present, surface damage is usually caused by the loss of cement matrix which, in its early stages, causes the aggregate to be exposed. Eventually the cement matrix may be weakened to the extent that the aggregate falls away, exposing a new surface which is likely to be more porous and thereby accelerating the deterioration.

General surface deterioration may result from exposure to aggressive agents over a wider area, or from inadequate specification and/or construction. It may also be caused by inadequate curing or exposure to frost, abrasion or aggressive agents before the concrete has adequately matured.

Repair techniques

Before repairs can be specified it is necessary first to understand and, if possible, remove the cause(s) of the problem. Where cracking is due to thermal or shrinkage stresses there is little that can be done except to fill them as described in Box 10.10. Where cracks are induced by the applied loadings (e.g. at the foot of the walls of a bund) it is essential to investigate whether they are structurally significant. If they are, it may be necessary to reduce future loadings or strengthen or prop the walls. If a bund floor is damaged as a result of inadequate or uneven support, the subgrade must be made good before it can be repaired, otherwise the problem will recur.

Symptoms and repair techniques are summarised in Box 10.10 for different types of damage. Repairs should be carried out precisely in accordance with the manufacturers' instructions. Failure to undertake the necessary preparatory work properly is a major cause of failure of concrete repairs.

Box 10.10 *Reinforced concrete bund defects and repairs*

Symptom	Investigation and repair
Thermal or shrinkage cracks	Ensure cracks are not structural (particularly in bund walls). Cracks > 0.3 mm: fill by resin injection (Appendix A8). Cracks < 0.3 mm: coat concrete with penetrating sealant.
Structural cracks	Determine whether structural weakening has occurred. Remove cause of distress if possible. Cracks > 0.3 mm: resin injection or patch repair (Appendix A8). Cracks < 0.3 mm: if structurally significant, patch repair. Otherwise coat surface with penetrating sealant.
Bund floor slab settlement	Remove affected area back to sound concrete, leaving clean vertical edge. Remove any weak subgrade material and replace with well-compacted granular material. Using suitable mix, cast new concrete on dpm to level of existing concrete. Use bonding agent to join new concrete to existing. Cure carefully to avoid shrinkage.
Bund floor or wall surface deterioration (local)	Remove affected area back to sound concrete and make good with suitable material. See Appendix A8 for typical situations.
Bund floor surface deterioration (general)	Replace or, if practicable, lay new slab or thin screed of suitable material on top of existing floor.

The techniques of resin injection, patch repairs and thin bonded repairs and a general guide to the suitability of repair materials for particular situations is given in Appendix A8.

10.4.8 Model designs and specification for reinforced concrete bunds

A model design and specification is presented at Appendix A9 for a range of small bunds constructed in reinforced concrete (and masonry, see Section 10.5). It covers bunds up to 7 m wide and in two height ranges; (a) up to 900 mm and (b) between 900 mm and 1 500 mm. Further information on the model design is included in CIRIA Report 163 (Mason *et al*, 1997).

The Institute of Petroleum (IP, 1996b) has also produced model designs for reinforced concrete bunds and these are reproduced at Appendix A10. The main difference between these and the model designs at Appendix A9 is that the latter assumes an integral reinforced concrete bund floor. The IP model designs are for free standing bund walls.

10.5 REINFORCED MASONRY BUNDS

This section gives guidance on designing bunds using reinforced masonry (blockwork and brickwork). Unreinforced masonry is not recommended for bund construction because of its susceptibility to thermal and shrinkage cracking and vulnerability to impact damage.

It is recommended that the use of reinforced blockwork is restricted to bunds for *Class 1* containment only. Reinforced brickwork (other than grouted cavity construction, see below) is not recommended for bunds.

10.5.1 Forms of construction

The forms of construction of reinforced masonry walls and their suitability or otherwise for use as bund walls are summarised in Box 10.11.

Appendix A9 includes a model design and specification for reinforced masonry bunds of structural form (a), i.e. concrete blockwork with reinforced and grouted cores.

Given the relatively large amount of thermal and drying shrinkage movement that occurs in blockwork walls it is necessary to incorporate movement joints at about 7 m intervals. Such movement joints are very difficult to seal effectively and blockwork is not therefore recommended other than for small bunds, i.e. where the walls will be less than 7 m long.

10.5.2 Outline specification for masonry bunds

A detailed specification is given in Appendix A9. Important specific points are summarised below:

(a) for blockwork bunds:

- hollow concrete blocks should have a minimum compressive strength of 10 N/mm^2
- mortar mix should be 'class (i)' to BS 5628 (BSI, 1992, 1985)
- cores of blocks should be thoroughly filled with high workability concrete

- reinforcement should be provided both in the blockwork cores and the bed joint (the latter is to reduce thermal and shrinkage cracking and susceptibility to impact damage)
- the need for movement and construction joints should be avoided wherever possible, even if that means restricting the size of individual bunds
- the inside face of the bund should be rendered with a dense sand/cement render.

Box 10.11 *Forms of masonry wall construction*

Form of construction	Suitability for bund wall construction
(a) Structural concrete blockwork with vertical reinforcement in filled cores	Recommended for Class 1 only. Susceptible to thermal and drying shrinkage.
(b) Concrete blockwork used as facing skin or cladding for reinforced concrete bund wall.	Suitable for all Classes provided that the reinforced concrete wall is properly designed and constructed. Provides protection to concrete in the event of fire.
(c) Structural brickwork with reinforcement in bed joints only.	Not recommended. Insufficient strength to withstand lateral loads from liquid in the event of an incident. Poor resistance to impact damage. Susceptible to thermal and shrinkage cracking.
(d) Brickwork as facing skin as in (b).	Suitable for all Classes as (b).
(e) Grouted cavity brickwork (brickwork acts as permanent shuttering for reinforced concrete, see Figs. 10.18 and 10.19)	Suitable for all Classes provided reinforced concrete in cavity is properly designed. Brickwork skin provides additional protection to concrete.
(f) Pocket wall, Quetta bond wall and vertical slot walls in blockwork or brickwork (see Fig. 10.18)	May be suitable subject to incorporation of sufficient reinforcement and provision of adequate concrete cover. Class 1 only.

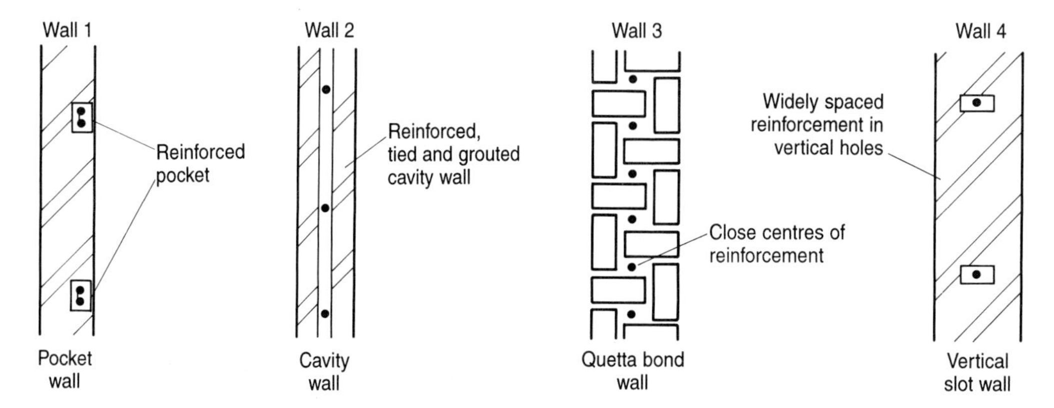

Figure 10.18 *Composite masonry and concrete wall construction*

Figure 10.19 *Grouted cavity brickwork storage tank*

(b) for brickwork bunds:

- bricks should be specified as clay, solid class B engineering bricks to BS 3921 (BSI, 1985c)
- mortars and renders as for blockwork
- avoid the need for joints, as for blockwork.

10.6 EARTH BUNDS

Earth bunds are frequently used on larger tank farm installations. Detailed structural design considerations are covered in Section 11 which deals with earth banked lagoons and other earthworks.

10.7 PREFABRICATED BUNDS

The performance criteria developed for bunds in this report (see Box 10.1) excludes double-skinned tanks and a range of products marketed as 'self-bunding tanks' and 'integrally bunded tanks', largely on the grounds that the inner and outer containers which make up the item are structurally joined or interdependent. Such products are considered briefly in Section 8.2.1, where they are defined as 'enhanced primary containment systems' as distinct from bunds. It should be stressed, however, that the failure of such products to satisfy the definition of a bund does not necessarily make them a less effective option than a conventionally bunded tank.

The following guidance is concerned solely with prefabricated bunds that conform with the performance criteria set out in Box 10.1.

10.7.1 General description

A prefabricated bund is a prefabricated tank, usually constructed from steel or plastics, inside which the primary container is placed. A typical example is shown at Figure 10.20. One-piece prefabricated bunds are available in capacities up to 100 m^3, the maximum size being dictated by the difficulties in transporting anything larger.

Larger bunds may be constructed on site by joining together two or more prefabricated sections. A lid may be provided to keep out rainwater.

Figure 10.20 *Prefabricated bund* (Reproduced courtesy of INTEG Ltd)

10.7.2 Specification and procurement

Prefabricated tanks are manufactured items and their detailed design is outside the scope of this report. The following comments are limited to general guidance.

Prefabricated bunds should comply with the capacity recommendations given in Sections 8, and the requirement for access for inspection, or alternatively provision for leakage detection, as described earlier in this section.

Prefabricated tanks used as bunds should be designed in accordance with the relevant material structural codes (where they exist) to withstand the actions defined in Section 10.3.10. Where there are no relevant structural codes it is recommended that a prospective purchaser should require the supplier to provide evidence, either in the form of independently certificated test results or analyses, that the product is capable of withstanding the actions defined in Section 10.3.10 with an expected service life of 50 years.

A prefabricated bund must not rely on structural linkage with the primary tank for its stability.

10.7.3 Installation

The installation requirements for prefabricated bunds are product- and site-specific. Suppliers should provide the necessary installation instructions to cover a specified range of site situations, and guidance on what to do if the site conditions fall outside that range.

Installation instructions should cover at least the following aspects of delivery and installation:

- loading and unloading
- support requirements
- any ancillary protection requirements
- health and safety requirements and need for notices
- commissioning.

10.7.4 Testing

Where prefabricated bunds are constructed under factory-controlled conditions in accordance with an appropriate British Standard or Code of Practice (for example BS 799: Part 5 (BSI, 1987d)) for tank manufacture, testing for leakage should be in accordance with those Standards or Codes. Where this does not apply, or where the relevant Standards or Codes do not include provision for leakage testing, prefabricated bunds should be tested in the manner described in Section 10.3.9 for *in-situ* bunds, although it is not necessary in this case to allow for a saturation period or to form a notch weir.

10.7.5 Maintenance

The supplier of a prefabricated bund should provide full instructions on inspection and maintenance covering at least the following:

- details of any finishes or other protective measures to be applied at time of installation
- frequency of inspection
- preventative routine maintenance requirements
- damage repair.

Prefabricated bunds should be inspected regularly for signs of damage, deterioration or general wear, and to ensure that nothing has collected in unroofed or uncovered bunds to reduce their effective capacity.

11 Earth banked containment basins (lagoons) and earth bunds

11.1 INTRODUCTION

Where the site topography and the ground and soil conditions are suitable, earth banked containment basins (in this report referred to as lagoons) can provide cost-effective *remote* secondary containment (Figure 11.1). Lagoons may be constructed either above or below the surrounding ground level and formation level is often determined by the economic advantage of balancing cut and fill. Depending on the soil type (particularly its impermeability and stability) and the environmental sensitivity of the site, a lagoon may need to be sealed using an impermeable membrane or other suitable liner to ensure that it does not leak in the event of an incident.

Earth banks can also be used to create bunds around an installation to provide *local* secondary containment (Figure 11.2). The recommendations and guidance that follows apply equally to lagoons and bunds. In the case of bunds, it is assumed that the floor of the bunded area is made adequately impermeable by adopting appropriate pavement and drainage design (see Section 13).

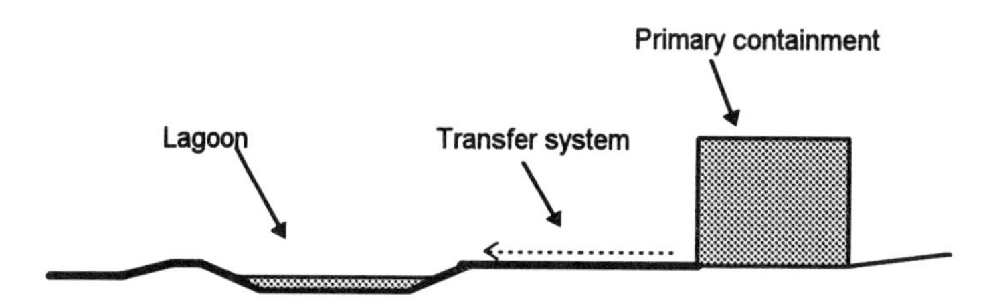

Figure 11.1 *Earth banked lagoon for 'remote' containment*

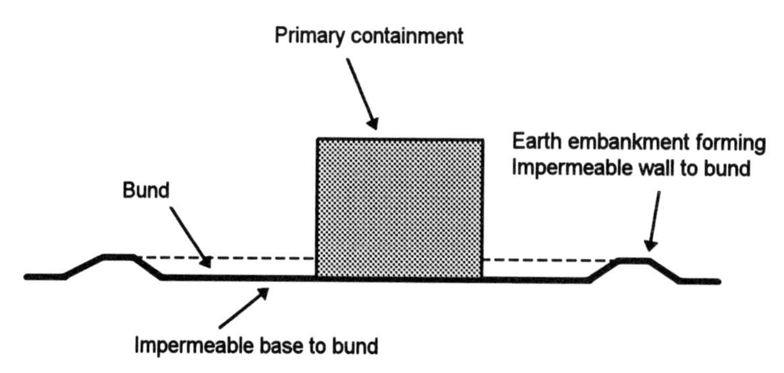

Figure 11.2 *Earth banked bunds for 'local' containment*

Given the variability of soils, even over a small site, and that soil characteristics can change significantly over time, particularly as a result of changes in moisture content, great care must be taken over the design and construction of lagoons and earth bunds to ensure adequate impermeability and reliability. It is strongly recommended, therefore, that design and supervision of the work on site is entrusted to a suitably qualified and experienced civil engineer.

This section of the report sets out design criteria for lagoon and earth bund construction, with the aim of achieving containment systems which are of an equivalent standard to those constructed using concrete and other conventional construction materials. In some situations this will require the use of an impermeable lining to ensure that the necessary level of impermeability is reliably achieved. Advice is given on selecting and installing liners.

The variability and relative unpredictability of soil as a construction material, and the absence of recognised design standards, makes differentiation between *Classes 1*, *2* and *3* construction more difficult than it is with other forms of construction. A simple approach to classification has therefore been adopted, requiring a more thorough soils' analysis and the use of separate impermeable liners for *Class 2* containment, backed up by leakage detection systems to allow periodic testing for *Class 3* containment.

The Reservoirs Act 1975 (HMSO, 1975) requires that reservoirs capable of holding more than 25 000 m^3 of water above the natural level of any part of the land adjoining the reservoir are designed by, and constructed under the supervision of, a qualified civil engineer on one of the Panels constituted under the Act. Periodic inspection by a Panel Engineer is also required. Where a lagoon holds more than 25 000 m^3 of water it may also fall within the provisions of the Act.

11.2 DESIGN CRITERIA FOR LAGOONS AND EARTH BUNDS

This section of the report sets out design criteria that the engineer should try to achieve. Advice on ways to satisfy these criteria is given in following sections. The essential requirements are an acceptable level of impermeability, stability and durability. Although containment basins and bunds should rarely be called upon to retain liquids for long durations, it is essential that the earth structure is sufficiently durable to perform reliably and effectively in the event of an incident occurring some considerable time after construction.

11.2.1 Soil permeability limits

It is not possible to achieve a completely impermeable soil structure and there will be some seepage from any unlined lagoon, however carefully constructed. The rate of seepage through a soil structure may be calculated using Darcy's Law. It can be shown, for example, that for a lagoon containing two metres depth of water, on a 600 mm thick stratum of clay with a permeability coefficient of 10^{-9} m per second, the seepage over a seven-day period would be approximately 2.5 litres per square metre of its surface area. In drawing up guidelines for the construction of *impermeable* containment lagoons it is therefore necessary first to consider the degree of impermeability that would be tolerable.

Current legislation in the United States requires surface impoundment lagoons to be constructed with a one-metre thick clay-type soil with a permeability not exceeding 10^{-9} m per second. Where hazardous liquids are stored long term, the US legislation requires an additional double liner system with provision for leakage detection (US Environmental Protection Agency, 1980).

In the UK, the Environment Agency/SEPA also recommend a minimum one-metre thickness of clay stratum with a maximum permeability of 10^{-9} m per second for natural soil barriers (NRA, 1989). This has become the adopted 'standard' within the UK for the design of landfill containment systems and farm waste lagoons. Even where this specification of clay lining is achieved, there is an increasing trend towards the incorporation of synthetic liners in the works to improve performance still further.

It is recommended that the same 'standard', i.e. a minimum of one metre of soil with a maximum permeability of 10^{-9} m per second, is adopted for the design of lagoons and earth bunds for secondary containment of industrial pollutants.

Typical values for the permeability of soils are shown in Table 11.1 and it is clear from this that only soils with a high clay content will give the required level of impermeability.

Table 11.1 *Permeability of soils by broad category*

Soil type	Coefficient of permeability (m/s)	Relative permeability
Coarse gravel	exceeds 10^{-3}	high
Sand	10^{-3} - 10^{-7}	medium to low
Silt	10^{-7} - 10^{-9}	very low
Clay	less than 10^{-9}	impervious

11.2.2 Stability

Lagoons and bund embankments should be designed and constructed to meet the following stability criteria:

- embankments to be accessible to, and capable of withstanding the loads from, machinery and vehicles used during maintenance and emptying operations
- embankments to remain stable during rapid draw-down and filling
- embankments to be capable of withstanding the erosion that would be associated with a 50-year return period rainstorm
- embankments to be capable of withstanding erosion by any fire fighting water likely to be used in the event of an incident, or wave action due to wind
- no reliance should be placed on short-life (defined here as less than 20 years) impermeable liners to provide or improve embankment stability.

11.2.3 Durability

Lagoons and earth bunds should be designed for a durability life of 20 years subject to normal routine maintenance. In assessing durability life, it should be assumed that the lagoon or earth bund remains empty for the whole period and that earthworks and any lining systems incorporated in the works will be exposed to the weather.

11.2.4 Classification of containment

As there are no British Standards or Codes of Practice covering earth structures for retaining liquids, it is not possible to define the differences between *Class 1*, *Class 2* and *Class 3* containment construction in terms of modifications to 'normal' standards. Nevertheless, it is still important to try to differentiate between the classes to reflect the range of environmental sensitivities that exist and thereby ensure that the most appropriate measures are put in place at each site.

In the following sections, advice is given first on constructing **unlined** lagoons and earth bunds, differentiating between *Class 1*, *Class 2* and *Class 3* containment in terms of soils' testing and the approach to embankment design. In addition, it is recommended that extra precautions are taken for *Class 2* and *Class 3* containment as follows:

- *Class 2* lagoon - the floor and banks should be lined with a suitable impermeable liner (see Section 11.8).
- *Class 3* lagoon - as *Class 2* lagoon except that, in addition, a suitable leakage detection system should be installed beneath the lagoon in order to allow periodic monitoring of impermeability (see Section 11.10).

These requirements are illustrated in Figure 11.3.

Class 1

Class 2

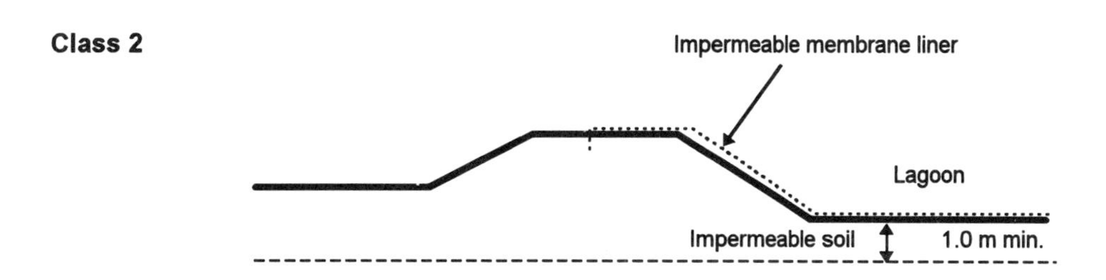

Class 3

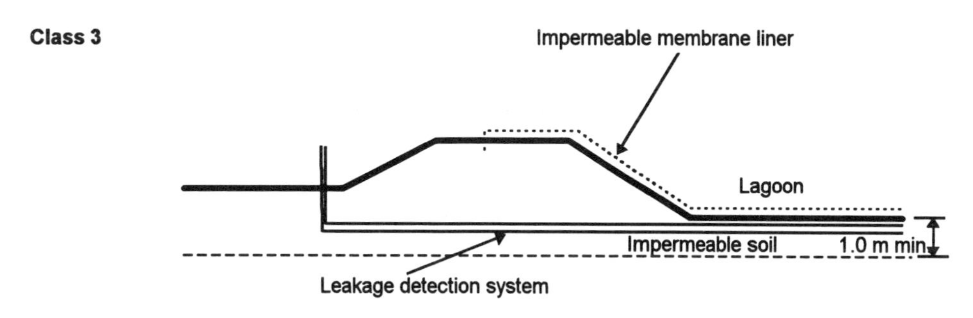

Figure 11.3 *Classification of lagoons*

It should be stressed that in all cases the ground and soil conditions should comply with the impermeability, stability and durability criteria set out above. Impermeable linings and leakage detection systems are required *in addition* to there being adequate ground and soil conditions, rather than to compensate for any inadequacies. If the recommended ground and soil conditions cannot be achieved, a lagoon should not be considered for secondary containment unless a properly designed structural liner is installed, in which case the lagoon becomes in effect a below-ground tank.

11.3 CAPACITY

The capacity of a lagoon or earth bund should be calculated in accordance with the recommendations given in Section 8.

11.4 LOCATION OF LAGOONS IN RELATION TO RECEPTORS

In the absence of specific controls governing the siting of containment lagoons in relation to controlled waters, it is recommended that subject to obtaining Environment Agency/ SEPA approval for a particular site, the requirements included in the Farm Waste Regulations (HMSO, 1991) are adopted. These Regulations require waste containment facilities to be sited no less than 10 m from any controlled water that an escaping pollutant could enter.

11.5 SITE INVESTIGATION

A detailed ground survey should be carried out in accordance with the recommendations of BS 5930 (BSI, 1981a) and BS 6031 (BSI, 1981b). Such a survey will normally involve a desk study, reconnaissance of the site and an investigation of the soils.

11.5.1 Desk study

The importance of carrying out a thorough desk study in the early stages of planning the works cannot be overstated. A lagoon or earth banked bund may cover a large site area and it is essential that any features which could affect performance are identified, for example drains or other services running beneath the site, and that appropriate design precautions are taken. A study of maps and plans of the site, particularly older documents, and discussions with the Local Authority and local utility providers may reveal features which are no longer apparent, such as infilled drains and wells, dried up streams and old building lines. Given their purpose, lagoons and earth bunds will often be constructed on, or close to, congested industrial sites which have supported a variety of different industries and processes, probably over many years.

In mining areas, enquiries should be made about past and present mining operations and any existing, or predicted, problems with subsidence locally.

11.5.2 Site reconnaissance

Site reconnaissance entails a methodical physical inspection of the site to identify features that could cause construction difficulties or longer-term reliability problems. As with the desk study, reconnaissance is particularly important for earthworks since their performance relies on the ground conditions across the site being as predicted. Topography is an important factor, not only in terms of facilitating drainage into the lagoon or bund, and other operational considerations, but also in relation to the possibility of soil creep, a process which can occur on slopes greater than about 1 in 10 and which may be evidenced by sloping walls, trees and fences.

Abrupt changes in local topography may indicate changes in ground type and consequential variations in soil settlement characteristics, strength and permeability across the site.

Vegetation is an important indicator of soil types and groundwater levels. Reeds and willows, for example, indicate a high water table (which could make a lined lagoon impracticable), whereas bracken and gorse usually indicate a well-drained soil with a low water table. As with topography, abrupt changes in vegetation may mean significant changes in ground characteristics and engineering properties.

11.5.3 Soils' investigation

Soil is the building element for lagoons and earth bunds. As with other structures which are built using, for example, concrete or steel, it is impossible to predict the performance of the completed structure without a thorough knowledge of the building element.

In the case of lagoons and earth bunds, two important characteristics of the soil govern whether *in-situ* soil can be used (with or without modification, e.g. by incorporation of admixtures, or compaction) or whether soil with different engineering properties will need to be imported. They are:

- **permeability** - can the recommended maximum permeability of 10^{-9} m per second be achieved economically over a one metre thick layer of soil?
- **stability** - will the soil remain stable under changing conditions (particularly moisture content changes) and can it be formed into a stable embankment that will satisfy the criteria set out at Section 11.2.2?

Assessment of these two factors requires careful analysis backed by judgement based on experience, and should therefore be entrusted only to a suitably qualified and experienced civil engineer or soils' expert. The following two sub-sections deal with permeability and stability (linked also to durability), make recommendations on methods for testing soils (differentiating between *Classes 1*, *2* and *3* containment) and summarise the test methods.

11.6 SOIL PERMEABILITY

11.6.1 Permeability assessment

There are several approaches to assessing soil permeability, ranging from quick and simple methods used in making approximate preliminary evaluations, through to more sophisticated laboratory and field testing methods. The method(s) chosen by the engineer will depend on the nature of the soil, the extent of the works and the sensitivity of the site in terms of pollution control. The engineer must take into account whether *in-situ* soils are to be used, and whether these soils will be remoulded during construction or, alternatively, whether soil with more suitable characteristics will need to be imported.

Laboratory tests are carried out on small samples of soil and the engineer must exercise skill and judgement in selecting a suitable number of representative samples. Guidance on sampling soils is given in BS 5930 (BSI, 1981a). It is essential also that the condition of the samples when tested, particularly their moisture content and consolidation state, is representative of the actual conditions on site.

Direct measurement of the permeability of clay soils with permeabilities down to 10^{-9} m per second is difficult, given the time it takes for water to percolate through the test sample. Permeability varies significantly with the degree of consolidation of the soil. If it is measured using laboratory samples, it is important to ensure that disturbance of the soil during sampling and sample preparation does not introduce errors. Some clay soils are susceptible to fissuring on drying and, where this happens, measurements of the impermeability of the in-tact soil will be of little value and may even be misleading.

The permeability of clay soils may be evaluated using long established empirical relationships between permeability and clay content. In situations where an approximate assessment is sufficient, this approach may be used instead of direct testing.

It has been found from experience that the most suitable soils for constructing impermeable embankments and lagoons contain generally between 20 and 30% clay, the remaining fraction being well-drained sand and gravel. Soils of this type are likely to remain stable even when subject to significant changes in moisture content. Soils with a clay content much below 20% are likely to exceed the recommended permeability limit of 10^{-9} m per second. On the other hand, soils with a clay content much higher than 30% are likely to be difficult to form into a stable embankment and they will have a greater tendency to shrink and crack on drying.

Empirical and direct assessments of permeability are both subject to error and uncertainty. It is therefore recommended that on moderate or high hazard/risk sites both methods are used together in making the assessment.

Recommendations on testing soils and assessing permeability are summarised in Table 11.2 which differentiates between the methods to be used for the three classes of containment. The methods are described in more detail in Appendices A11 to A14.

Table 11.2 *Recommendations for assessing soil permeability*

Containment Class	Method for assessing soil permeability	Acceptance criteria
1	(a) make preliminary assessment using *settlement* or *hand-texturing* methods (Appendices A11, A12), and	clay content* in the range 10% to 35%
	(b) classify soils according to the British Soil Classification System (BSCS) (Appendix A13).	soils to be in Casagrande groups CL, CI or CH
2	as (a) and (b) above and, in addition:	as above
	(c) direct measurement using *falling-head cell* method ensuring that samples reflect conditions of soil in completed structure (Appendix A14).	permeability coefficient $< 10^{-9}$ m per second
3	as (a), (b) and (c) above and, in addition:	as above
	(d) where floor of a lagoon is to comprise *in-situ* soil, field measurement of permeability using *falling head cell* method (Appendix A14).	permeability coefficient $< 10^{-9}$ m per second

* *less than 0.002 mm particle size.*

Where the permeability of the soil on site is found to be too high it may be possible, depending on the type of soil, to reduce it to a satisfactory level by consolidation (see Section 11.6.2). Alternatively, it may be possible to modify the naturally occurring soil by blending it with imported clay-rich soils or bentonite. However, this can be costly to carry out and, in circumstances where this is seen as the only way to achieve the required impermeability, a lagoon or earth bund may not be the most economical solution.

11.6.2 Effect of soil consolidation

The impermeability (and shear strength) of a clay soil can be improved by consolidation. During consolidation, the voids between the particles in the soil mass are reduced in size, making it more difficult for water to percolate. The maximum density that can be achieved through consolidation is related to the moisture content of the soil and compactive effort applied. In practice the compactive effort relates to the type and weight of compaction machinery used, the thickness of the layers in which the soil is placed, and the number of passes of the compacting machinery. For each soil and level of compactive effort there is an optimum moisture content for the achievement of maximum density and a slightly higher moisture content for the achievement of minimum permeability. At moisture contents below the optimum, a clay soil becomes increasingly stiff and an increasing amount of compactive effort is required to break down the soil structure. Conversely, at moisture contents above the optimum, a clay soil becomes more difficult to work owing to the build-up of pressure in the water-filled pores.

The optimum moisture content for a particular soil is established in the laboratory using the Proctor compaction test described in BS 1377 (BSI, 1990b). Figure 11.4 illustrates the relationships between soil moisture content and compactive effort, and the resultant soil density and permeability.

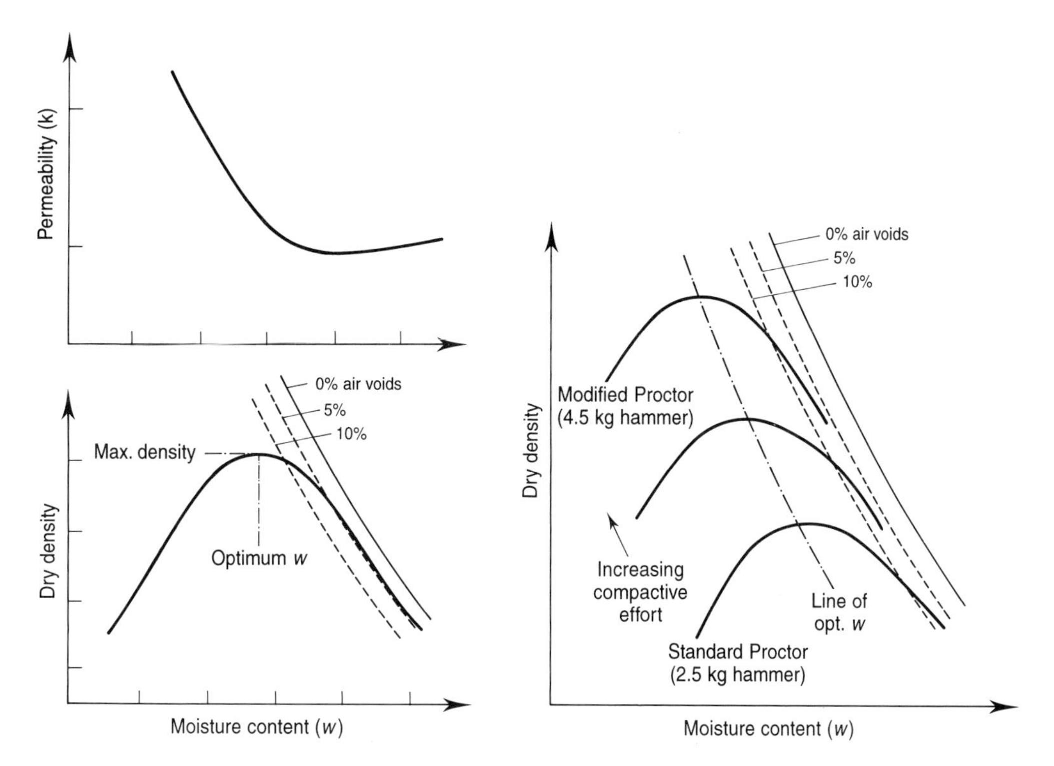

Figure 11.4 *Relationship between soil moisture content and compactive effort, and soil density and permeability*

Where it is necessary to consolidate *in-situ* soils to achieve the required level of impermeability, Proctor tests should be carried out in accordance with BS 1377 to determine either (a) the minimum permeability that can be achieved given the natural moisture content of the soil on site, or (b) the extent to which the moisture content of the soil on site must be changed in order to be able to achieve the required level of impermeability.

11.7 DESIGN AND CONSTRUCTION

11.7.1 Design

It is recommended that earth embankments forming part of a *Class 3* lagoon or bund are designed in accordance with the stability requirements specified in BS 6031, Section 6 ('Cuttings') or Section 7 ('Embankments and general filling') (BSI, 1981b), as appropriate.

Section 6 of BS 6031 covers in some detail:

- factors governing the stability of slopes in cuttings
- the behaviour of cohesive soils
- the selection of parameters of soils for assessment of slope stability
- modes of failure of cuttings and natural slopes
- methods of analysis of stability of slopes
- design.

Section 7 covers:

- general factors affecting the design of embankments
- strength and deformation characteristics of fill material
- special site conditions affecting embankment design, including soft ground and sloping sites
- suitability of materials.

BS 6031 is concerned with the design of earthworks in general, rather than liquid retaining structures, and the engineer must make appropriate allowances for the hydraulic loads which will arise in the event that a lagoon or earth bund fills with liquid (perhaps very rapidly) as a result of an incident.

As an alternative to BS 6031, embankments up to a maximum height of 5 m above lagoon or bund floor level forming part of a *Class 1* or *2* containment system may be constructed in accordance with the cross-sections shown in Figure 11.5 and embankment widths shown in Table 11.3, and the recommendations on construction given in Section 11.7.2.

Table 11.3 *Embankment widths for Class 1 and 2 containment structures*

Bank height (m)	Minimum top width (m)
2	2.5
3	2.75
4	3.0
5	3.25

The embankment profiles shown in Figure 11.5 are suitable for lagoons or bunds which are partly excavated below the surrounding ground level, giving economies in construction and a reduced external embankment height.

The following assumptions are implicit in the designs shown in Figure 11.5:

- the site is level
- the subgrade is sufficiently strong to support and prevent displacement of the embankment
- embankment and foundation soils are homogeneous throughout and have the required strength and impermeabilty

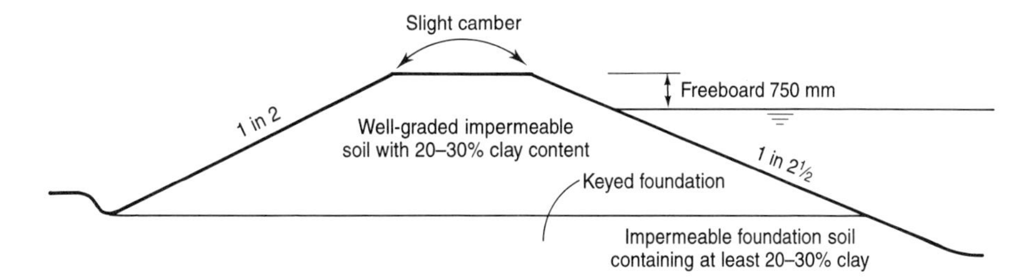

(a) Embankment on impermeable soil

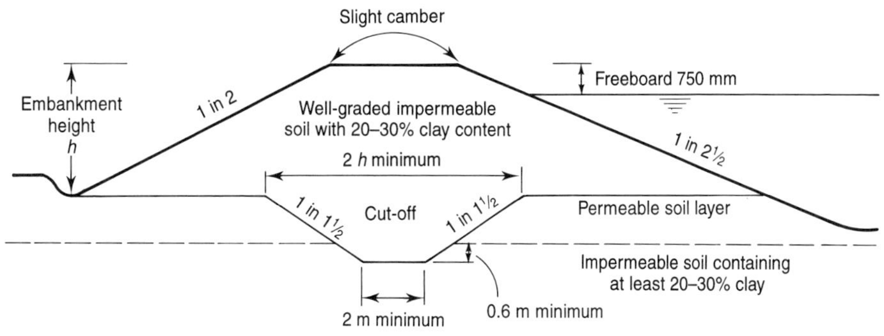

(b) Embankment on permeable soil with 'cut-off'

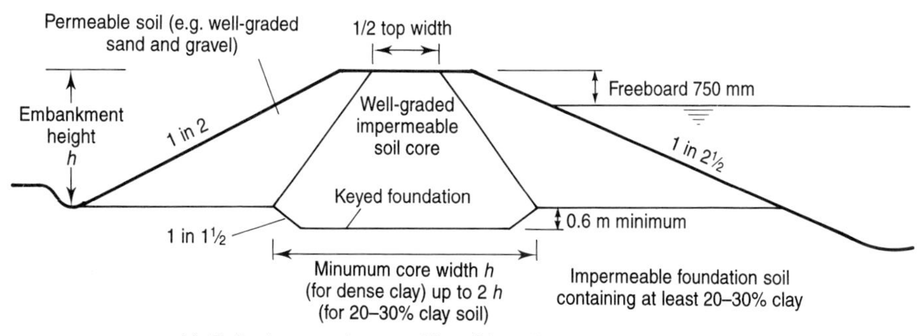

(c) Embankment on impermeable soil (zoned construction)

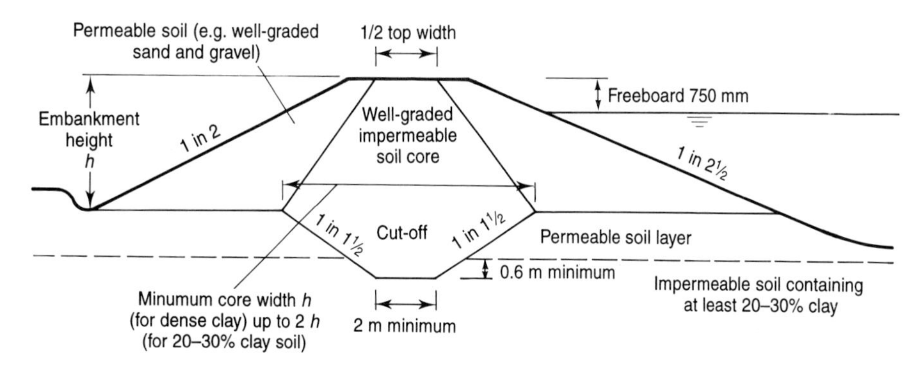

(d) Embankment on permeable soil with 'cut-off' (zoned construction)

Figure 11.5 *Embankment construction for Class 1 and 2 containment*

- the compacted soils making up the embankment have a dry density of at least 18 kN/m^3, an undrained shear strength of at least 50 kN/m^2 and an effective angle of shearing resistance of at least 28 degrees
- there is a freeboard of not less than 750 mm below the top of the embankment
- regular maintenance is carried out, including attention to surface cracking, erosion and damage by burrowing animals etc.

11.7.2 Construction

Site preparation

For lagoons:

- sites should be carefully cleared of all debris, vegetation and top soil, the latter being set aside for re-use
- soft spots or pockets where the soil type or condition have the potential to create areas of higher permeability should be excavated and filled with soil of the same parameters as elsewhere on the site
- drains which pass under the site or within 10 m of it should be either stopped and sealed, or diverted. Culverting drains which pass beneath a site is not recommended as an acceptable alternative.

For earth bund embankments:

- the site of the embankment should be cleared and prepared as for lagoons
- unless cut-off trenches are to be constructed, the embankment site should be loosened to a depth of approximately 300 mm before the first layer of embankment fill is placed.

Placing and compaction of impermeable fill material

The engineer should specify the required moisture content of the fill material, both for embankment construction and to provide the lagoon base, and the method of compaction to achieve the necessary impermeability. The Department of Transport *Manual of Contract Documents for Highway Works* (HMSO, 1991) provides comprehensive guidance on placing and compacting a range of different soil types.

So far as possible, each layer of soil should be placed and compacted along the entire length of the embankment or section of lagoon base, in one continuous process. This is to avoid creating discontinuities which could lead to differential settlement and areas of weakness and potential leakage.

Embankment and lagoon base protection

Unless they are to be covered with a hard pavement or a separate membrane lining, embankment sides and tops should be covered evenly with 150 mm of topsoil and seeded with grass to prevent erosion. Creeping bent and rough stalk meadow grasses are suitable for this purpose.

On no account should shrubs or trees be allowed to establish on embankments, where they would seriously impair stability, or lagoon floors where they could create a leakage path.

The banks and base of a lagoon should be adequately protected against scouring at the points where site drains discharge and where, in an emergency, fire fighting water would be likely to enter.

11.8 EMBANKMENT AND LAGOON LININGS

One of the main differences in the recommendations for *Class 1* and *Classes 2* and *3* lagoon and bund construction is that the latter should both incorporate impermeable membrane linings. This section provides advice on the selection and installation of linings.

11.8.1 Types of liner and lining systems

The various types of liner and lining system include:

- liners of natural clay, or a mix of sandy soil and bentonite
- asphaltic concrete and asphalt
- reinforced bitumen membranes and geotextiles
- geosynthetic clay liners (GCL)
- geomembranes and geomembrane/geotextile composites.

Compacted natural clay liners and soils with bentonite admixtures

In deciding the overall geometry of the lagoon and the embankment slopes, consideration should be given to the strength and stability requirements of the liner and natural soil interface. The coefficient of permeability of the compacted clay liner should not exceed 10^{-9} m per second.

Sodium montmorillonite is the most common form of bentonite. Bentonite has hydrophilic properties and in contact with water it swells to give a low strength impermeable clay. Bentonite may be mixed or blended *in-situ* with the naturally occurring soil. Mixing may be carried out using a rotovator, but a cement mixer usually provides better mixing and consistency. In order to achieve the required impermeability, well graded soils require about 10% sodium bentonite by dry weight rising to 15% or more for uniformly graded sands.

The admixed soil is placed in a uniform layer over the lagoon or embankment to form a natural clay lining system.

Close control over mixing, placing and compaction is essential to achieving an effective impermeable lining using natural clays and bentonite/soil mixtures. Proper placing and compaction of clay liners may be particularly difficult on lagoon embankments. For these reasons, geosynthetic clay liners are more commonly used.

Asphaltic concrete and asphalt

Asphalt is not commonly used for lining lagoons. Its use as a surfacing material on hardstandings and trafficked areas to provide impermeable catchment areas, is discussed in Section 13.5.

Reinforced bitumen membranes

Bitumen reinforced with a nylon fabric provides a high strength membrane. The product is easy to repair and join. It also has good UV resistance, thereby reducing the need for UV protection by covering with soil or other material.

Geosynthetic clay liners

Geosynthetic clay liners (GCLs) are composite liners made from a geotextile or geomembrane which is bonded with a layer of unhydrated bentonite or polymer. GCLs are manufactured in panels measuring approximately 5 m by 30 m which may be effectively joined on site by lapping to make larger sheets.

GCLs are normally covered with a layer of overburden, typically 300 mm of soil or granular fill, to prevent desiccation and to provide protection from mechanical damage and UV deterioration. The relatively low shear strength of unhydrated bentonite limits the use of GCLs on embankments to slopes not exceeding about 1 in 3. Some of the stated advantages of GCL over natural clay liners include:

- factory controlled manufacture
- relatively simple and rapid installation
- less affected by poor weather during construction
- they exhibit some self-sealing properties.

Geotextiles used in GCLs include woven and non-woven synthetics, and woven natural materials such as jute. An additional geomembrane is also incorporated into some bentonite GCLs. The terms 'reinforced' and 'non-reinforced' which are applied to GCLs refer to the method of bonding. In the former, the bentonite is mechanically bonded by stitching or needle punching. In the latter it is bonded using an adhesive. The bentonite content is usually about 4 to 5 kg/m^2.

Geomembranes

Geomembranes are manufactured from either thermoplastic or thermosetting polymers or combinations of the two. Commonly used geomembranes include: polyethylene (VLDPE, (i.e. very low density etc.), LDPE, MDPE, HDPE); polyvinyl chloride (PVC); chlorinated polyethylene (CPE); polyamides (a form of thermoplastic); butyl or isoprene isobutylene (IIR), ethylene propylenediene monomer (EPDM) and polychloroprene (neoprene) - which are all thermosetting polymers; and butyl rubber-PVC-nitrile rubber (PE-EPDM), PVC-ethyl vinyl acetate, cross linked CPE, and chlorosulphonated polyethylene (CSPE) - which are all combined polymers. The recommended minimum thickness of these unreinforced membranes is 0.75 mm.

Box 11.1 summarises the characteristics and performance of commonly used geomembranes.

Geomembranes and their jointing methods must be sufficiently robust to prevent rupture or puncture when in use. Soft ground, voids, gaps and sharp angularities in the sub-base may cause the geomembrane to deform and to rupture or puncture when under load from, for example, a head of water in a lagoon. For this reason geomembranes should be laid slack to help prevent puncture when loaded. Soils eroded by rain during construction, forming soft areas at the base of slopes, can be a potential cause of failure of geomembranes if not rectified.

Box 11.1 *Typical performance characteristics of commonly used geomembranes*

Description	Performance and characteristics
Polyethylene (LDPE and HDPE)	Resistant to many chemicals, good strength, seams heat welded, good durability, good flexibility, poor puncture resistance, needs protection from UV - carbon black additive helps performance. Minimum recommended thickness for seaming 1.5 mm. (HDPE most commonly used geomembrane).
Polyvinyl chloride (PVC)	Seams solvent welded, high strength but poor resistance to UV, weather, ozone, and sulphides. Affected by temperature. Easy to work.
Chlorinated polyethylene (CPE)	Strength good, poor resistance to chemicals and oil, poor seam performance, good performance at low temperature, reasonably resistant to UV and weather.
Ethylene propylenediene monomer (EPDM)	Reasonable UV and weather resistance, high strength, poor resistance to hydrocarbons and solvents, good low-temperature resistance, poor seam performance.
Butyl rubber	Low strength, affected by hydrocarbons e.g. oils and petroleum, resistant to UV and weather, difficult to seam, temperature performance good.
Chlorosulphonated polyethylene (CSPE)	Fairly resistant to acids, oils, some chemicals and bacteria, some resistance to UV, and weather, reasonable performance at low temperature, difficult to seam, low strength.

Compared with geotextiles, most geomembranes have relatively poor strength characteristics. Both, however, exhibit visco-elastic properties and are relatively inelastic in engineering terms. Loading can produce permanent deformations, a reduction in membrane thickness and a consequent reduction in performance. Owing to their relatively high strength, geotextiles are commonly used with geomembranes to provide reinforcement, anchorage, soil stabilisation and additional puncture resistance, and to help provide a filtration mechanism in leakage detection systems. Geotextiles are an essential component of many lining systems.

The principal properties of geomembranes and geotextiles that need to be considered when selecting a suitable liner are considered in Appendix A15. BS 6906 (BSI, 1987-1991) Parts 1 to 8, describes the test methods which are used to determine the strength and other engineering properties of geotextiles.

A British Standard covering geotextile design is currently being developed. A Eurocode, EC7 (Building Research Establishment, 1995), which will include geotechnical design and general rules, is also in course of development. EC7 will categorise geotechnical structures into three classes which are expected to be comparable in principle with the three classifications proposed in Section 7 of this report.

Pending publication of these standards, it is proposed that the method presented in Appendix A16 is adopted for the design and use of geosynthetic impermeable liners where empirical rules may not be applicable. Further guidance is given in (Jewell, 1996).

11.8.2 Anchorage and protection

The need to anchor membranes securely and to protect them from mechanical and UV damage has been mentioned above. Figure 11.6 illustrates a suitable method.

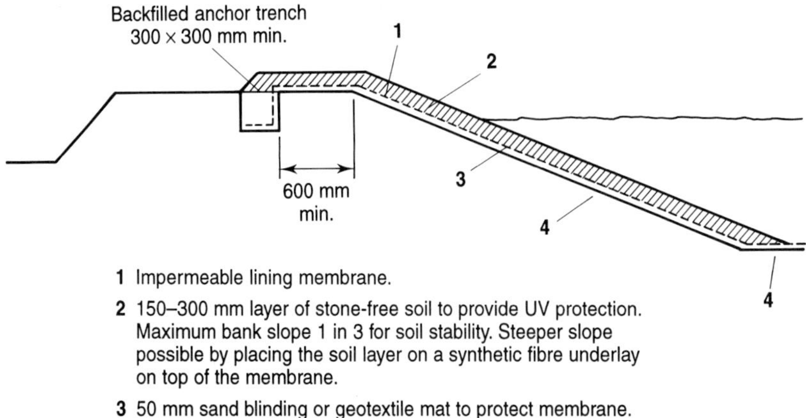

1 Impermeable lining membrane.
2 150–300 mm layer of stone-free soil to provide UV protection. Maximum bank slope 1 in 3 for soil stability. Steeper slope possible by placing the soil layer on a synthetic fibre underlay on top of the membrane.
3 50 mm sand blinding or geotextile mat to protect membrane.
4 Embankment and base soil treated to prevent weed growth.

Figure 11.6 *Typical arrangement for protecting and anchoring a membrane liner*

11.9 PIPE ENTRIES THROUGH EMBANKMENTS

Unless it is unavoidable, embankments should not be penetrated below the design liquid surface level. Where it is necessary for a pipe to penetrate an embankment, anti-seepage collars should be provided at a spacing of not more than 10 times the pipe diameter. The collar increases the length of the potential seepage pathway and thereby minimises any leakage.

Unavoidable pipe entries through liners should be properly made and positioned so that they can be readily maintained (see Figure 11.7).

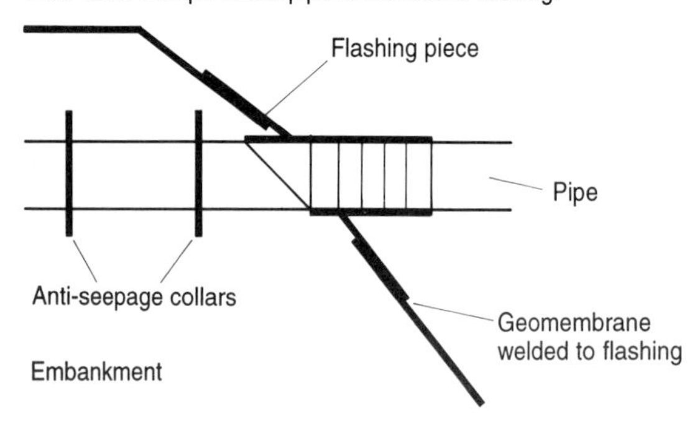

Figure 11.7 *Arrangement for sealing pipe penetrating a geomembrane liner*

11.10 LEAKAGE DETECTION FOR *CLASS 3* LAGOONS AND BUNDS

Leakage detection is a recommended requirement for *Class 3* containment lagoons so that they may be routinely tested for impermeability.

Several systems are currently used for leakage detection including the following:

- underdrains
- findrains
- groundwater monitoring by boreholes
- electronic groundwater monitoring by conductivity measurement.

Figure 11.8 shows schematically the arrangements for a detection system installed beneath an impermeable liner. Any leakage through the liner is intercepted by the underlying sand/ gravel layer and runs through the drains in the free-draining layer to monitoring points.

11.11 MAINTENANCE

Embankments must to be sufficiently large and structurally stable to allow for movement of mechanical excavators which may be used for maintenance of the banks. Concrete or hard-core ramps may be required for access into bunded areas and concrete headwalls may be needed for pipe or channel inlets and outfalls. The construction of these elements must be carried out carefully to protect the integrity of impermeable liners, particularly at the joints between concrete and liner. Granular overburden is often provided both to protect a liner and to allow access for machinery.

Regular inspection and maintenance helps prevent damage to liners from rodents and other burrowing animals. High tensile steel wire reinforced geosynthetics are available for protection against such attack. Alternatively, fine mesh wire netting may be used to protect the liner.

Maintenance of erosion damage caused by weathering and surface wave action may be minimised by the use of revetments or *rip rap* placed on the embankment slope.

Damage to liners, usually in the form of puncturing, bursting or tearing, can be minimised by ensuring that the site is properly prepared and that the liner is placed carefully. All sharp objects must be removed from the sub-base prior to laying the liner. Sufficient slack should be allowed for ground deformations and the loading conditions. Liners should be weighted to prevent billowing. Seams should preferably be laid in the direction of the slope of an embankment. In order to prevent UV deterioration of sensitive geotextiles and membranes, overburden should be placed on top of the liner as soon as possible after laying.

Depending on the nature and extent of any site damage, and the type of liner, it may be possible to make site repairs by patching. Instructions on repairs should be provided by the liner manufacturer or supplier. On sites with a high hazard or risk rating it is recommended that the entire damaged section is replaced with undamaged material. Where a liner has deteriorated generally through age, it should be replaced entirely.

Systematic regular inspections should be carried out. Vulnerable areas requiring special attention during maintenance inspections include:

- pipe entries, particularly if they penetrate the embankment
- the foot of embankments
- joints and discontinuities, particularly between dissimilar materials.

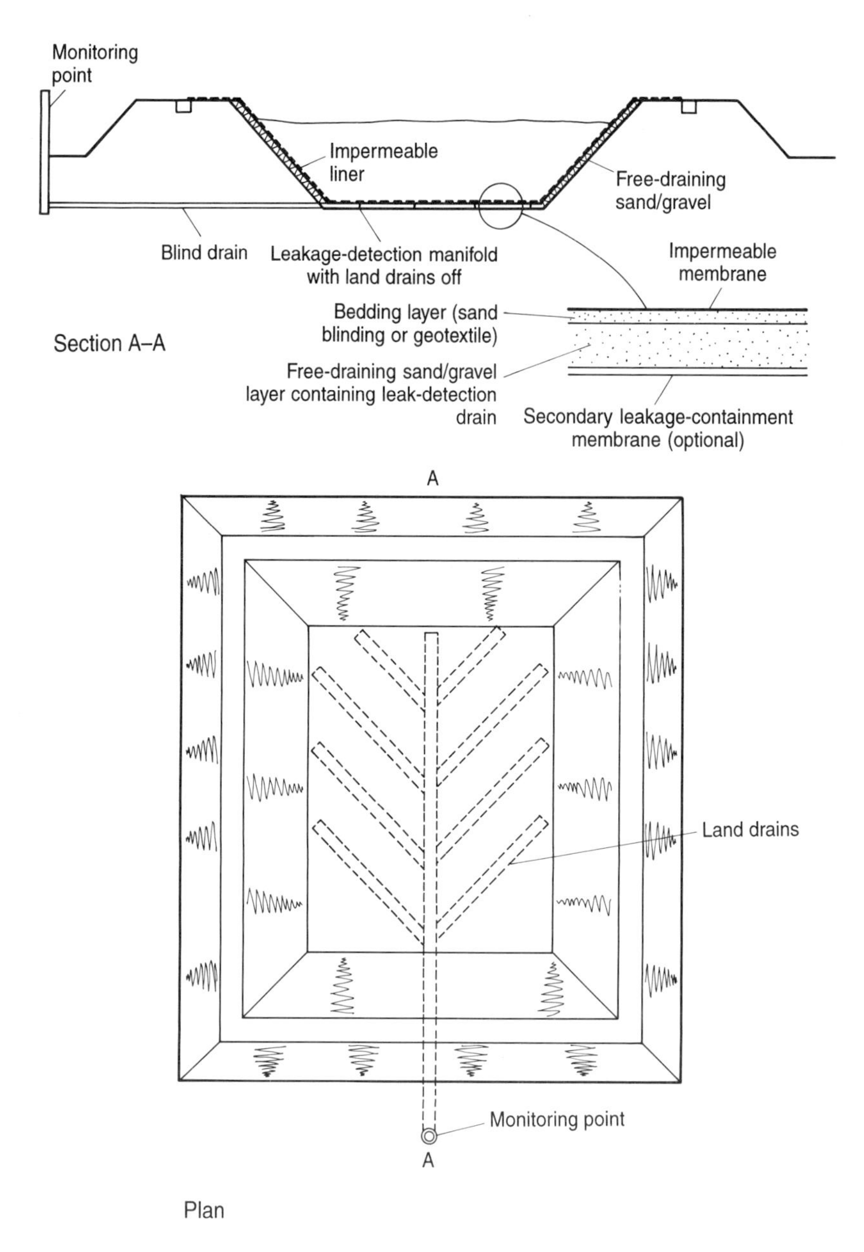

Figure 11.8 *Leakage detection system for a lined lagoon*

11.12 SAFETY

Safety measures should include:

- notices warning of the danger and type of hazard presented by the lagoon and the substances it may contain
- secure fencing and other barriers to prevent unauthorised entry
- means of escape and rescue of persons.

Signs should be erected prominently along each side of the lagoon, and conform to the Safety Signs Regulations (HMSO, 1980) and Hazardous Substances Regulations (HMSO, 1988). Fences should be sufficiently high and constructed to prevent people from climbing over. Gates should be lockable.

12 Secondary containment tanks

12.1 INTRODUCTION

Most tank design standards do not deal specifically with the containment of industrial or trade effluents or secondary containment. The few exceptions include tanks for agricultural wastes and sewage. However, the UK standards for liquid storage tanks and vessels are high and many of these tanks are eminently suitable for use as secondary containment. This section summarises the principal design features and specifications and their relevance to containment use.

Secondary containment in the form of cavity wall and base construction (i.e. double-skin) is sometimes built into primary containment steel tanks and vessels (e.g. to BS 5500 (BSI,1996)) to control leakage, and for safety reasons. The false bottom of such tanks is fitted with leakproof inspection hatches enabling access for maintenance purposes. Each skin of a double-skin tank should be designed to withstand the same loading as a single-skin tank, with consideration given to additional pressures which may arise from a sudden rupture of the internal skin. Other construction systems include specially designed impermeable foundations which provide for leakage monitoring, interception and collection.

Factors to be taken into account in selecting a secondary containment tank include:

- ground conditions
- topography
- access provision
- site location
- site classification
- proximity to ground or surface water
- pumping and drainage requirements
- pathway catchment characteristics
- health and safety requirements.

Above-ground tanks may be chosen in preference to below-ground tanks where limited space is available, or where the presence of groundwater or the ground conditions make a below-ground tank uneconomical. Above-ground tanks may be used at many installations. Figure 12.1 illustrates typical arrangements for above-ground tanks supplied either by pumping or by gravity.

Given the wide range of materials that tanks may be required to contain, it is not possible to generalise on appropriate internal or external protective systems. These will be subject to the specific circumstances of the site. The site classification, the retention time, quantity and nature of the material stored will all influence selection of the tank type, size, design standards and protective finishes.

Freeboard

It is recommended that a freeboard of not less than 300 mm is provided as buffer capacity. No overflows should be permitted within the freeboard depth. Tanks that

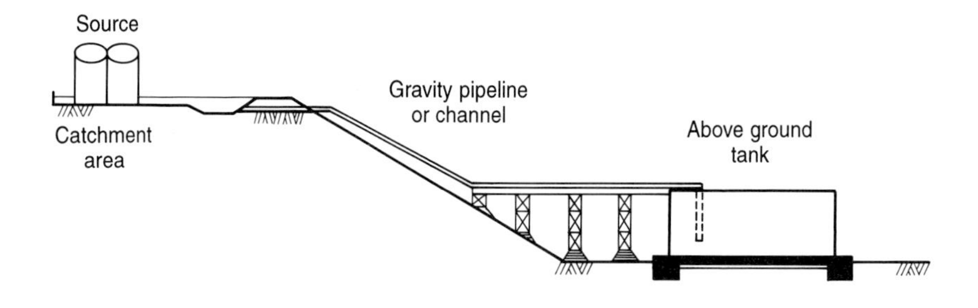

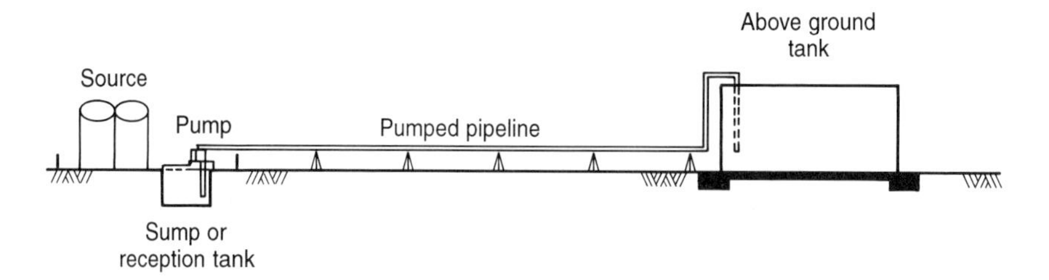

Figure 12.1 *Typical arrangements for above-ground tank systems*

may take in rainwater require regular monitoring and emptying to maintain the necessary secondary containment capacity.

The most onerous loading condition should be considered for design purposes. The overall depth of the tank including freeboard should be taken as the minimum design depth when assessing the static head of contained liquids.

Fire fighting water reservoirs

Compartmention will be required where collected rainwater is stored for use as fire fighting water. It should be noted that mandatory standards are set for fire fighting water reserve tanks supplied directly from the water mains and it may not be possible to consider these for doubling-up as a secondary containment facility because of the potential risk to potable supplies.

12.2 TANKS CONSTRUCTED ABOVE GROUND

12.2.1 Suitable tank systems

The majority of large capacity above-ground containment tanks are assembled on site from prefabricated components, although *in-situ* reinforced concrete construction is sometimes used where a characteristic such as robustness is a particularly important factor.

The principal categories of tanks that may be considered suitable for above-ground secondary containment of potentially polluting substances include:

- proprietary cylindrical tanks as used for agricultural wastes
- welded steel tanks as used for oil, petroleum and other liquid products
- sectional steel rectangular liquid storage tanks
- reinforced plastics tanks

- reinforced concrete tanks
- reinforced concrete/masonry tanks.

Other materials sometimes used for above-ground tanks include stainless steel, aluminium and plastics. Aluminium and plastics are disadvantaged by their relatively poor resistance to fire.

Protection from corrosion and aggressive conditions may be provided by a range of coatings including bitumastic paints, epoxy coatings, and rubber and glass linings. These are specified according to the contained substance and other corrosive influences.

BS 5502 (BSI, 1993c), the agricultural storage tanks standard, provides for a unique design classification system according to the intended design or durability life of the structure. A structure or tank constructed in accordance with Class 1 of BS 5502 is equivalent to Building Regulations' standards and therefore equivalent to the proposed *Class 1* category defined in this report. **(Tanks to BS 5502 Classes 2, 3 or 4 (i.e. other than Class 1) are designed to lower standards than those proposed in this report and are not therefore acceptable for secondary containment on industrial sites.)**

To provide secure containment, tank component design and construction must be properly specified. The various aspects of design, specification, fabrication and site works for above-ground tanks considered suitable for containment, together with the relevant British Standards and codes of good practice, are summarised in Table 12.1 and in the following sections.

Table 12.1 *Design standards for common forms of above-ground tank construction*

Tank construction	British Standard	Nominal capacity range
Pressed steel sectional rectangular tanks, founded at ground level or on higher level support structure, static head. Recommended maximum depth 4.88 m. (See Section 12.2.2)	BS 1564	7 to approx. 920 m^3
Welded steel cylindrical tanks, petroleum industry standards, static head and pressure, set on foundations at ground level. (See Section 12.2.3)	BS 2654	7 to approx. 183 000 m^3
Specification for oil storage tanks, oil burning equipment to BS 799 Part 5, rectangular and cylindrical, horizontal and vertical. (See Section 12.2.4)	BS 799	Up to 150 m^3
Cylindrical tanks as used in agriculture, founded at ground level on concrete base: (a) lapped and bolted vitreous enamelled steel sheets, (b) precast concrete panels ('staves') held together by external hoops, (c) corrugated galvanised steel panels, and (d) sectional precast concrete. (See Section 12.2.5)	BS 5502, Pt 50 Design to BS 5502 class 1 standards (static head only)	Max circa 4 000 m^3
Reinforced plastic tanks (used for containment of various substances). (See Section 12.2.6)	BS 4994	Small to medium
In-situ and precast reinforced concrete (may require waterproof tanking and special joints). (See Section 12.2.7)	BS 8110	Not restricted.
In-situ reinforced concrete (structures for retaining aqueous liquids). (See Section 12.2.7)	BS 8007	Not restricted

12.2.2 Pressed steel sectional rectangular tanks

Pressed steel rectangular tanks are generally assembled on site from prefabricated sectional components. These are proprietary systems and are available from several manufacturers. The elements are easily transported, making the system particularly suited for confined sites and sites with poor access. It is possible to design tanks for future increases in volume and, depending on specification, the tanks may be used to store a variety of liquids. Specifications should comply with BS 1564 (BSI, 1983a) which was developed from the earlier (1949) Standard. Their use is limited to above-ground installations because of strength considerations, and to internal static head pressures. Figure 12.2 shows the general arrangement for this type of tank which is fitted, as is usual, with internal stays.

Tanks are designed to BS 449 (BSI,1969) using steel to BS 4360 and BS 1449. Minimum plate thicknesses are given. For example, for storage of cold liquids with a density not exceeding 1.0, plate thickness is 5 mm for the sides and bottom where depths do not exceed 3.6 m, and 6 mm for the bottom and first tier sides for depths of 4.8 m. For hot liquids the thickness is required to be at least 6 mm throughout. Denser liquids may require increased plate thickness.

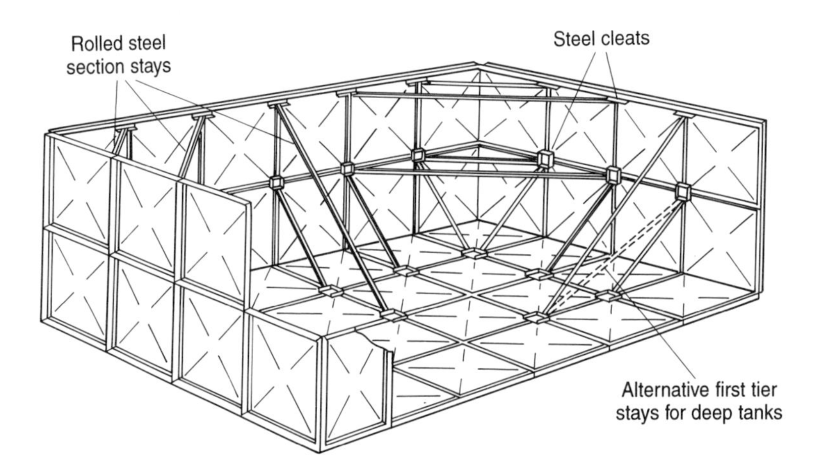

Figure 12.2 *Pressed steel sectional rectangular tank* (Source: BSI)

The plates are bolted together and the Standard requires the use of jointing compounds which are insoluble in the stored liquid. Protective coatings and jointing materials must be specified appropriate to the aggressive nature of the stored liquid.

Bases may be configured with internally flanged plates allowing even contact with the foundation. A sand/ bitumen bed is recommended in such cases. Alternatively, tanks may be supported on a suitably designed steel grillage. Both methods allow leakage monitoring by observation of the base/ foundation interface. A leakage test, using water, should be carried out prior to commissioning the tank.

(Note: the specification of rectangular sectional tanks for oil storage having a capacity of not more than 150 m^3 is described in BS 799 which is discussed later in this section.)

12.2.3 Vertical cylindrical welded steel tanks for the petroleum industry

Welded steel non-refrigerated cylindrical tanks to BS 2654 (BSI, 1989b) may be grouped according to their design pressures as follows:

- non-pressure tanks. These may be open or closed at the top, but designed for an internal pressure of 7.5 mbar and a vacuum of 2.5 mbar
- low pressure tanks. Designed for an internal pressure of 20 mbar and a vacuum of 6 mbar
- high pressure tanks. Designed for an internal pressure of 56 mbar and a vacuum of 6 mbar.

A typical example of a vertical cylindrical welded steel tank is shown in Figure 12.3.

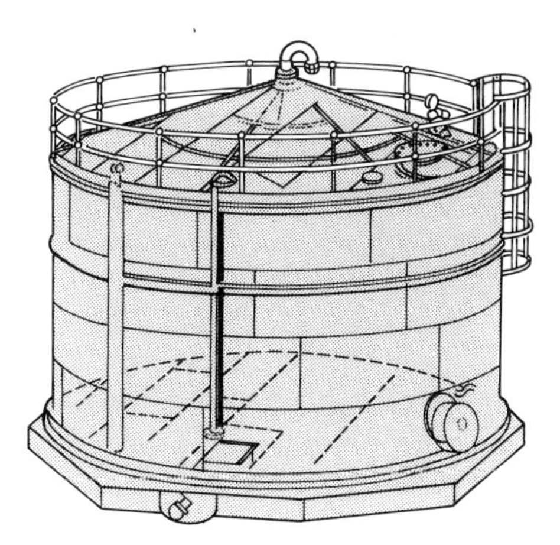

Figure 12.3 *Vertical cylindrical welded-steel tank*

The design relative density of the stored product should be taken as not less than 1.0 for calculating loading. The tank shell thickness is calculated on the assumption that the tank is full to the top of the shell or to a purpose-designed overflow.

The formula used to calculate the minimum shell plate thickness t (in mm) is as follows:

$$t = \frac{(98wD(H-03)) + p}{20S + C}$$

where

H	=	distance from the bottom of the course under consideration to maximum fluid height (m)
D	=	tank diameter (m)
w	=	maximum density of the contained liquid (not less than 1.0) (g/ml)
S	=	allowable design stress (N/mm^2)
p	=	design pressure (zero for unpressurised tanks) (mbar)
c	=	corrosion allowance (mm).

Other design considerations include steel suitability, stress analysis for dead and imposed loads, standards for tank bottoms, anchorage against uplift, roof design where applicable, weld design and testing. Open-top tanks require primary stiffening rings at or near the rim of the tank to withstand wind loads that would otherwise cause out-of-round distortion. The minimum required section modulus of the primary stiffening ring is calculated using the expression:

$$Z = 0.058D^2H$$

where

Z	=	section modulus (cm^3)
D	=	tank diameter (m)
H	=	overall height including freeboard (m).

Corrosion protection may be provided by various systems including paints, epoxy coatings, glass liners and rubber liners, which must be specified according to the nature of the stored product and required durability.

It is recommended that tanks are tested by filling with water before commissioning, and that this should be done under controlled conditions so that the performance of the tank and its foundations can be monitored during the test.

Large diameter open-top tanks are susceptible to damage as a result of foundation settlement. Since tanks are not designed to accommodate differential settlement, the foundations should be designed accordingly.

Tanks of this type are usually built on a 50 mm layer of bitumen/ sand mix. The purpose of this layer is to retard corrosion, to provide a clean working surface for welding the bottom plates, and for weatherproofing where compacted granular material is used as a tank foundation. This layer is also recommended where rigid concrete foundations are used, although it is omitted where cathodic protection is adopted. A suitable bitumen/ sand mix consists of the following proportions by mass:

- 9% non-toxic cutback bitumen
- 10% filler, either limestone dust (passing 75 μm sieve) or Ordinary Portland Cement
- 81% clean and dry washed sand (table 2 BS882, 1201).

75 mm PVC pipes, set around the perimeter at 5 m centres, extending into the granular fill sub-base and protruding through the bitumen sand are commonly used to detect bottom plate leakage, but this is not required where the tanks are to be used only for secondary containment.

12.2.4 Oil storage tanks to BS 799 Part 5

Subject to meeting certain durability criteria, tanks constructed to BS 799 (BSI, 1987d) may be suitable either for short-term containment prior to transfer to a larger facility, or at sites requiring only minimal containment capacity. The Standard includes design and fabrication specifications for cylindrical tanks in horizontal or vertical configuration, oval tanks and rectangular tanks. Tanks exceeding 3.5 m^3 capacity are designated according to their design head capability as shown in Table 12.2.

Table 12.2 *Tank designations according to design pressure* (Source: BS 799)

Arrangement	Type	Capacity (m^3)	Design head above top of tank (m)
Horizontal cylindrical	A (1)	≤ 60	≤ 0.5
	B (2)	≤ 60	≤ 4.5
	C (2)	≤ 60	> 4.5
Vertical cylindrical	D (2)	≤ 65	≤ 0.5
	E (2)	≤ 145	≤ 7.5
	F (2)	> 65	> 7.5
	G (3)	≤ 145	≤ 0.5
Rectangular	J	≤ 150	≤ 0.5
	K	≤ 150	≤ 7.5
Rectangular Sectional	L	≤ 150	≤ 7.5

Notes : (1) dished or flat ends (2) dished end (3) flat bottom plates

12.2.5 Cylindrical tanks to BS 5502 Part 50 (for agricultural waste containment)

BS 5502 Part 50 (BSI, 1993c) gives recommendations for the design and use of agricultural storage tanks and reception pits for liquid waste.

Design wind loads are assessed in accordance with CP3: ChapterV: Part 2 (BSI, 1972). For a full tank the maximum suction value of Cp (the wind pressure coefficient) is taken as 1.3. For an empty tank the maximum inward pressure value of Cp is taken as either 1.5 or 1.8 depending on the height to diameter ratio of the tank. The engineer would need to consider the adequacy of these loading parameters for specific installations.

The Standard specifies a minimum durability life of 10 years , subject to routine maintenance.

The four most common forms of construction for cylindrical tanks to BS 5502 are:

- lapped and bolted vitreous enamelled steel sheets
- corrugated galvanised steel panels
- sectional precast concrete
- precast concrete panels ('staves') held together by external steel hoops.

These four types are illustrated in Figure 12.4.

The walls are usually manufactured components of steel or concrete which are then site assembled on an *in-situ* reinforced base which forms the tank floor. Walls are designed primarily to withstand circumferential hoop stresses, assuming no fixity at the base. Figure 12.5 shows a section through typical baseworks. Figure 12.6 illustrates a number of methods of forming wall to base junctions. Containment integrity is heavily dependent on the wall to floor joint and on the impermeability of the *in-situ* concrete floor.

To achieve *Class 1* or a higher classification, the design for the tank base and floor should conform to BS 8007, *the Code of Practice for design of concrete structures for retaining aqueous liquids* (BSI, 1987b).

Figure 12.4 *Examples of liquid waste storage tanks to BS 5502*

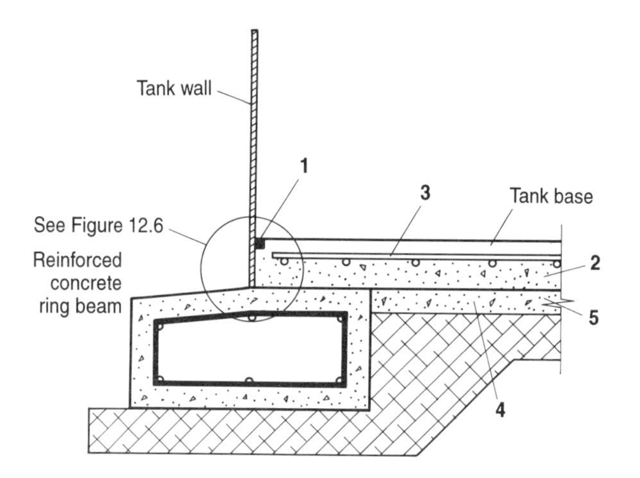

NOTES:

1 Flexible joint sealant between wall and base.

2 Concrete floor designed to BS 8007 for "severe" or "very severe" exposure.

3 For floor slab without contraction joints, provide reinforcement in both directions in top face in accordance with BS 8007. Minimum of 40 mm cover required. Provide reinforcement in bottom face also for slabs more than 300 mm thick.

4 BS 8007 requires a blinding layer of a minimum of 75 mm of C20 (equivalent to Designated mix GEN3 or Standard mix ST4) concrete.

5 For a non-restrained floor (i.e. with contraction joints) provide a 1000-gauge polyethylene membrane. For a fully restrained floor, cast directly onto blinding concrete.

Figure 12.5 *Section through typical baseworks*

Further design guidance on these types of tanks is contained in CIRIA Report 126 (Mason, 1992) which provides information on site selection, baseworks, tank design, construction, durability and maintenance.

12.2.6 Reinforced plastics tanks

The most common types of reinforced plastics tanks are made from glassfibre resin laminates. The tanks are factory manufactured under carefully controlled conditions, using resins such as polyesters, epoxy and furane, which are specified according to the intended use. The choice of resin affects chemical resistance properties and heat distortion temperature. Glass fibres must be compatible with the resin and should comply with BS 3396, BS 3496, BS 3691 or BS 3749. Tanks are categorised into three classes according to use and specification and these are fully described in BS 4994 (BSI, 1987c).

In view of the difficulty in achieving proper quality controls, site jointing using glassfibre resin mixes is not normal practice. For this reason the capacity of single tanks is limited by transportation considerations, and 100 m^3 is the usual maximum. It is possible to increase capacity by connecting a number of tanks together. A number of

proprietary systems have been developed using tanks formed from composites of rectangular GRP panels and steel and also vertical cylindrical GRP multi-straked tanks.

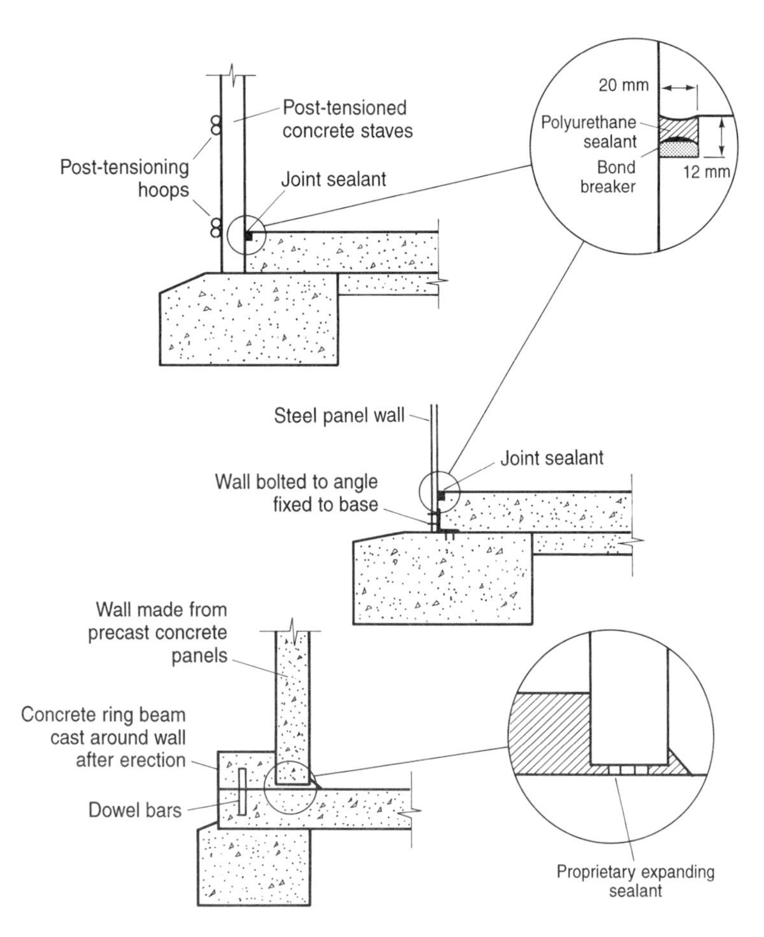

Figure 12.6 *Wall/ base junction details*

Thermoplastics such as PVC, PP, PVDF, ECTFE, CPVC and FEP are used as lining systems to increase resistance to aggressive chemicals. Although GRP tanks may be manufactured to tolerate highly aggressive chemicals and effluents, it should be noted that when used above ground they are more susceptible to fire damage than steel or concrete tanks. GRP tanks are used extensively for below-ground storage of fuels and effluents.

GRP is an anisotropic non-ductile material. For this reason BS 4994 recommends that designs are based on factored ultimate unit strength, rather than permissible stress theory. The ultimate unit strength of the laminate is factored according to a number of considerations including for various criteria such as manufacture, chemical environment, temperature, cyclic loading and curing, to obtain the allowable unit loading. Other considerations include debonding of the laminate under strain, and long-term deterioration. A minimum design safety factor of 8 is recommended.

Tank loading criteria and specifications must be agreed with the manufacturer. This information is discussed in detail in Section 4 of BS 4994 which requires tanks to be designed to withstand the anticipated internal and external dead and imposed loadings, including internal pressures, wind and snow. In addition, buried tanks must be designed to withstand loads due to surcharge, construction, and soil and groundwater.

12.2.7 *In-situ* reinforced concrete tanks

The majority of proprietary tank components are manufactured under factory controlled conditions where quality procedures can be readily maintained. *In-situ* concrete work is affected by site conditions, including the weather, and quality control is much more difficult.

Concrete tanks and bases may be designed to either BS 8110 or BS 8007.

Figure 12.7 shows a typical example of an underground *in-situ* reinforced concrete tank built in compliance with BS 8007. Figure 12.8 shows a construction detail at the junction between the wall and the floor, including a waterstop and polyurethane sealant to ensure a leakproof join.

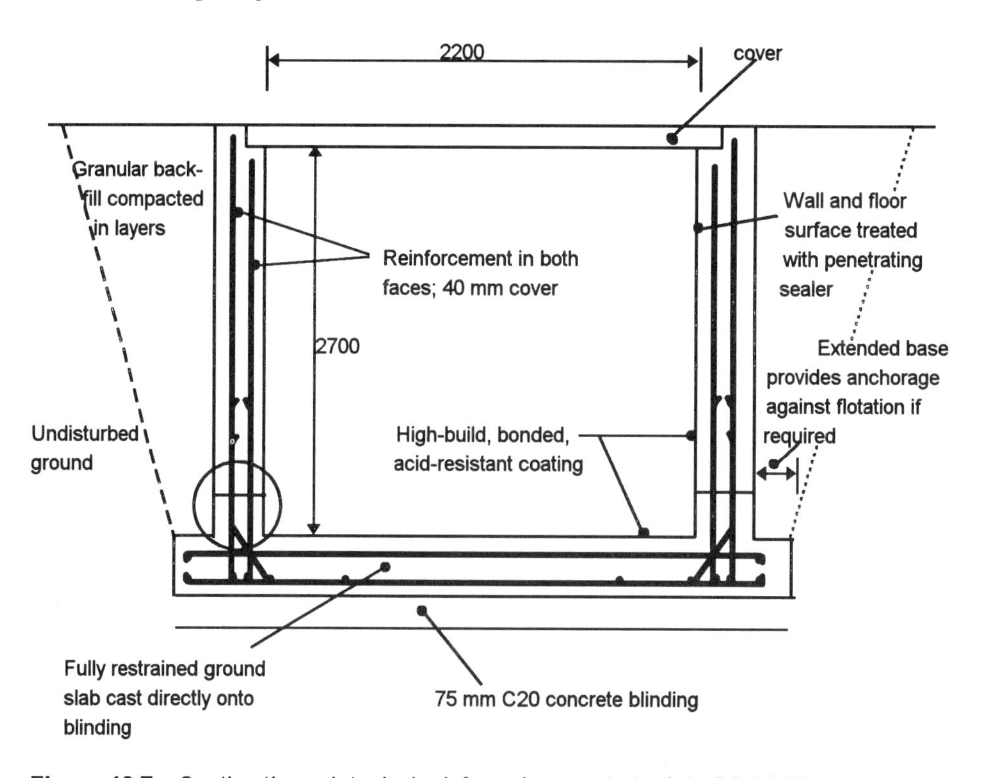

Figure 12.7 *Section through typical reinforced concrete tank to BS 8007*

The recognised Standard for liquid retaining structures is BS 8007 and tanks built to this Standard can be expected to be adequately impermeable for secondary containment. This is achieved by specifying a mix of concrete that can be well compacted, resulting in low permeability, and by specifying details that control

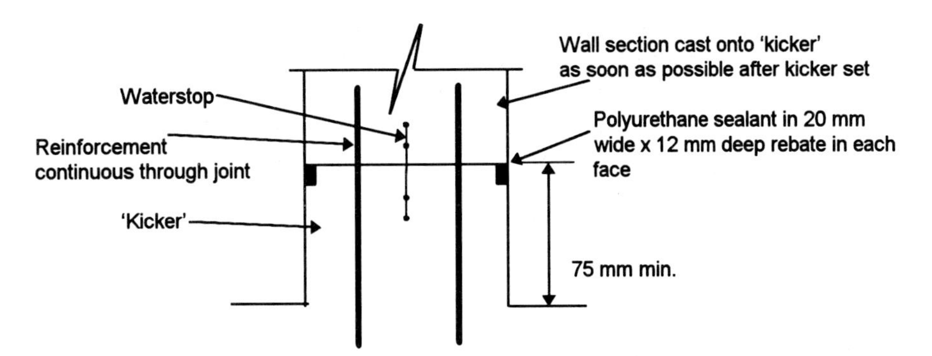

Figure 12.8 *Typical detail at wall/floor junction in reinforced concrete tank to BS 8007*

structural cracking. The Standard is specifically concerned with water containment. Where the contained products are likely to have a deleterious effect, e.g. where pH < 7, protective coatings may be needed.

Tanks built to BS 8110 cannot be expected to achieve the same degree of impermeability as BS 8007 structures since it includes less stringent controls on measures to prevent leakage. BS 8007 also sets standards of construction which ensure a high durability structure.

A summary of the main provisions of BS 8007 is given in Section 10.4.1.

BS 8007 requires that concrete in containment structures is designed to the 'severe' exposure rating as defined in clause 3.3.4 of BS 8110 Part 1, with a nominal cover to all steel reinforcement of 40 mm. Crack widths in concrete must not exceed 0.2 mm. BS 8007 specifies a concrete mix of C35A or better, together with a minimum cement content.

BS 8007 is the relevant standard for many containment structures. The following two sections provide a number of design examples to BS 8007 for comparison with the design procedures of BS 8110.

12.2.8 BS 8007 approach to reducing thermal and other shrinkage cracking in concrete walls and slabs

Thermal and other shrinkage effects are controlled by the design of the size of panel, and by the amount of reinforcement. Table 12.3, which reproduces an extract from BS 8007, indicates three options for crack control by specifying the spacing and minimum amount of reinforcement in a section.

The reinforcement provides additional tensile restraint against thermal movement. The heat of hydration within the first three days of concrete placing is considered in BS 8007 to be the dominant effect for thermal shrinkage. In some instances crack control reinforcement may be found to meet the design limit-state requirement. Typical values of fall in temperature (T_1) between the hydration peak and ambient are given in BS 8007. Further information is contained in CIRIA Report 91 (Harrison, 1992).

The section of the panel is divided into surface zones, according to thickness, in order to calculate the required minimum steel area. The criteria for surface zones for walls and slabs are set out in Figures A.1 and A.2 in Appendix A of BS 8007.

Table 12.3 *Table 5.1 BS 8007*

Option	Type of construction and method of control	Movement joint spacing
1	Continuous: full restraint	No joints other than expansion joints at wide spacings
2	Semi-continuous: partial restraint	a) complete joints ≤ 15 m b) alternate partial and complete joints ≤ 11.25 m c) partial joints ≤ 7.5 m.
3	Close joint spacing: freedom of movement	a) complete joints ≤ 4.8 m + w_{max}/ϵ b) alternate partial and complete joints ≤ 0.5 s_{max} + 2.4 m + w_{max}/ϵ c) partial joints ≤ $s_{max} + w_{max}/\epsilon$

It may be demonstrated that the required reinforcement ratio (ρ) is given as follows:

Options 1 and 2 $\quad \rho = \leq \rho_{crit}$ 1

or $\rho = (fct / fb)(\epsilon\phi / 2w_{max})$ 2

whichever is the greater

Option 3 $\quad \rho = 2/3\rho_{crit}$ 3

where

w = design surface crack width in concrete section

fct/fb = ratio of tensile strength of the concrete to the average bond strength between concrete and steel (taken as 1.0 for plain

bars

and 2/3 for deformed bars)

ϕ = bar diameter

s_{max} = the limiting crack width

ρ_{crit} = fct/fy

ρ_{crit} = critical steel ratio (minimum ratio of steel area to gross area

of

concrete section)

fct = direct tensile stress of the immature concrete at three days

(1.6 N/mm^2) fy = characteristic yield strength

ϵ = concrete strain

For concrete grade C35 and Grade 460 steel:

ρ_{crit} = 0.0035 (Table A.1 BS 8007)

s_{max} is the estimated likely maximum crack spacing

ϵ (the effective strain) $= 0.5\alpha\,(T1 + T2)$ for Option 1

$= 0.5\alpha\,T1$ for Option 2

where

α = coefficient of thermal expansion

$T1$ = initial temperature fall from hydration peak to ambient (Table A.2BS 8007)

$T2$ = subsequent fall in temperature due to seasonal variations.

A worked example is given in Box 12.1.

Box 12.1 *Example calculation of the required reinforcement ratio*

Determine the reinforcement required for a 300 mm thick ground slab (assume polythene under-membrane and freedom to slide). Concrete grade is C35, with OPC content 350 kg/m^3. Concrete placing temperature is 20°C and ambient temperature during construction is 15°C. Maximum crack width w_{max} is 0.2 mm.

Figure A2 BS 8007: h < 500.

Therefore top zone thickness = 300/2 = 150 mm and thickness of bottom zone may be ignored, i.e. no bottom reinforcement required for shrinkage crack control.

Option 1: Continuous construction:

From Table A.2 BS 8007:

for a concrete placing temperature of 20°C and a 300 mm ground slab with 350 kg/m^3 OPC:

$T1$ = 17° C

α = 12×10^{-6} per degree C

$\in$ = $0.5\ \alpha\,(T1 + T2)$

= $0.5 \times 12 \times 10^{-6} \times (17+15)$

= 192×10^{-6}

Use ρ_{crit} = 0.0035

if ρ_{crit} is greater than $\rho = \in\phi/3\ w_{max}$

(derived from eqn. 2 where fct/fb = 2/3 ....deformed bars grade 460)

$\rho = 3.2 \times 10^{-6}$ but $\rho_{crit} = 0.0035$ and $\phi \leq 12$

similarly - Option 2 (semi continuous partial restraint)

Option 3: Complete close movement joints

Required % steel area (Eqn 3):

$\rho \geq 2/3\rho_{crit}$ for grade 460 (ρ_{crit} = 0.0035)

$\in$ = $0.5\alpha\ T1$

= $0.5 \times 12 \times 10^{-6} \times T1$

= 102×10^{-6}

Maximum joint spacing (m) $\leq 4.8 + w_{max} / \in$

= $4.8 + 0.0002/ 102 \times 10^{-6}$

= <u>6.8 m spacing</u>

Summary:

Option 1 use A 393 fabric or equivalent in top zone only

Option 2 use A 393 fabric or equivalent in top zone only

Option 3 use A 252 fabric with complete joints at 6.8 m maximum spacing.

12.2.9 Cylindrical concrete tanks

A cylindrical tank provides the optimum structural arrangement and in comparison to an equivalent capacity rectangular tank demonstrates marginal economies in the use of concrete, although construction costs tend to be higher. The circular plan may also be more difficult to fit on to an existing site.

As with rectangular tanks, the design analysis depends on the structural configuration. Walls are primarily designed to withstand the tensile forces resulting from hoop tension. For a given wall thickness, it can be demonstrated that for a thin-walled cylinder the maximum hoop stress is given by the expression:

$F_t = C\rho L_v r$

where

F_t	=	*hoop stress at depth* L_v
C	=	loading coefficient
ρ	=	density of liquid
r	=	radius of tank

Flexural and shear stresses must be considered, taking account of the dimensions of the tank, the loading condition and the degree of fixity at the base.

Table 12.4 lists the coefficients for determining maximum hoop tension stresses, flexural stresses and shear stresses for a range of cylindrical tank sizes and loading conditions. Box 12.2 gives provides a worked design example.

12.3 TANKS CONSTRUCTED BELOW GROUND LEVEL

12.3.1 Introduction

The design of below-ground structures requires information on the topography, the nature of the ground, the water table and any other factor that may influence the integrity of the construction.

When designing to *Class 2* and better, the engineer should make a detailed geotechnical investigation of the site. Where sufficient information is available, e.g. from trial holes, site inspection or previous site history, it will be possible by reference to standards and Codes of Practice, to make assumptions about the characteristics of a site and its soil properties. This will often be the case when assessing soil properties for *Class 1* structures. Since floatation is a possible cause of failure it is important to establish the levels and fluctuations of the water table.

Two British Standards provide information on the characteristics of soils and the loads that may be imposed upon below ground level tanks and structures:

- BS 8002 (1994b) - *Code of practice for earth retaining structures*
- BS 5502 Part 50 (1993c) - *Buildings and structures for agriculture, code of practice for the design, construction and use of storage tanks and reception pits for livestock slurry.*

Both Standards provide methods for determining the physical parameters of soils where detailed tests are unavailable.

The following list summarises the main contents of BS 8002:

- information on site, features, groundwater and soil characteristics
- detailed methods for selection of soil parameter values in absence of tests
- minimum load values/parameters, for example:
  - minimum surcharge value of 10 kN/m^2 (c.f. BS 5502 which gives specific circumstances for surcharge load without minimum requirement)
  - design depth of retained height to include possible additional excavation depth of 0.5 m in front of wall, or 10% total retained height in the case of an unpropped cantilever wall, (not stated in BS 5502)
  - water/ liquid pressure to be taken as most onerous condition

- limit state design methods
- total and effective stress parameters
- equilibrium calculation
- design situations
- design to structural codes
- specific information on various earth retaining structures
- technical information for design use.

Table 12.4 *Maximum load coefficients for cylindrical tanks* (Source: Portland Cement Association)

Table a *Tension in cylindrical tank*
(Triangular load, fixed base, free top)

$L_v^2/2rh$	Depth from top of tank	Coef. for maximum value
O.4	$0.0L_v$	+0.149
0.8	$0.0L_v$	+0.263
1.2	$0.0L_v$	+0.283
1.6	$0.1L_v$	+0.268
2.0	$0.3L_v$ $0.4L_v$	+0.285
3.0	$0.5L_v$	+0.362
4.0	$0.5L_v$	+0.429
5.0	$0.5L_v$	+0.477
6.0	$0.6L_v$	+0.514
8.0	$0.6L_v$	+0.575
10.0	$0.6L_v$	+0.608

Table b *Tension in cylindrical tank*
(Triangular load, hinged base, free top)

$L_v^2/2rh$	Depth from top of tank	Coef. for maximum value
O.4	$0.0L_v$	+0.474
0.8	$0.0L_v$	+0.423
1.2	$0.3L_v$	+0.362
1.6	$0.4L_v$ $0.5L_v$	+0.385
2.0	$0.5L_v$	+0.434
3.0	$0.6L_v$	+0.519
4.0	$0.6L_v$	+0.579
5.0	$0.6L_v$	+0.617
6.0	$0.7L_v$	+0.643
8.0	$0.7L_v$	+0.697
10.0	$0.7L_v$	+0.730

Table c *Moments in cylindrical wall*
(Triangular load, fixed base, free top)

$L_v^2/2rh$	Depth from top of tank	Coef. for max. value
0.4	$0.3L_v$	+0.0021
	$1.0L_v$	-0.1205
0.8	$0.4L_v$	+0.008
	$1.0L_v$	-0.0795
1.2	$0.5L_v$	+0.0112
	$1.0L_v$	-0.0602
1.6	$0.5L_v$	+0.0121
	$1.0L_v$	-0.0505
2.0	$0.5L_v$	+0.0120
	$1.0L_v$	-0.0436
3.0	$0.6L_v$	+0.0097
	$1.0L_v$	-0.0333
4.0	$0.6L_v$	+0.0077
	$1.0L_v$	-0.0268
5.0	$0.6L_v$	+0.0059
	$1.0L_v$	-0.0222
6.00	$0.7L_v$	+0.0051
6.0	$1.0L_v$	-0.0187
8.0	$0.7L_v$	+0.0038
	$1.0L_v$	-0.0146
10.0	$0.7L_v$	+0.0029
	$1.0L_v$	-0.0122

Max. BM = coef. x ρL_v^3 (kNm per m)
+ sign denotes tension on outside face

Table d *Shear at base of cylindrical wall*

$L_v^2/2rh$	Triangular load fixed base
0.4	+0.436
0.8	+0.374
1.2	+0.339
1.6	+0.317
2.0	+0.299
3.0	+0.262
4.0	+0.236
5.0	+0.213
6.0	+0.197
8.0	+0.174
10.0	+0.158

V = coef. x ρL_v^2 (kN)

Box 12.2 *Example of cylindrical concrete tank design*

A reinforced concrete cylindrical tank, 5 m deep and 14 m diameter is to hold an aqueous liquid. Assume the walls are C35 grade concrete with a uniform thickness of 200 mm. The base and walls are contiguous and fixity may be assumed. Calculate reinforcement required to limit crack width to 0.1 mm and compliance with BS 8007.

$L_v^2 / 2rh = 5 \times 5/(2 \times 7 \times 0.2) = 8.93$

Interpolation of Table 12.4a gives maximum hoop tension coefficient = 0.59
Maximum hoop tension F_t = coef. x ρ L_v r
= 0.59 x 9.8 x 5 x 7
= <u>202.3 kN/m</u>

Interpolation of Table 12.4c gives maximum bending moment coefficients:
at bottom of wall = 0.0133
at top of wall = 0.0034
Maximum BM at bottom of wall = coef. x ρL_v^3
= $0.0133 \times 9.8 \times 5^3$
= <u>16.3 kNm/m</u>

Maximum BM at top of wall = $0.0034 \times 9.8 \times 5^3$
= <u>4.2 kNm/m</u>

Tensile reinforcement for circumferential hoop stress:

SLS (serviceability limit state) tensile stress analysis to BS 8007 indicates that for 200 mm section (40 mm minimum cover) and a limiting crack width of 0.1mm, 12 mm diameter grade 460 reinforcement at 200 centres set in each face will have a tensile strength of 272 kN, at a bar stress of 240 N/mm^2.

With a maximum hoop stress of 202 kN the bar stress will be 179 N/mm^2 (i.e. satisfactory).

The anti-crack reinforcement must be checked for a limiting crack width of 0.1 mm since this may be the governing requirement.

Flexural reinforcement requirement:

Maximum vertical BM = 16.3 kNm/m

A 200 mm section with a bar cover of 52 mm and a limiting crack width of 0.1 mm when designed to BS 8007 stress limits indicates that 12 mm diameter grade 460 reinforcement at 125 mm centres will provide a SLS moment of resistance of 17.4 kNm/m, which is therefore satisfactory.

Shear check:

From Table 12.4d, V = $0.164 \times 9.8 \times 5^2 \times 1.4$
= <u>56.3 kN/m</u>

d = 200 -50 - 6 = 144 mm
v = $(56.3 \times 10^3)/(10^3 \times 144)$
= 0.39 N/mm^2

With 12 mm diam bars at 125 mm crs 100As/ bd = (100 x 905)/(1000 x 144)
= <u>0.64 N/mm^2</u>

v_c = 0.78 N/mm^2 (i.e. satisfactory).

12.3.2 Tank systems for use below ground

Many above-ground tank systems may be designed and adapted for below-ground installation. In many cases it will be necessary to check the manufacturer's specification to determine the suitability for below-ground use.

Small capacity, welded steel and GRP cylindrical tanks are commonly used for below-ground storage of chemicals, fuel, oil, sewage effluent.

The most common types of structure for large capacity below ground tanks are *in-situ* reinforced concrete, sheet piled walls and deep shaft construction.

12.3.3 Reinforced masonry tanks

Reinforced masonry is occasionally used for small capacity tanks. The difficulties inherent in the provision of watertight joints and waterproof rendering to withstand high static pressures, tends to exclude the use of reinforced masonry for large capacity structures. The most reliable structural system involves reinforced concrete filled cavity wall construction where jointing techniques similar to BS 8007 are employed. This technique essentially provides a concrete wall surrounded on each side by masonry. Although the system may involve a relatively high initial capital cost, it might be favourably considered where concrete structures need to be protected against fire or aggressive chemicals and groundwater, since the masonry skin will help protect the concrete from spalling and deterioration.

12.3.4 *In-situ* reinforced concrete below-ground tanks

In-situ concrete may be used to provide structural support for tanks made from other materials such as small capacity welded steel or GRP tanks. This can serve a double purpose by acting as additional back-up containment which will have particular relevance at a *Class 2* site. At particularly sensitive locations, a further containment system may be required in the form, for example, of a double geomembrane/ geotextile incorporating a leakage detection facility that surrounds the enveloping concrete, thus giving an overall *Class 3* rating.

To prevent ground contamination and or groundwater pollution, tanks formed from *in-situ* concrete must be watertight. The external walls and base of the tank will be inaccessible for regular inspection and maintenance. This constraint means that the stringent standards of BS 8007 will be more appropriate than BS 8110 for constructing below-ground tanks.

Designs should allow for the additional loadings imposed by below-ground conditions including soil pressures, groundwater and surcharge loading. Consideration needs to be given to the possibility of flotation whilst the tank is empty where groundwater is present. The potential threat of pollution means that it may not be possible to use pressure relief valves which could allow ingress of groundwater. No allowance should be made for external passive restraint of the retained soil when designing the tank walls for internal loads.

The use of waterstops is mandatory for construction in accordance with BS 8007. In order to be effective, they must be properly specified, carefully located and joined using the specified method.

12.3.5 Deep shaft tanks

The 'bored shafts' technique describes the method of excavating and forming a vertical cylindrical shaft or tank to depths of around 60 m, the circumference of which is usually concrete lined (Figure 12.9). After excavation the base of the tank is concreted to provide a watertight seal. Bored shafts are commonly up to 5 m in diameter, although it is possible to go to 9 m. Deep shafts were first used in the UK in the 1980s for sewage sludge treatment and aeration.

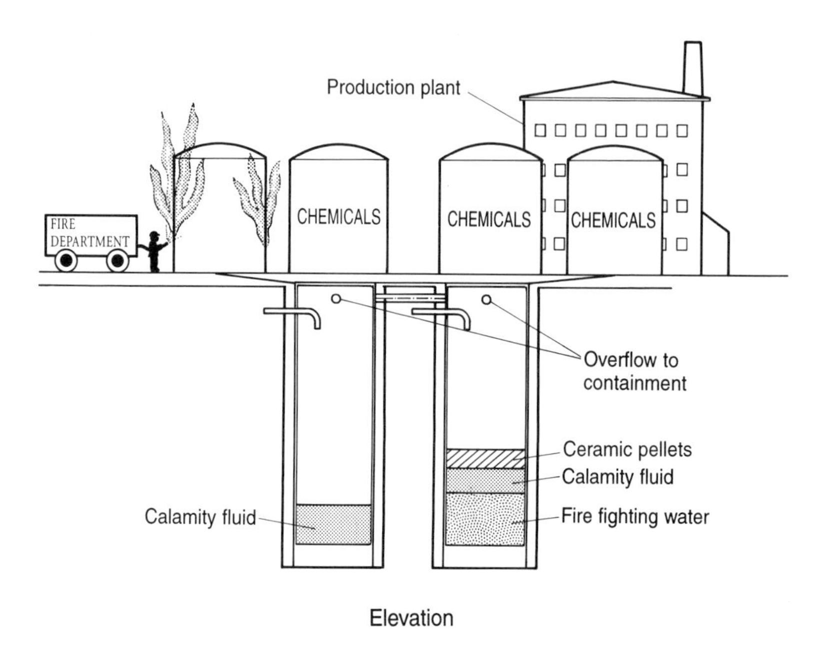

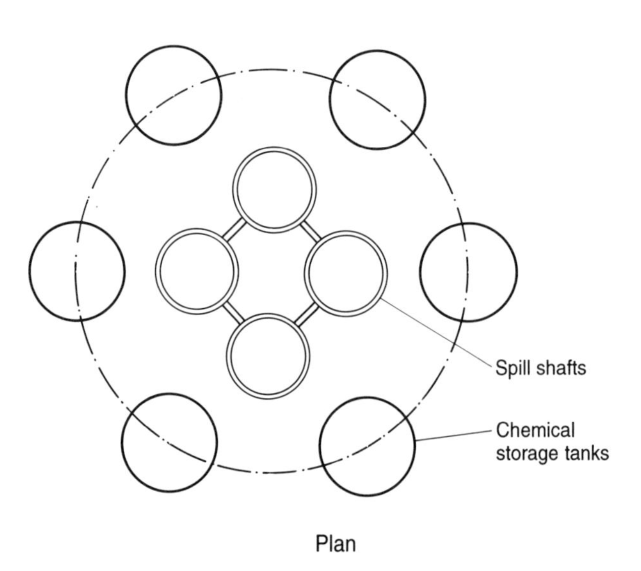

Figure 12.9 *Use of multiple deep shaft tanks to provide containment adjacent to the source*

The bored shaft technique has been developed as a proprietary system by a Dutch company who are now marketing it throughout Europe. Potential uses include containment tanks for a variety of substances, water and fire water reservoirs, abstraction well heads, sewage treatment, balancing tanks etc. The small plan area of the tanks gives a number of potential advantages, including minimal site area requirement and siting of full containment close to the source. The small effective diameter makes fire control easier within the tank. The system cannot be used in all ground conditions and the emptying costs are relatively greater, and emptying is more difficult than for a shallow system.

Bentonite is normally used to lower ground friction whilst boring the shaft and also to maintain watertightness until replaced by a cementitious grout. The company have developed a fire control system using carbon dioxide and ceramic pellets which float on the surface of any spilled liquid (labelled 'calamity fluid' in Figure 12.9) to suppress flames.

12.3.6 Tanks formed with embedded walls of steel sheet piling

Continuous steel sheet piling is a common method for retaining wall construction and is used to form open storage reservoirs in impermeable soils. Where the site soils are permeable, bentonite may be used together with concrete to make the walls and floor sufficiently impermeable. In many parts of Europe, secondary containment tanks are constructed using sheet steel piles and concrete floors. Normal steel sheet piling may be used either as cantilever wall or as an anchored or propped cantilever. The maximum retained height of cantilever walls is usually restricted to about 5 m although this may need to be less where soft ground is encountered in front of the wall. The effective retained height of an anchored wall will depend on the degree of anchorage and the ground conditions at the foot of the wall. The height restrictions influence the required area and effective capacity of the tank.

The various design methods are discussed in some detail in CIRIA Report 104 (Padfield and Mair, 1984).

12.3.7 Compartmentation of tanks

Compartmentation or multiple containment may be necessary for various reasons including:

- fire control
- rainwater/ fire water storage
- separation and handling of contents
- maintenance
- duplication and multiple use requiring different capacities
- site conditions
- maintaining reserve capacity whilst emptying, etc.
- safety.

12.4 MAINTENANCE

All tanks should be inspected at regular intervals, preferably at least once a year externally and every two years internally. It will not be possible to inspect readily the exterior of below-ground tanks. Where leakage detection monitoring is not installed, periodic watertightness tests are recommended.

Items subject to deterioration and requiring particularly detailed inspection and regular maintenance include:

- welded joints
- bolted and riveted joints
- joint sealants

- laps and seams
- tank bottoms
- protective coatings
- pipe and other connections
- valves
- access hatches
- surfaces subject to corrosion by weathering or aggressive attack.

Welded joints have a tendency to corrode more rapidly than the parent metal, particularly at the base of the tank and at all positions subject to exposure and aggressive conditions. Ultrasonic and X-ray techniques should be used to check weld integrity. Site welding repairs should be carried out to the same standards as the original tank specification.

A wide range of sealants and adhesives is available to rectify joints between the various materials and components used in tank construction. Table 4 and sections six and seven of BS 6213 (BSI, 1992b) provide guidance on the selection and suitability of sealants and the method of application for repairing concrete and other structural materials. The tank manufacturer should be consulted to ensure compatibility of a sealant intended for use with proprietary components.

Similarly, a wide variety of protective and remedial coatings is available. The selection and application should be compatible with the structural requirement and the aggressive nature of the contained substance.

12.5 SAFETY

The Construction (Design and Management) Regulations 1994 apply to all containment systems, including tanks. Tank maintenance operations require strict controls because of the additional hazards imposed by the need to access confined spaces where hazardous gases and substances may be present. All tank maintenance operatives should be fully trained and be competent to deal with emergencies. Washdown showers, breathing apparatus, protective clothing, first aid and other safety measures should be available as required under HSE recommendations.

Operational procedures during transfer of tank contents or while emptying tanks for maintenance purposes must be safe for operatives and without risk to other persons or property.

Operatives should not enter tanks without the means to communicate with persons who can give immediate assistance in the event of an emergency. Safety equipment and safety lines should be used and escape ladders should be fixed within tanks. Scaffolding and other temporary works should conform to HSE requirements.

Notices, complying with statutory requirements, clearly stating the hazards and characteristics of any dangerous substances should be prominently sited adjacent to tanks.

Permanent and adequate security fencing with lockable gates should be provided around the site perimeter.

Fire is a major hazard at many tank sites. The site should be provided with facilities that conform to the Fire Service's requirements and with apparatus that can be deployed in the event of fire within the containment installation.

13 Transfer systems

13.1 INTRODUCTION

The term transfer system is used to describe the means for collecting and conveying spillage and contaminated water from the primary containment area to the facility designated for secondary containment. With *local* secondary containment systems, i.e. bunds, the transfer system is effectively the bund floor and this is covered in Section 10 which deals in detail with bund design. This section is concerned with transfer systems that comprise part of *remote* and *combined* secondary containment systems.

Transfer systems comprise the following elements:

- catchment areas in the immediate vicinity of the primary containment to control and channel any polluting materials ready for transfer
- conveyance systems to transfer the material from the catchment area to the secondary containment.

Catchment areas may be purpose designed and built with the sole function of intercepting polluting materials resulting from an incident. Alternatively, they may be areas such as roadways, hardstandings and paved areas which double as catchment areas in addition to performing other functions.

Conveyance systems may take the form of pipe networks, open channels or culverts. As with catchments, these may be designed for the sole purpose of dealing with spillages and other incidents or they may serve other purposes as well as the routine drainage of contaminated water from a site. In some cases, on-site roadways may be modified or adapted to provide a safe means of transfer.

The essential requirements of any transfer system are that it must be:

- leakproof
- sufficiently strong and durable to perform adequately for the duration of its design life with only routine maintenance
- resistant to fire
- resistant to attack from the materials that may be released from the primary containment
- of sufficient capacity to cope with the worst flow scenario without overflowing.

Material collected by the transfer system must be stored, where appropriate treated, and disposed of in a safe manner. In order not to overload storage and treatment systems designed specifically for dealing with contaminated water, it is usually beneficial on larger sites to segregate catchment areas into 'clean' and 'dirty' zones so that the runoff from each may be dealt with separately. This avoids the unnecessary costs of storing and treating 'clean' water as if it was contaminated and, conversely, the risk of contaminated water reaching, for example, a consented outfall.

13.2 CATEGORISING CATCHMENT AREAS

13.2.1 Categories of waste

Drainage on most sites can be divided into three categories, commonly referred to as:

- storm water drainage
- foul drainage
- trade waste drainage.

Storm water drainage

This is designed essentially for 'clean' water and most storm water drainage outfalls to a watercourse or soakaway, usually under an Environment Agency/ SEPA discharge consent. With such outfalls it is essential that inlets to all storm water drains are protected against ingress by foul and trade wastes.

On sites which are designated a moderate or high hazard or risk rating (i.e. Class 2 or 3) it is recommended that all storm water drainage is designed to outfall initially to a holding facility (e.g. lagoon, reception tank, interceptor - depending on the capacity required) so that any unplanned entry of polluting material into the system may be dealt with safely.

Foul water drainage

This deals with sewage and normally outfalls to a Water Utility's sewage treatment works. It is important that trade waste is prevented from entering foul water drainage systems since it may cause long-lasting damage to the treatment works process which could in turn result in the discharge of untreated sewage.

Trade waste

Trade wastes may be subdivided into those resulting from planned operations (e.g. washing out vessels used to store polluting materials) and those resulting from unplanned incidents. The latter may range from, for example, a minor spill in a loading bay through to a major incident involving the loss of a large inventory of harmful materials coupled with the application of large amounts of fire fighting water.

Some trade wastes may be consented to go to a Water Utility's sewage treatment works but, more usually (particularly on larger sites) they will go to the discharger's own on-site treatment works. Where large volumes of material have to be dealt with, holding tanks will be required. In the event of a major incident the polluting material may be held in large containment tanks and taken off-site for treatment prior to disposal.

This report is concerned primarily with unplanned incidents.

13.2.2 Site zoning

In planning a new transfer system, or assessing the adequacy of existing drainage systems for dealing with unplanned incidents, it is recommended that the site is divided into catchment areas, each designated as either 'storm water' or 'trade waste' or a combination of both. Catchment areas may be hardstandings, roadways, floors of buildings, roofs of buildings or any other areas that may be subject to runoff of water or other material.

An assessment should be made of the amount of runoff likely to occur from any catchments designated as trade waste or a combination of trade waste and storm water. The approach described in Section 8 should be used to estimate the total quantity of material (i.e. material in primary storage, rainwater, fire fighting and cooling water) likely to be released from a catchment area. The rate of release, which is the factor that affects the design of the transfer system, must be assessed by means of a detailed consideration of all of the events that could cause a release to each catchment area.

Great care must be exercised before designating any catchment area as solely 'storm water'. To be so designated, there must be no possibility of the catchment becoming contaminated with trade waste during the course of an incident, when operating procedures for ensuring that trade wastes and storm water are normally kept apart, may break down. Even where catchment areas are modified using kerbs, gullies, low height compartmentation walls etc. in order to improve the separation of storm water and trade waste, the possibility and consequences of inundation in the event of a serious incident should be considered in consultation with the Regulators. In general, the higher the site's hazard or risk rating, the more onerous will be the safeguards required to support the designation of a catchment as storm water only.

Where a zone is designated as 'trade waste', it is important to identify the types of material likely to be released from it since the characteristics of the material, as well as its rate of release, affect drainage design. Where more than one material is involved, consideration should be given to the characteristics of the cocktails that could result. Important characteristics in terms of drainage design are:

- corrosive effects on materials used for constructing the drainage system
- density (in particular whether lighter than water)
- flammability (liquid and/or vapour)
- flow characteristics (viscosity and changes in viscosity according to temperature and mixing with other substances or fire fighting water)
- tumescence and possible solidification on burning.

It is recommended that the characteristics of all catchment areas on a site are tabulated in a form along the lines shown in the sample pro forma shown at Appendix A17, as an aid both to planning new and to assessing existing systems. This approach recognises that in some situations it may not be possible to design catchments or conveyance systems for all contingencies (for example the very rapid release of materials stored in primary containment) but requires that any shortfall in, say, drainage capacity, is made explicit and that the consequences are understood and agreed with the Regulators.

The output from the catchment assessment is used to design the conveyance system.

13.3 GRAVITY AND PUMPED TRANSFER SYSTEMS

Transfer systems may be designed to operate entirely by gravity or they may be pumped, depending on the layout of the plant and the topography of the site, particularly the location of the primary storage areas in relation to the secondary containment. There are advantages and disadvantages associated with both types of system as outlined below.

13.3.1 Gravity systems

Advantages

- simple and relatively inexpensive
- little to go wrong
- do not rely on operator intervention or automatic controls to activate.

Disadvantages

- difficult to control flow rate (necessary, for example, for input to treatment works)
- pipework usually necessarily underground, making monitoring, inspection and maintenance more difficult
- site layout and topography may make it impracticable.

It should be stressed that the above refers to transfer systems which are part of a *remote* or *combined* secondary containment system, and not to bunds where gravity discharge arrangements are strongly recommended against.

13.3.2 Pumped systems

Advantages

- more flexibility (can pump up hill)
- makes above-ground pipework feasible
- more control over transfer rates.

Disadvantages

- more complicated, therefore more to go wrong
- require reliable power supply and controls
- may require back-up system
- more maintenance required.

Many transfer systems are part gravity and part pumped. Ideally a transfer system would be gravity operated but with pipework suspended above ground to enable effective monitoring and maintenance. In practice, the benefits of the inherent reliability of gravity systems has usually to be balanced against the convenience and accessibility of pumped systems.

13.4 TRANSFER SYSTEM CLASSIFICATION

Overall containment system classification is dealt with in Section 8. Table 13.1 summarises the performance requirements that components of a transfer system should meet in order to satisfy the overall system classification.

Table 13.1 *Performance requirements for transfer system components*

Main System component	*Class 1*	*Class 2*	*Class 3*
Catchments (refer to Table 13.2 for detail)	Impermeability and durability consistent with life of primary installation. Designed and constructed to normal standards using materials unaffected by fire or damaging substances. Adequate safety margins including safe collection and transfer of substances. Normal levels of supervision.	As *Class 1* but to the enhanced standards described in Section 7. Built-in redundancies e.g. impermeable coatings, underlays, jointing. Leakage interception/ monitoring system under primary tanks. Enhanced regular site inspections - condition, efficacy of system and state of impermeability.	As *Class2* but with the additional requirements described in Section 7.6. Additional redundancies including duplication of bypass channels, pipes, basins etc. Telemetry and/ or CCTV etc. together with continual monitoring and action procedures for diversion of flow etc.
Pipes/ channels (refer to Tables 13.3 13.4 for additional information)	As above. Design and construction generally to BS 8031. No detectable leakage through pipes, joints and ancillary items such as manholes, sumps, gullies etc. Pumped and gravity pipes, inlets and outfalls to be capable of dealing with spate flow. Dual purpose drainage systems (i.e. normal use - site and surface water drainage; emergency use - transfer system drainage) must be designed to prevent pollution of controlled waters (N.B. not permissible for *Class 2* or *3*).	As *Class 1* but to the enhanced standards of Section 7.6. Subsidence and movement joints to be fully watertight. Additional site specific redundancies which may include sleeved pipes, dual pipe systems each capable of full flow capacity in event of pipe failure, pipe bunds etc. More rigorous operational and management procedures including systems' maintenance.	As *Class 2* but to the enhanced standards of Section 7.6. Duplicate facilities separate and additional to *Class 2* requirement. Telemetry, including flow metering of all pipe runs to permit leakage detection. Regular CCTV inspections of below-ground installations. Permanent CCTV surveillance of above-ground pipe networks. Alarm systems.
Pumps* * see note below	As above. Since pump failure may cause a major incident it is recommended that all sites should have back- up provisions in the form of a holding tank of sufficient capacity or an emergency pump. All ancillary items e.g. sumps, detention tanks, etc. to be built to *Class 1* requirements.	As *Class 1* but to the enhanced standards of Section 7.6. Dual pumps and controls required. Redundancies to be designed into the sump or pump chamber all to meet *Class 2* requirement. More rigorous monitoring and operational controls to deal with emergencies as they occur.	As *Class 2* but to the enhanced standards of Section 7.6. In addition to *Class 2* requirement duplicate facilities to be provided. Flow monitoring and leakage detection telemetry provided at all times together with emergency operational procedures. Regular CCTV inspections with permanent CCTV installations at appropriate locations.

* **Note**: where the system is totally reliant on a working pump, the pump and its controls should be designed to minimise any risk of failure and be accessible for repair. This is particularly relevant where fire damage is a possibility. The pump and its controls may be best sited away from the bund or sump so that it can be maintained at all times during an incident. Dual pumps should be considered. Alternatively, or additionally, portable pumps with a rating equivalent to the fixed pump should be available in the event of an exceptional emergency or a failure of the main system.

13.5 CATCHMENT AREA DESIGN

In the absence of a bund, most primary containment areas require an impermeable catchment to intercept spills and other unintended discharges. The area beneath and surrounding the installation must be impermeable and contoured, or kerbed, to collect the spillage prior to its transfer by gravity or pumping to the remote secondary containment. Catchments may be constructed using earth, if suitably impermeable, but more usually they are constructed in concrete. The effectiveness of a catchment area relies heavily on the ability of the downstream drainage system to transfer the spillage away from the catchment area at a sufficient rate to prevent overflow to surrounding areas which may not be impermeable.

Catchment areas sometimes serve a dual purpose, e.g. to provide a hardstanding or stable working area. The catchment function may be secondary to the main function, as in the case of roadways which may double as catchment areas. With roadways, additional contouring and kerbing is likely to be required to ensure that spillages do not overflow. Open channels may be provided along the sides of such roadways for this purpose.

For safety reasons, roadways normally have provision for drainage of storm water and it would be difficult in the event of a serious incident to ensure that pollutants do not enter the storm water system. For this reason, dual purpose catchment areas such as roadways cannot be recommended for high hazard or risk situations unless storm water is routinely collected in holding tanks for sampling prior to treatment (if necessary) and discharge.

Oil separators are an essential requirement for most catchments that are used as hardstandings for vehicles (see later in this section).

The two main criteria to be satisfied when designing a catchment area are:

- adequate impermeability
- sufficiently large (plan area and capacity) to prevent the escape of surge flows of primary materials or fire fighting water while not compromising unduly the access arrangements required for day-to-day operations of the plant.

Forms of catchment construction are summarised in Table 13.2.

One of the most common and reliable forms of construction for catchment areas is an impermeable *in-situ* reinforced concrete base surrounded by a reinforced concrete upstand which forms a shallow retaining kerb. The most appropriate design standard for this type of construction is BS 8007. Guidance on the design and construction of impermeable works in accordance with BS 8007 is provided in Section 10.

Where drain outlets can be properly sealed, impermeable catchment areas may be used for detention storage.

Table 13.2 *Summary of common forms of catchment area construction*

Form of construction	Comment on construction	Characteristics	Notes on possible Classification
Catchments with an earth base	The impermeability of some soils may be increased by reworking and compaction, and by blending with other soils and admixtures. Further information is given in Section 11	Runoff increases with decreasing permeability and increasing slope. Conversely, infiltration increases with increasing permeability and decreasing slope. Cultivation, saturation and vegetation influence runoff characteristic	Soils having an impermeability of 10^{-9} m/s or less may be suitable for *Class 1*, *2* or *3*. Suitability of soil with a higher permeability depends on any additional precautions, e.g. liners, and the expected retention time. Gravels sands and silts with high to medium permeabilities (exceeding 10^{-7} m/s) generally unsuitable for catchments. Section 11 gives characteristic permeabilities and tests for various soils
Modular paving units, non-reinforced concrete, hardcore and similar surfaces	Modular units generally hand laid on a compacted hard core sub-base. An underlying impermeable membrane, which must be able to accommodate deformation and be unaffected by deleterious materials, may be specified	Hardcore and similar surfacing has a relatively high permeability due to the nature of its grading. The joints of modular units must be sealed. Impermeability of modular units improved by laying on a reinforced concrete base	Non-reinforced concrete and modular units may meet *Class 1, 2* or *3* requirements if laid on an impermeable liner over soil as described above
Reinforced concrete to BS 8110	Concrete mix design should conform to BS 5328, design and construction to BS 8110	BS 8110 design permits crack width of up to 0.3 mm which may allow leakage in some circumstances. Detail design recommendations in the Standard not aimed specifically at achieving impermeability, e.g. no requirement for waterstops in joints. Range of concrete mixes and durability standards	Given the lack of specific requirements for watertightness in the provisions of BS 8110, it is recommended only for *Class 1* situations. Where a higher classification is required additional measures will be necessary including waterproof lining systems, and design and construction standards equal at least to the 'severe' categories described in Tables 3.2 & 3.4 of BS 8110 Part 1
Reinforced concrete structures designed to BS 8007	BS 8007 sets out specific requirements for the design and construction of watertight concrete structures	The design standard specifies waterstops in joints, specified concrete grades and cement contents to minimise cracking and anti-crack reinforcement to limit crack widths. Principal differences between BS 8007 and BS 8110 summarised in Section 10	All reinforced concrete containment structures, including catchments, to meet *Class 2* or *3* requirements, designed and constructed to comply with the requirements of BS 8007. Structural details of reinforced concrete roads are given in DoT publication
'Black top' surfaces - flexible and semi rigid paving to DoT specification	Asphalts and asphaltic concretes, tar and bituminous macadams. Hot rolled asphalt (HRA) most common for heavily trafficked areas. Light traffic bitmac, tarmacadam, cold asphalts etc. Construction specification depends on CBR of formation	Flexible pavements tend to deform under concentrated loads. Black top on concrete sub-base demonstrates superior performance for industrial use. Black top surfaces are susceptible to damage by many solvents and by fire. Tar surfacing less prone to damage from diesel spills	A Class 2 black top paving suitable for up to 700 commercial vehicles per day on a formation 5%< CBR ≤ 15% comprises typically: *40 to 50 mm HRA to DoT C1.911* *60mm base course of asphalt to C1.905 or bitmac to C1.903 or tarmacadam to C1.902.* *At least 150 mm of C15 wet lean-mix concrete.* *150 mm capping layer of compacted granular Type 1 fill.* Concentrated loads may require base of reinforced concrete. Suitability depends on assessment of the likelihood and nature of fire and other damage

13.6 OPEN CHANNEL DESIGN

Open channels are used to drain large areas or for dealing with large flows. On many sites the channels may be formed by contouring and kerbing suitable catchment areas such as roadways. The kerb height and width and slope of the catchment determine the flow capacity. Where unprotected earth is used to construct drainage channels, flow velocities should be limited to 0.5 to 0.8 m/sec to prevent scour of fine materials.

Mannings formula is the most commonly used method for determining open channel flow. The flow capacity Q of a channel (typically in the context of this report the channel may take the form of a kerbed roadway) is given by the expression:

$$Q = (A/n)m^{0.67}\,I^{0.5}$$

where

Q	=	discharge (m^3)
A	=	wetted perimeter (m)
m	=	hydraulic mean depth
I	=	slope (mm/metre)
n	=	Mannings coefficient (ranges from 0.015 for smooth concrete to 0.12 for an earth channel overgrown with weeds, with an average value of 0.030 for a channel lined with short grass).

13.6.1 Design flow in channels

The worst scenario is represented by the simultaneous occurrence of the following events:

- the instantaneous release of the contents of the primary containment
- the return period rainfall
- application of peak flow fire fighting water
- aggravating occurrences, e.g. drain blockage etc.

The last item should be considered as part of a HAZOP or similar assessment. When carrying out such an assessment it should be recognised that aggravating occurrences such as blocked drains may not be independent from the first and third items since the released materials, or fire, may themselves encourage drain blockage.

Normal catchment and drainage design techniques (see later in this section) may be used to calculate the design requirements for the second and third items above.

The instantaneous release of the contents of the primary containment is site specific and depending on the anticipated nature and volume of the release it may be necessary to mitigate against overflow of the catchment. Attenuation of a surge flow can be achieved by a variety of means including:

- increasing the catchment rugosity
- provision of barriers e.g. open fences, humps, kerbs etc.
- provision of sumps, interception channels etc.
- increasing the height of the kerb etc.
- increasing the catchment drainage capability (ideally so that it discharges at a similar rate to which material enters the catchment).

Although it may often be impracticable to design a channel to cope with the highest possible flow rate that could occur on a site, it is important at least to recognise those scenarios where the design flows may be exceeded, particularly on sites with a high hazard or risk rating.

13.7 PIPEWORK FOR TRANSFER SYSTEMS

13.7.1 Design capacity

It is recommended that transfer pipe networks and, where appropriate, pump sizes are designed to cope with the simultaneous occurrence of the following:

- the flow arising from an instantaneous release from primary containment (subject to any attenuation provided by catchment areas or local containment)
- the flow resulting from 50 mm of rainfall falling on relevant catchment areas during a period of 1 hour. At the designer's discretion, this requirement may be relaxed for roofed installations
- the equivalent of a 90-minute duration fire fighting water load, quantified as described in Section 9. At the designer's discretion, this requirement may be relaxed where non-flammable substances are involved.

It is recommended that the design flow rate is calculated on the basis that the pipe or pipes are three quarters full at the maximum release rate of flow.

Allowance should be made for any change in state in the released substance in the event of fire etc. For example, water applied to some flammable substances considerably increases their viscosity (see Allied Colloids case study - Section 6).

Design of the pipework should take into account the effect that items such as flame traps, siphons, restricting valves and meters may have on flow capacity.

A number of software packages are available to assist in drainage design, some of which (e.g. WALRUS (ENDS, 1993a)) include designs for detention and catchment storage.

Further information on the design of piped drainage systems is given in Appendix A18.

13.7.2 Pipework materials and construction

Table 13.3 summarises the materials and properties of commonly used effluent pipes. Table 13.4 indicates the resistance of a range of pipe and jointing materials to trade effluents.

13.8 ASSESSMENT AND IMPROVEMENT OF EXISTING DRAINAGE SYSTEMS

Many existing site drainage systems are inadequate to deal with the effects of an incident.

Table 13.3 *Pipework materials, properties and application*

Component material	Description and design references	Characteristic properties	Suggested Classification
Clayware	Manufactured diameters up to 1 m. Three strength classes given in BS 65: *standard*, *extra* and *super*. Design data including design loads, imposed traffic loads and construction of trench bedding is given in CPDA publication, *Design tables for determining the bedding construction of vitrified clay pipelines.*	Suitable for non-pressure application and use below-ground. Good internal and external resistance to aggressive chemicals. Buried clayware pipes prone to damage which is not readily detected below ground. Ground movement may affect joint integrity.	Buried clayware pipes to BS 8301. Disadvantages, including potential loss of joint integrity, brittleness etc., limit recommendation to *Class 1* only.
Plastic	Two main categories:(a) thermosetting resins, and (b) thermoplastics. (a) reinforced thermosetting resin pipes include glass reinforced plastics (GRP) and unreinforced sand and resin (RPM). GRP pipes available in diameters up to 4 m and for high pressure use. (b) polyolefins: polyethylene (PE), polypropylene (PP), polybutylene (PB). Vinyls: polyvinylchloride (PVC), acrylonitrile butadiene styrene (ABS). Extruded (PE) pipes made in diameters up to 3 m in welded form. Made in different densities, e.g. MDPE, HDPE, suitable for pressures up to 12 bar. ABS limited to about 300 mm maximum diameter. PP pipes made in diameters up to about 1.2 m. PB pipes available in diameters up to about 600 mm. Detail design and specification data available from pipe manufacturers.	Strength reduces as temperature increases (thermoplastics show significant loss). PP and PB better high performance temperature than PE. Normal maximum operating temperature of most plastics about 20°C. Categories (a) and (b) have good chemical resistance which may reduce under stress or strain. Polyolefins can be heat welded. UPVC to BS 4772 widely used for utility pipelines. UPVC pipes suitable for low pressure installations; careful design critical for high pressure pipelines. Vinyls may be solvent welded. Pitch fibre pipes which do not fall within categories (a) and (b) are limited to maximum diameters of 200 mm.	Depending on the nature of the effluent and site conditions, plastic pipes suitable for many above-and below-ground applications, and for some pressurised pipelines. Consideration must be given to potential effects of fire. Depending on various factors, plastic pipes may be considered for *Classes 1* and *2*. *Class 3* use depends on redundancies and other precautionary measures e.g. dual pipes, sleeves, pipe bunds, and monitoring controls.
Metallic	Non-ferrous metals not commonly used for effluent pipelines. Welded steel and stainless steel pipes used throughout industry for conveying fuels, oil, foodstuffs and many chemicals. With adequate protection may be used for some harmful effluents. Steel pipes are fabricated in all diameters and welded steel is commonly used to form double-skinned pipes and independent sleeves for all types of pipe. Various British Standards, e.g. BS 5500 (BSI, 1996), provide guidance. Cast iron to BS 437 (BSI, 1978) used for small diameter effluent pipework. Ductile iron is most commonly used metal for effluent pipes. Pipes available up to 1.8 m diameter, with standard bends and fittings to ISO 2531 up to 1 m diameter. Pipe lengths from 6 to 8 m. Ductile iron pipes for effluents should be specified to BS EN 598 (BSI, 1995). Detail design data provided by manufacturers.	Suitability of steel for welding, and high tensile stress, allows fabrication of continuous above- and below-ground pipelines capable of withstanding high pressures. Typical specification for ductile iron pipe for materials with pH in range 4 to 12 would consist of a pipe to BS EN 598 with an external coating of 200g/m^2 zinc to ISO 8179, overlaid with a 250 micron epoxy seal coat. An additional outer skin of polyethylene would help combat aggressive soil conditions. Tape wrapping provides further protection. Internally, pipes typically coated with epoxy and lined with high alumina cement. Exposed end surfaces coated with epoxy. Pipes may be protected internally with polyurethane lining for materials with pH< 4. High alumina cement liners offer good abrasion resistance at flow rates up to 7 m/sec. Various proprietary joining systems available; specified according to use. Ductile iron pipes may be used above or below ground and for gravity or pressurised pipelines.	Subject to design, specification and construction standards, ferrous pipes are recommended for *Class 2*. This applies to above- and below-ground installation and to gravity and pressurised networks. *Class 3* installations require redundancies such as system duplication, together with full monitoring and control of leakage. In *Class 3* situations, pipes should be above ground level and within sleeves or ducts. Alternatively they should be within concrete ducts to BS 8007, where located below ground level. Sleeves and ducts to conform to *Class 3*. These provisions allow for easier inspection and maintenance.

Table 13.3 (contd.)

Component material	Description and design references	Characteristic properties	Suggested Classification
Cements	Concrete effluent pipes may be unreinforced, steel or fibre reinforced, or prestressed. Principal design performance standards are BS 5911(BSI, 1992) and BS 4625 (BSI, 1972). Unreinforced pipes, made up to 1.4 m diameter, suited only to gravity flows. Reinforced pipes are up to 3 m diameter and used in some low pressure systems. Prestressed pipes also made in large diameters and used in pressurised systems. Number of proprietary pipe systems manufactured from composites of concrete and reinforced plastic. Technical Bulletin No 6 (CPA, 1981) provides guidance on installation of concrete pipes.	Concrete is robust but prone to attack from sulphates in groundwater, acid effluents, sugars and bacteria. Expensive measures needed for protection and these tend to offset initial cost advantages of concrete. Pipe abrasion and erosion governed by material flow and nature. Pipe linings usually applied after manufacture; epoxy or polymer resins commonly used. Linings cannot be prestressed.	Unreinforced concrete pipes recommended for *Class 1* only. Reinforced and prestressed pipes recommended for *Class 2*. Specifications and site installation controls must be high to warrant *Class 2*. This recommendation applies to pipe components. Additional safeguards will be essential for class3.
Joint materials	Many joints systems are proprietary or have been developed over the years by various bodies such as The Water Research Centres. A number of methods are also described in the BSs for pipes.	Most flexible joints rely on a form of spigot and socket. Other methods include double collared joints, sleeved joints and flanged joints. Welding is used for thermoplastics, steel and some non-ferrous metals. Flexible joints rely on an elastomeric sealing ring or gasket. Ethylene-propylene rubber and styrene butadiene rubber used in preference to natural rubber for normal sewage effluent, since natural rubber subject to biodegradation in sewage. Some aggressive materials may require use of other synthetic polymers.	Since many jointing systems are proprietary, necessary to confirm the classification and agree the specification details with the manufacturer.

Table 13.4 *Chemical resistance of pipework materials* (Source: BS 8301)

Group	BS no.	Material and applications	Normal domestic sewage	Trade effluent								
				At normal temperature		Organic solvents	Containing oils and fats		At sustained high temperature		Soil environment containing	
				Acids	Alkalis		Vegetable	Mineral	Acids	Alkalis	Sulphates	Acids
Ceramics	65,1196	Clayware pipes and fittings	A	S	S	S	S	S	A	A	S	S
	3921	Bricks and blocks of fired brick-earth, clay or shale	A	S	S	S	S	S	A	A	S	S
Concrete	5911	Concrete:										
		ordinary Portland cement	A	E	A	A	E	A	E	A	E	E
		sulphate-resisting Portland cement	A	E	A	A	E	A	E	A	A	E
Asbestos cement	486	Asbestos cement pressure pipes and joints										
	3656	Asbestos cement pipes, joints and fittings (gravity) for sewerage and drainage	A	E	A	A	E	A	E	A	A	E
Metals	534	Steel pipes and fittings	A	E	A	A	A	A	E	E	E	E
	437	Grey iron pipes and fittings (gravity) }										
	4622	Grey iron pipes and fittings (pressure) }	A	E	A	A	A	A	E	E	E	E
	4772	Ductile iron pipes and fittings }										
Pitch fibre	2760	Pitch fibre	A	A	A	E	E	E	E	E	S	S
Plastics	4660, 5481,	uPVC -gravity drain and sewer pressure	A	S	S	E	A	A	E	E	S	S
	3506, 5480	GRP - gravity drain and sewer pressure	A	A	A	E	A	A	E	E	A	A
Jointing materials (other than Portland cement and iron and steel components (see under pipes))	2494	Rubber										
		natural (NR)	A	A	A	E	E	E	E	A	A	A
		chloroprene (CR)	A	S	A	E	A	A	A	A	S	S
		BUTYL (IIR)	A	S	S	A	A	A	A	A	S	S
		styrene-butadiene (SBR)	A	A	A	E	E	E	E	A	A	A
		nitrile-butadiene (NBR)	A	A	A	E	A	A	E	A	A	A
		ethylene-propylene terpolymer (EPDM)	A	A	A	E	A	E	A	A	A	A
		ethylene-propylene copolymer (EPM)	A	A	A	E	A	E	A	A	A	A
		*Bituminous compositions	A	A	A	E	E	E	E	E	A	A
		*Polyester resin	A	S	A	E	A	A	E	E	S	S
		Polyethylene:	A	A	A	A	S	A	A	A	S	S
	1972	(Type 32) cold water services										
		(Type 32) general purposes including chemical and food										
	3284	(Type 50) cold water services										
	6437	(Type 50) general purposes including chemical and food										
		*Polyurethane	A	S	S	A	S	A	A	A	S	S
		*Polypropylene	A	S	S	A	S	A	A	A	S	S
		*Epoxy	A	A	S	E	E	A	E	S	S	A

A - normally suitable E- need expert advice, each case to be considered on its own merits S - specially suitable

Note: It is important to take account of **quantities and concentrations** of all types of chemical likely to be encountered.

*There are no relevant British Standards for these jointing materials. The classification given to them in the table assumes that the formulations and methods of curing employed are appropriate for pipe jointing.

Where there is any uncertainty about the layout and condition of drainage systems for trade waste it is recommended that a survey is carried out. Potential hazards in terms of the system's ability to cope with major incidents should be identified. Typical hazards may include:

- combined surface and trade waste systems, especially where provided with storm overflows
- direct runoff or infiltration of contaminants through drains to surface and groundwater
- unprotected clean water gullies and other devices for collecting storm water from areas likely to become contaminated
- unprotected roof rainwater pipes on hazardous buildings which, if damaged (typically in a fire) could allow pollutants to enter storm drainage
- inadequate or poorly maintained valves and other devices used for flow control
- defective pipes or joints
- inadequately sized systems
- inadequate robustness or durability.

Simple improvements, such as providing gully surrounds or re-contouring catchment areas, can overcome many minor defects for relatively little cost. Where it is concluded that replacement or additional drainage is required this should be designed to cope with all foreseeable future requirements. The Environment Agency recommends that separate drainage systems are identified by colour coding drainage installations, including access chamber covers and pipe runs, wherever feasible. Red is suggested for effluents and blue for clean water.

Internal surveys by CCTV may be necessary to identify leakage pathways. Defective and inadequate pipelines should be replaced or, if feasible, repaired or upgraded. Repair techniques have been investigated extensively by the WRc (1983). The following categories describe the principal techniques used for pipeline renovation:

- glassfibre concrete and glass reinforced plastic linings which are bonded to the interior of existing pipes
- non-bonded lining systems such as PE pipes or thermosetting resin and other plastic liners inserted within the old pipe
- lightweight GRP or resin plastic liners used as annular formwork for injected grout.

It is important to institute an ongoing maintenance programme to minimise blockages and to keep drainage systems fully serviceable.

13.9 UPGRADING TRANSFER PIPEWORK CLASS RATING

There are a number of ways in which pipes and pipe networks may be upgraded to a higher classification. These include coatings, linings, composite pipes, special joints, protective covers or sleeves and pipe bunds for above- and below-ground drains. Most of the methods provide a measure of inbuilt redundancy. Bunding of pipework prevents the escape of any leakage and permits visible inspection of below-ground pipelines. Below-ground reinforced concrete ducts or bunds should be constructed to BS 8007 requirements. Above-ground pipe bunds should comply with the relevant structural codes and also be of watertight construction. Ventilated roofs and covers may be needed to prevent rainwater collecting in the pipe bund.

13.10 FLAME ARRESTORS

Traps should be provided in gullies, inspection chambers and tanks etc. to prevent the spread of fire.

13.11 SYSTEM TESTING

BS 8301 Section 5 describes procedures for air and water testing of drains and ancillary works for gravity pipe networks. It is recommended that all below-ground pipelines are tested before and after backfill. Acceptable head losses are given for testing gravity pipes to BS 8301.

Test procedures are also set out for pressure pipelines. Test pressures are given at up to twice the maximum flow head pressure of the pump, plus an appropriate allowance where surge conditions can occur.

13.12 TANKER OFFLOADING AND LOADING FACILITIES

Catchment areas should be adequately sized to contain at least the maximum contents of one compartment of a road tanker and in any case not less than 1 000 litres. Additional provision should be made for dealing with rainwater. Ventilated below-ground impermeable sumps or tanks may be used to increase the capacity of the catchment area to the required volume. Collection tank alarms and automatic shut-off valves at loading and unloading points are desirable on installations with a high hazard or risk rating.

13.13 DETENTION TANKS, SUMPS AND PUMPING STATIONS

Detention tanks and sumps are used primarily as balancing tanks prior to transfer to containment. They may also be used for dosing or neutralising harmful substances, for example by pH adjustment. The capacity of a detention tank, and any pump balancing tank, together with its catchment, should be sufficient to prevent the peak flow of drainage systems being exceeded and sufficient to allow time to deploy pumps, or repair or replace failed pumps. Designs must accommodate normal effluent flows, spills, potential contaminated rainwater and potential fire fighting water runoff. Where detention tanks, sumps and pumping stations form an integral part of a catchment or bunded area the detention requirement should be determined cumulatively.

Drainage and pump capability will determine the most effective method for coping with a 90-minute duration fire water runoff and a 60-minute duration rainfall. Where detention tanks are to be used for sedimentation of materials, they should be designed for non-turbulent flow conditions. Long shallow tanks are better than short deep tanks for the sedimentation of discrete particles, with a design flow rate through the tank of 30 to 40 m^3 per day per m^2 cross sectional area, when based on two hours' retention time.

For pumped systems, wet-well or dry-well pumping stations may be used.

14 Sacrificial areas and temporary containment

14.1 INTRODUCTION

The Environmental Protection Act 1990, sub section 33, makes it an offence to deposit, knowingly cause or knowingly permit controlled wastes to be disposed in or on any land, except under, and in accordance with, a waste management licence. Controls are exercised by the Waste Regulation Authority (WRA) and the Environment Agency/SEPA. The penalties for non compliance are severe and similar to those set for contraventions of the Water Resources Act 1991.

Waste management licences do not include discharges of trade and sewage effluent and it is improbable that any other form of consent would be issued. A possible defence with respect to water pollution may be argued under Section 89 of the Water Act 1991 if an unauthorised discharge was made in an emergency in order to avoid damage to life or health.

Where sacrificial areas form part of an emergency plan the sacrificial soil or media used to retain the pollutant should be contained within a barrier to prevent the spread of contaminants. The contaminated soil or media should be disposed to a licensed disposal site as soon as possible after the incident.

This section illustrates examples of 'sacrificial' and temporary methods which may be used to contain the spread of pollutants.

14.2 EXAMPLES OF SACRIFICIAL AREAS

14.2.1 Car parks, sports fields and other landscaped areas

The method relies on interception of spills at the source and conveying the contaminant runoff to a remote area designated as a sacrificial site area. Sites which may be designated for this purpose include car parks, landscaped areas, sports fields etc. The sacrificial area soaks up the contaminant, containing the spill within a depth of permeable soil or porous media. The contaminant should be prevented from dispersing into other strata or groundwater by an impermeable barrier of clay or by means of some other impermeable lining system. The storm water drainage serving the area must be capable of being shut off quickly and effectively during an incident and until the contaminant is dealt with. After contamination the permeable material should be excavated and removed to disposal or for recycling treatment. The method may not be suitable for sites with a high water table.

The sacrificial area should be designed to allow infiltration and to prevent run off. The infiltration capacity should be sufficient for storage of the contaminants, and other infiltrates, prior to disposal. This requires that the area is provided with adequate under-drainage to cope with percolation, i.e. the normal vertical flow of water through the soil. The outfall of the percolation drains must be capable of immediate closure at the time of an incident.

Figure 14.1 illustrates a parking or landscaped area used as a sacrificial area. The source is surrounded by an impermeable catchment for directing spills and rainwater towards the sacrificial area. The porous medium (a) allows infiltration to the percolation drains (b). In this example, a concrete lined open channel (e) is used to intercept the outlets of the drainage pipes (b). A sluice (d) is used to close off the outlet of the open channel during emergencies. An impermeable liner and leakage detection system (c) is set below the pervious medium of granular fill (a).

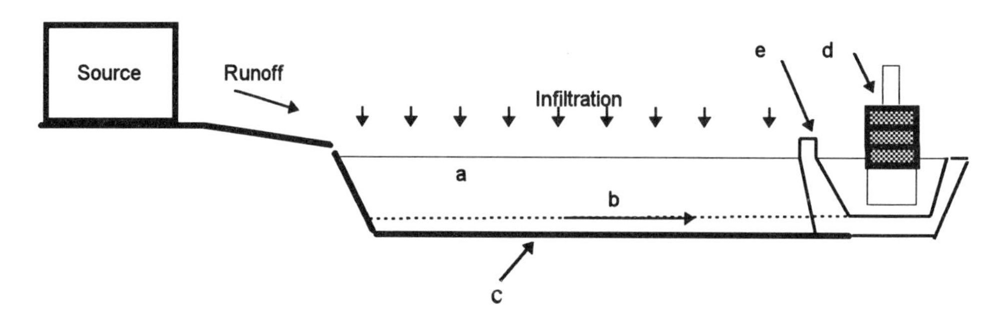

Figure 14.1 *Longitudinal section through a typical surface area using sacrificial media for containment*

Where layers of geotextile are used to reinforce the granular fill it is be important to ensure that the infiltration capacity is not impeded by the geotextile. This can be determined using Darcy's law:

$v = k\,(dh/dl)$

where

v	=	velocity of flow per unit area
k	=	permitivity
dh/dl	=	hydraulic gradient (head/ thickness of geotextile).

To maintain effective filtration and to minimise the migration of small particles, the O_{90} size of the geotextile should be about 5 D_{90} for non-wovens and 2.5 D_{90} for woven fabrics. D_{90} refers to percentage grading classification of the soil/ media.

The spacing of the percolating drains to give a maximum head (h) may be determined theoretically as follows:

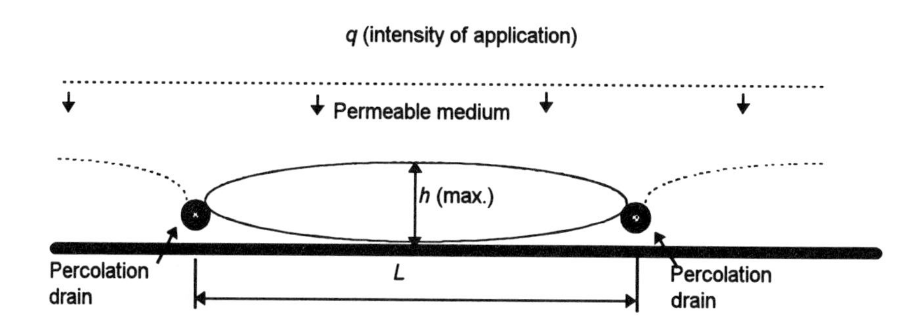

$L = 2h(\text{max})(\,k/q\,)^{0.5}$

from which $h(\text{max}) = (q/k)^{0.5}\,(L/2)$

Knowing the permeability and voids ratio of the permeable medium, the drain spacing and flow capacity, it is possible to determine the maximum head *h* and the theoretical capacity of the sacrificial medium.

It is important to maintain the flow capacity of the percolating drains to provide a fully effective system.

Retaining gabions or mattresses and impermeable soil or lined open channels may be used as alternative systems.

14.2.2 Areas surrounded by vertical cut-off walls

The most common method in the UK for constructing a vertical barrier wall is by excavating a trench under bentonite/cementitious slurry (see Figure 14.2). The slurry hardens within about 24 hours to form a low permeability (10^{-8} to 10^{-10} m/sec) barrier with the strength of a stiff clay. This is known as the 'single phase' method. Other methods involve excavation under a bentonite slurry followed by injection with a bentonite/cementitious grout.

Geomembranes, placed vertically, may be provided for additional integrity. The membranes, commonly HDPE, are installed with the help of a temporary metal frame following trench excavation, in panels usually 2 to 6 m wide and up to 30 m deep, with purpose-made high integrity inter-panel joints.

A detailed site investigation should be carried out prior to excavation to determine the presence and extent of the aquiclude and its level and dip, and also the presence of any geological faults, fissures or discontinuities. The investigation should determine the hydrogeology of the site and any physiochemical characteristics that may adversely affect the performance of the barrier wall.

Vertical barrier walls may be used to enclose the circumference of a site. Full containment is provided where the wall is keyed into underlying impermeable strata. In the absence of an aquiclude, a low permeability horizontal layer can be created by grout injection. When used to control groundwater flow, it is not always necessary for the walls to extend down to the aquiclude (see Figure 14.3).

Trenches are normally about 600 mm wide; widths over 1 m are unusual. Hydraulic back-acter excavators may be used at depths to about 12 m in most ground conditions. Diaphragm wall grabs or cutters are used at depths below 12 m. The walls should be capped with 0.5 m clay soon after completion, to prevent desiccation and cracking.

The walls prevent pollutants from migrating beyond the boundaries of the enclosure. In the event of an incident, any contaminated soils will require appropriate treatment or safe disposal.

Cement/ bentonite slurry mix materials should conform to the following specification:

- sodium exchanged bentonite (Engineering grade bentonite)
- Ordinary Portland Cement (BS 12)
- ground granulated blast furnace slag (ggbs) (BS 6699)
- pulverised fuel ash (BS 3892 Pt 1and Pt 2).

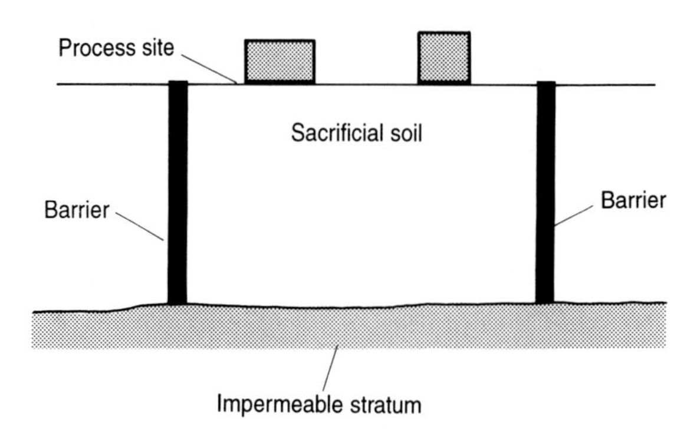

Figure 14.2 *Section through barrier wall containment system*

Cementitious materials are usually a blend of OPC and 70 to 80% ggbs. Permeabilities of around 10^{-9} m/sec can be achieved using these proportions at total cementitious content of 90 to 150 kg/m^3 of slurry.

Barrier wall construction requires specialised techniques. In the UK, the method is used for landfill leachate containment and for encapsulation of contaminants, or where additional fail-safe containment is required at particularly sensitive installations. It may also be used to contain incidental contaminants, preventing migration in groundwater.

Barrier walls are used in the USA to control groundwater flow. The US Environmental Protection Agency demonstrates a number of techniques that may be used for this purpose. An example of a barrier wall used to contain, and extract, leaked fuels is illustrated in Figure 14. 3.

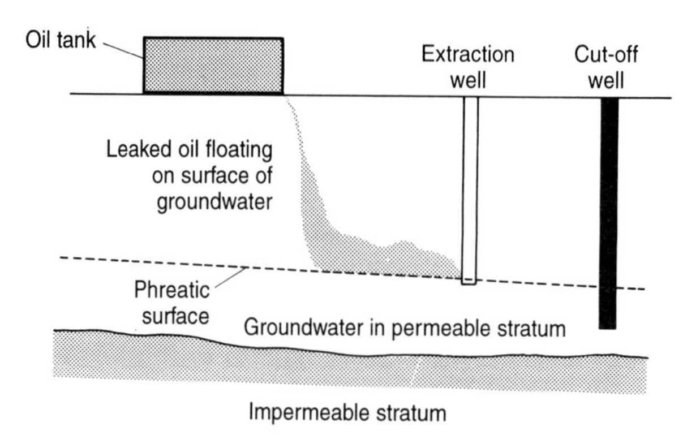

Figure 14.3 *Cut-off wall used to contain oil spillage within permeable stratum*

14.3 EMERGENCY PLANS FOR SACRIFICIAL AREAS

The relevant regulatory bodies should be consulted on the use of sacrificial areas for dealing with major spills. Sacrificial areas should not be located where they might endanger life or property in the event of an incident. Where dual purpose areas are used for containment and these areas are provided with clean storm water drainage facilities, it is essential that all operatives, the Fire Service and any other emergency service, are made aware of the correct procedures for using the areas, including the diversion or closure of drains.

14.4 EMERGENCY AND TEMPORARY CONTAINMENT MEASURES

Permanent containment facilities will be provided at most sites as part of emergency planning for controlling hazardous substances. The plan should also include incident response strategies and preventative measures for dealing with exceptional events that cannot be dealt with by the permanent facility. Many sites having the potential to cause pollution may not require permanent containment facilities since, although they may be handling polluting substances, only small quantities are involved. These sites may rely on the use of suitable emergency pollution control materials and equipment to prevent pollution. As part of their environmental control duties, the Environment Agency/ SEPA have developed many emergency procedures for containing pollutants to prevent and minimise water pollution. A number of these methods are briefly described in this section. It is recommended that the Environment Agency/ SEPA are consulted on suitability where these methods are to form part of an emergency plan.

14.5 EXAMPLES OF TEMPORARY CONTAINMENT MEASURES AND EMERGENCY MATERIALS AND EQUIPMENT

Temporary bunding of vehicular parking areas and other hardstandings

Impermeable yards, roads and parking areas can be converted to temporary lagoons using sandbags, suitable excavated soils or sand from emergency spoil heaps to form perimeter bunds. This is shown schematically in Figure 14.4.

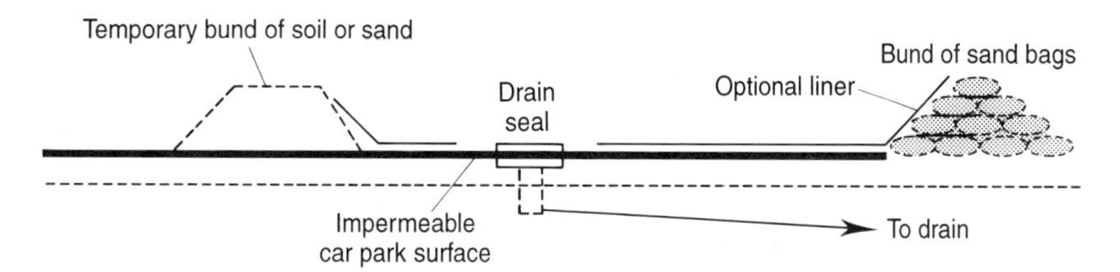

Figure 14.4 *Sacrificial car park area showing temporary bunds and drain seals*

All drain inlets such as gullies and manhole covers within the area must be sealed to prevent the escape of pollutants. Proprietary equipment is available from several manufacturers for this purpose. Liners may be used to help protect the bunding material from contamination and to help improve the impermeability of the land surface. Where soil or sand used for bunding becomes contaminated it must be properly disposed of as sacrificial material. Pits can be excavated and lined to form temporary sumps for collecting and pumping pollutants.

Interception/detention pits and trenches

Interception trenches should be used only where other methods have failed and where it is essential to protect life, or where the risk of damage to the environment, or property, outweighs the threat of ground or groundwater contamination. Where the natural ground is relatively impermeable, contamination will be minimised by removing the pollutant as soon as possible after the incident. Liners can be used as temporary barriers to improve impermeability although these cannot be used in fires or for deleterious substances. After the event, contaminated earth should be removed to a disposal site. Pits and trenches may help prevent the spread of burning substances. The pits may also be used to add reagents for neutralising harmful substances.

Portable tanks and tankers

Portable tanks made from synthetic rubber, polymers and other materials can be used to contain small spills. In most cases a sump or pit is needed to collect and pump the pollutant into the tank. Vacuum or similar mobile tankers may also be used for collecting and containing small spills. Where tanks and tankers are part of an emergency plan, these should be in a readily accessible location and maintained in a serviceable condition.

Drain pipe seals

Pipe seals may be used within a drainage system, providing significant storage volumes within the pipe network. Seals should be kept in a readily accessible location close to vulnerable drainage runs. Care should be taken in their installation to avoid exposure to hazardous conditions within the drainage system and to ensure that the contained liquid does not overflow into other gullies or drainage systems. Special arrangements will be necessary to empty the system after use.

Absorbents

Absorbent materials, such as sand and sawdust may be used to soak up small spillages. Wet or moist absorbents are unsuitable in situations where they are likely to come into contact with chemicals that react with water. Stocks of suitable absorbents should be available adjacent to potential spillage sites. The Environment Agency is able to advise on suitable absorbents, including proprietary absorbents.

Adsorbents

Granular activated carbon (GAC) is used extensively throughout industry to remove contaminants from water, and to treat and purify a wide range of liquids and gases. The pores within GAC have an extremely large internal surface area giving GACs the ability to adsorb a very large amount of substance by molecular attraction. For some, but not all contaminants, GAC may be successfully used to 'dose' detention tanks or pits and trenches, and lower pollutant concentrations towards acceptable handling standards. The spent GAC is afterwards removed for recycling or disposal. Although GAC is not in itself a containment system it may be possible in certain instances to use this method to reduce the capacity requirement for total containment.

Booms

Floating booms are used to contain spills of oil and similar substances that float on water, to prevent migration and to facilitate clearing operations. These are beyond the scope of this project. A number of proprietary inflatable plastic booms have been used as temporary devices to contain pollutants on land although this use is not widespread.

A1 References to existing guidance listed by topic area

Topic area	Reference number (see reference list)
Bunds	4, 6, 54, 81, 90, 97, 108, 115, 116, 117, 121, 127, 129, 155
Capacity (contaminants)	6, 90, 115, 116, 117, 121, 129
Capacity (fire fighting water)	6, 38, 90, 93, 99, 102, 116,121, 129, 150
Capacity (rainwater)	115, 116, 121, 129
Chemical Stores	1, 54, 80, 99, 121, 125, 129, 150, 156,
Concrete	3, 9, 10, 15, 16, 17, 19, 21, 27, 41, 55, 82, 109, 116, 117, 137, 141, 142
Drainage	9, 12, 22, 26, 31, 34, 35, 40, 42, 59, 60, 79, 101, 102, 117, 152, 154
Earthworks	14, 33, 51, 63, 115, 120
Emergency planning	65, 93, 94, 99, 105, 106, 129
Fire fighting and fire water	38, 79, 80, 91, 93, 99, 117, 129, 150, 153
Flammable liquid stores	80, 81, 84, 87, 88, 99, 105, 106, 129, 153
Foundations	25, 33, 117
Geotechnical investigation	57, 118
Geotextiles	14, 30, 103, 11, 118, 135, 136
Impermeable liners	103, 117, 118, 135, 140, 149
Incidents/hazardous substances	1, 5, 65, 66, 68, 69, 72, 73, 74, 75, 76, 77, 78, 80, 92, 114, 120, 121,.125, 129, 131, 134, 138, 139, 143, 151
Lagoons and reservoirs	14, 95, 112, 115, 135
Maintenance of structures	114, 115, 116
Masonry	23, 45, 62, 83, 115, 135
Oil separators	123, 124
Oil storage	2, 29, 62, 98, 105, 116, 122, 124
Paved areas	43, 46, 47, 115
Sealants and joints	3, 44, 46, 58, 112, 116
Site investigation	10, 14, 39, 104, 120
Soil classification	14, 39
Soil stability	39, 113
Structural steelwork	2, 7, 24, 32, 145
Tanks (concrete)	49, 115, 116, 133
Tanks (GRP)	28, 116
Tanks (steel)	2, 18, 29, 36, 48, 49, 115, 116, 133, 145
Water pollution controls	6, 56, 61, 64, 65, 66, 67, 68, 69, 70, 84, 85, 86, 87, 88, 89, 90, 91, 93, 94, 95, 97, 98, 117, 119, 144

A2 Examples and uses of existing hazard and risk assessment methodologies

In some cases it will be obvious that a particular hazard, e.g. an unbunded stock tank of a toxic chemical on the bank of a good quality stream, requires mitigation, and there will be no need to carry out long and detailed risk assessments. Equally there will be cases, particularly at large chemical complexes, where the situations of highest risks are not obvious and cannot be ascertained without detailed investigation by experts in safety and reliability techniques.

The following sections describe some of the hazard and risk assessment procedures that are available to the risk assessor. They concentrate on practical issues which need to be taken into account by the site environment adviser, the safety and reliability engineer or the design engineer, and on how the results of the assessment can be used as an input in decision making.

A2.1 QUALITATIVE HAZARD ASSESSMENT

(Commonly referred to as end-of-pipe approach)

Description:

- calculate the concentration at the discharge point to the river, of materials resulting from total loss of containment, assuming a direct pathway to the river
- assume spill takes place over a fixed period, say 1 hour
- allow for dilution in the site effluent
- the river provides a minimum dilution of 50-fold: assume good mixing in the river
- there are fish and other aquatic life in the river and it is decided that the appropriate trigger level for environmental damage is 10 times the EQS[1]
- compare concentration of diluted effluent with the pre-set trigger levels
- method is qualitative and simple to use.

---

[1] *The EQS (environmental quality standard) is often derived from laboratory toxicity tests on aquatic life. It is an average value, typically set at one hundredth of the concentration in water which kills 50% of the exposed aquatic organisms within a given time period, usually 96 hours (LC50). For short-term duration exposure the EA may require compliance with a maximum allowable concentration of 10 times the EQS. In reality the same hazard ranking is obtained whatever trigger threshold is chosen for environmental damage, but the absolute hazards differ. Drinking water standards may need to be considered.*

Output:

- water polluting materials in use at a site can be ranked in terms of their damage potential
- the method can be used to calculate the size of spill which would trigger an incident.

Example:

Three substances are in use at a chemical works. Loss of containment would result in each chemical flowing into the adjacent river from the one consented outfall. Concentrations in the river can be calculated and compared with trigger levels as shown in Table A2.1.

Table A2.1 *Comparison between in-river concentrations after releases and environmental trigger concentrations*

	Chemical		
	Vinyl acetate	**Methylene chloride**	**Ferric chloride (5% solution)**
Maximum potential release (tons)	5	2	1
Flow at outfall (m^3/h)	45	45	45
Dilution in site effluent	10 x	23.5 x	46 x
Concentration at outfall (mg/l)	100 000	42 500	370 (as Fe)
Concentration after dilution in river (mg/l) - A	2 000	850	7 (as Fe)
EQS (mg/l)	0.2	2	1
Trigger level (mg/l) - B	2	20	10
A/B	1 000	42	0.7

The methodology indicates that:

1. Loss of containment of vinyl acetate would generate a significant environmental impact and hence that its storage, transfer or use requires secondary containment.
2. The loss of containment of methylene chloride would also produce adverse environmental damage, although much less than a spill of vinyl acetate, and that secondary containment is necessary.
3. Loss of containment of the solution of ferric chloride would not produce an incident and hence secondary containment is not necessary.

Comment:

The method can be refined by taking different dilutions depending on the flow of receiving water, by allowing for reactions in the river, by using different trigger levels for different levels of risk, by taking different sizes of spills etc. (see example 2 below).

A2.2 SEMI-QUANTITATIVE SYSTEMS

Description:

Phase 1

- define incidents in terms of environmental impact; e.g. low, moderate, high (see example)
- define spill duration, say 1 hour

- allow for dilution in other site effluents, if any
- obtain maximum dilution in receiving water, e.g. use dry weather flow (obtainable from the EA) to simulate worst scenario and take volume of water which flows past discharge point in 1 hour
- calculate size of potential releases which would generate the three categories of incident
- decide if quantities in use at site can generate incidents.

Phase 2

- if incidents could be generated, carry out a HAZOP at site to determine likely frequency of events, such as failure of plant and equipment, or fire, which could cause incidents in receiving water
- specify tolerable frequency of incidents; e.g. 1 per 10^2, 10^3, 10^4 years for low, medium and high risk respectively.

Output:

- reasons for potential loss of containment are identified and quantified
- frequency of events/ incidents are evaluated, and hence the risk is quantified
- frequency of events/ incidents can be compared with tolerable frequency
- focused action can be taken to increase integrity of containment if frequency too high.

Example:

A factory uses a range of chemicals in varying quantities. There are fish in the river close to the factory's outfalls and trigger concentrations are specified as 10, 100 and 1000 times the EQS for low environmental damage (category 1), moderate damage (category 2) and high damage (category 3) impact. From a knowledge of the spill duration, dilution in the site effluent, dilution in the receiving water, and the EQS values for the chemicals concerned, the quantities of chemicals spilled which will trigger the three incident categories can be calculated as shown in Table A2.2.

Table A2.2 *Size of release to trigger incident categories*

Substance	Maximum potential release (tons)	Size of release to trigger incident categories		
		Cat 1	Cat 2	Cat 3
Phenol	0.2	>0.005	>0.05	NI
HCl	2	NI	NI	NI
NaOH	10	>2.5	NI	NI
Ammonia	0.5	>0.002	>0.02	>0.2
Methanol	0.5	>0.3	NI	NI

Where there is not sufficient material to trigger an incident the term NI (no incident) is used.

In the case of methanol the effect of a spill is based on its potential to deplete the river water of oxygen (rather than directly on aquatic toxicity).

Phase 1 of the procedure indicates that only ammonia is stored in sufficient quantities to generate severe incidents if the maximum potential release takes place; that loss of phenol can trigger an incident of moderate severity; that loss of caustic soda and methanol can generate low severity incidents; but that hydrochloric acid cannot be released in sufficient quantity from this site to generate an incident.

A HAZOP study is carried out to assess the likely frequency of incidents, as shown in Table A2.3

Table A2.3 *Likely frequency of incidents*

Substance	Frequency of event triggering an incident category (years)		
	Cat 1	Cat 2	Cat 3
Phenol	1/100	1/100	NI
NaOH	1/10 000	NI	NI
Ammonia	1/10	1/100	1/100
Methanol	1/10	NI	NI
Tolerable frequency (years)	1/100	1/1000	1/10 000

In the case of phenol, the HAZOP shows that the same event triggers both category 1 and 2 incidents; in the case of ammonia, one event triggers a category 1 incident, another event triggers both category 2 and 3 incidents. The probability of an incident is higher than the tolerable frequency with ammonia (category 3 and below), with phenol (category 2) and methanol (category 1). The highest priority should be given to improving the containment of ammonia since the ratio of the frequency of a category 3 spill occurring compared with the tolerable frequency of such a spill is ten times greater than that of a category 2 spill with phenol. The company concerned would need to take a decision, perhaps in conjunction with the statutory authority, on whether phenol or methanol containment is the higher priority, since a spill of methanol has less impact than that of phenol but is more likely to occur. The chance of a spill of caustic soda causing a category 1 incident is lower than the tolerable frequency and containment is probably not necessary.

Comment:

The method is difficult to use because:

- modelling the flow of the pollutant material from source of spill to water and its subsequent environmental impact can be very complex
- it may be necessary to allow for attenuation of the polluting properties of the material, on site, off site and in the receiving water through, for example, such effects as volatility, absorption, reaction or biodegradation
- defining environmental incidents and tolerable frequencies is contentious - in particular the choice of trigger concentrations for incidents will depend on the dose/response relationship of the aquatic species targeted
- a detailed study of all aspects of the plant and equipment at the site has to be carried out by safety and reliability engineers.

The effect of dispersion in the river and variable river flow can be built into the model using more complex procedures such as those incorporated into the PRAIRIE model used by the Environment Agency to simulate the effect of spills in the River Dee catchment and other rivers.

Although the two foregoing methods relate to the effect of pollutants on fish, they can be modified to predict the effect on humans. The potential abstraction risk index, for example, was developed in order to assist the Water Authorities to assess the effect of spills of toxic substances into rivers used for potable supply. A tolerable level of the pollutant in the water supply is derived and the ratio of the predicted concentration to the tolerable concentration is interpreted in risk potential terms; a ratio of up to 10, 100 and 1000 representing low, high and catastrophic scenarios respectively.

There is no need to carry out the HAZOP if incidents as defined by pre-set criteria cannot be generated.

The method can be applied to substances which float or sink in water provided the nature of the environmental impacts can be quantified.

The first two examples assume that hazardous materials are spilled as isolated events from, say, a storage tank over a one-hour period. They can equally be used for assessing the risk of a very large amount of contaminated fire fighting water entering a watercourse as a result, say, of a fire at a warehouse. In the latter case the spill duration could be several hours. The main difference between the fire risk at a warehouse and the examples given is that there would be only one event to consider, but this would involve the simultaneous discharge of a large number of polluting materials in a very large amount of diluting fire fighting water.

A2.3 MATRIX SCORING SYSTEMS

Description:

Sites may be ranked subjectively on the basis of the environmental risks posed by the operations taking place there. Pritchard (1994) has reviewed various models of which the following is an example.

Example:

A risk management procedure for underground storage tanks (UST) has been developed in the USA, where tens of thousands of such tanks have been found to be leaking. A rating system ranks the frequency of a release from an UST and its environmental consequences into broad risk category bands - low, moderate and high risk.

Various key factors are included in the assessment of the release potential as shown in Table A2.4.

Table A2.4 *Release Potential Factors for underground storage tanks*

Features of tank	Release Potential Factor		
	Low risk	Moderate risk	High risk
Age of tank and associated equipment (years)	0-10	11-20	>21
Corrosion protection	Tank and pipework protected	Tank or pipework protected	None
Leak detection	Continuous	Occasional	None
Soil corrosivity	pH > 7	pH 5.5-7	pH < 5.5

No points are allocated for criteria falling in the low risk band, one point in medium risk and two points in high risk.

The key factors in the assessment of damage potential (consequence) are tabled in similar fashion as shown in Table A2.5. The damage is measured in terms of impact on drinking water quality, i.e. human health.

The same scoring system for the release potential is used. Thus the range of available scores for each factor is 0 - 8.

Table A2.5 *Damage Potential Factors for underground storage tanks*

Water supply/use factors	Damage potential factor		
	Low risk	**Moderate risk**	**High risk**
Depth to water table (ft)	>50	15-50	<15
Soil hydraulic conductivity (gallons/day/ft^2)	<1 (e.g. clay)	$1\text{-}10^3$ (e.g. sand)	$>10^3$ (e.g. gravel)
Population served by ground water supply	$<10^3$	$10^3 - 10^4$	$>10^4$
Distance to surface water used for potable supply (miles)	>1	0.1-1	<0.1

Output:

The values for the two factors are placed in a risk rating matrix as shown in Figure A2.1

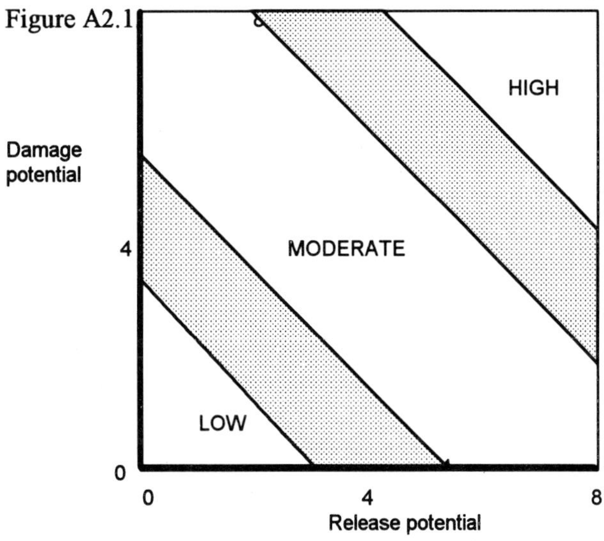

Figure A2.1 *Matrix scoring method: risk rating chart*

Comment:

- method is simple to use but qualitative and subjective
- it can be used for storage of materials in tanks
- as many release and damage factors can be built in as required, tailored to the environmental targets
- method can be applied to the release of contaminated fire fighting water.

A2.4 PRESCRIPTIVE APPROACH (GERMAN WGK SYSTEM)

Description:

Many thousands of chemicals have been classified according to their ecotoxicity properties and their potential to endanger water:

WGK class 0 (very low), class 1 (slight), class 2 (average), class 3 (high).

Containment measures are prescribed for the storage, transfer and loading of substances according to the WGK class.

Since 1986 plants manufacturing, treating or using water-endangering substances are also included, with the result that practically all industrial plants in Germany are controlled in this way.

Output:

Prescribed containment measures for certain activities depending on WGK class.

Example:

A simplified version of the German method for the loading and off-loading of materials is shown in Table A2.6 below.

Table A2.6 *Protective measures for the loading/off-loading of materials according to WGK class and vessel size*

Filling/emptying moveable vessels (capacity ranges)	Automatic safety devices	WGK 0	WGK 1	WGK 2	WGK 3
up to 1000 litres	No	A1+B1	C1+A2+B3	C2+A5+B5	C2+A5+B5
	Yes	None	C1+A1+B3	C2+A3+B5	C2+A3+B5
greater than 1000 litres	No	A1+B1	C1+A4+B3	C2+A5+B5	C2+A5+B5
	Yes	None	C1+A3+B3	C2+A3+B5	C2+A3+B5

Notes:
Various automatic safety devices are specified for the two size categories of vessels.
A relates to retention capacity and varies in size from A1 to A5:
A1 - retention of small amounts of liquid, e.g. a drip tray under a pump
A5 - retention for entire contents of vessel from which filling takes place
B relates to measures for disposing of contaminated rainwater and snow, increasing in integrity from B1 to B5:
B1 - connection, without additional precautions, to the foul sewer
B5 - connection to an appropriate internal effluent treatment plant
C relates to the impermeability of the ground area and has two categories C1 and C2:
C1 - impervious paving (e.g. concrete, asphalt)
C2 - as in C1 but with certification of resistance

Comments:

By its very nature the system makes no allowance for specific site or local environmental factors.

The method is based on source classification, i.e. on the ecotoxicity of hazardous substances in water. **It is not a risk assessment method.**

The method is representative of the German precautionary principle, i.e. if a material is dangerous to water it must be contained in a specified manner.

The system is applied to new and existing installations.

A3 Lists of hazardous substances

Lists of hazardous substances are available from many sources, including UK and EC legislation, international conventions, national priority pollutants from non-EC countries etc. The source documents generally aim to protect surface or groundwater, fish or shellfish, or water for human consumption.

Other legislation has generated lists of hazardous chemicals; for example, the CIMAH regulations or the classification, packaging and labelling regulations. Their prime aim may be health and safety, consumer protection, or transportation rather than environmental protection. It is beyond the scope of this report to comment upon them.

Although all substances which are hazardous in the aquatic environment must be fully considered in any risk assessment procedure, it is likely that the statutory authorities will be particularly interested in measures to reduce the likelihood of spills of the most dangerous materials. It is generally agreed that the most dangerous substances are those on the EC Directive 76/464 Black List and the UK Red List, which are reproduced in
Boxes A.3.1 and A.3.2 respectively.

Box A.3.1 *EC Black List substances*

Aldrin, Dieldrin, Endrin & Isodrin
Cadmium and its compounds
Carbon tetrachloride
Chloroform
DDT (all isomers)
Hexachlorobenzene
Hexachlorobutadiene
Hexachlorocyclohexane (all isomers)
Mercury and its compounds
Pentachlorophenol
1,2-Dichloroethane
Trichloroethylene
Tetrachloroethylene
1,2,4-Trichlorobenzene

Box A.3.2 *UK Red List substances*

Aldrin, Dieldrin, Endrin	Atrazine
Azinphos-methyl	Cadmium and its compounds
DDT	Dichlorvos
1,2-Dichloroethane	Fenitrothion
Endosulphan	Hexachlorobutadiene
Hexachlorobenzene	Malathion
c-hexachlorocyclohexane	Polychlorinated biphenyls
Mercury and its compounds	Simazine
Pentachlorophenol	Trifluralin
Trichlorobenzene (all isomers)	Triorganotin compounds

EC Directive 76/464 also contains a grey list (list II) of water-endangering substances requiring control. Environmental Quality Standards (maximum concentrations in the aquatic environment) for a number of grey list substances have been selected or proposed in the UK. The substances are listed in Box A3.3.

Box A3.3 *Substances for which UK water quality standards have been selected or proposed*

Copper	Tributyltin compounds
Zinc	Triphenyltin compounds
Lead	Ammonia
Arsenic	Sulphide
Chromium	Benzene
Nickel	Chlordane
Boron	Heptachlor
Iron	Chlorobenzene
Vanadium	Dichlorobenzenes
Aluminium	Xylene
Tin (inorganic)	Toluene
A range of mothproofing agents	

Other lists of substances or generic groups of substances are contained in:

- List I (Black List) and List II (Grey List) of EC Directive 76/464
- List I and list II of the Oslo and Paris Conventions
- List I and list II of the Groundwater Directive
- EC priority candidate list
- First priority candidate red list
- North Sea Conference list of banned or restricted pesticides
- North Sea Conference list of priority hazardous substances
- North Sea Conference reference list of substances
- Schedule 5 of SI 1991 No 472 (prescribed substances to water).

Dangerous Substances in Water: A Practical Guide (Edwards, 1992) contains lists of the specific substances from the above sources. The International Maritime Organisation has published a list of marine pollutants and the German water hazard class system (WGK) is an equivalent listing and, as mentioned above, many other lists of dangerous chemicals have been produced for a variety of purposes.

Most of the substances in the above lists were selected on the basis of specific selection criteria. The most commonly used criteria are:

- toxicity
- persistence
- bioaccumulation
- carcinogenicity.

The above lists indicate those substances which have long-term, widespread effects in water and target organisms including man, e.g. heavy metal and chlorinated hydrocarbons, or short-term acute effects, such as highly toxic biologically active pesticides. They do not include the very wide range of materials which are likely to have equally severe effects at least in the short term and in the vicinity of the discharge point. An indication of the broad classes into which these materials fall is given in Box A3.4.

Box A3.4 *Broad classes of materials with the potential to pollute water*

acids and alkalis
oxidising and reducing agents
corrosive materials
inert solids
other inorganic materials (ammonia, chlorine, sulphide, metal salts)
dyes, colours, pigments
detergents
organic solvents
oils, fuels, fats and waxes
flammable materials
other organic compounds (including foodstuffs)

A4 Estuarial quality classification scheme

Table A4.2 shows the allocation of points under the Estuarial Quality Classification Scheme. The points awarded to each area under the headings of biological, aesthetic and water quality should be summed, and the areas classified according to the scale shown in Table A4.1.

Table A4.1 *Estuarial Quality Classification Scheme*

Classification	Number of Points	Description
Class A	30-24	Good Quality
Class B	23-16	Fair Quality
Class C	15-9	Poor Quality
Class D	8-0	Bad Quality

Table A4.2 Allocation of points for the Estuarial Quality Classification Scheme

Description	Points awarded if estuary meets this description
Biological Quality (Scores under a, b, c & d to be summed)	
(a) Allows the passage to and from freshwater of all relevant species of migratory fish, when this is not prevented by physical barriers. Relevant species include salmonids, eels, flounders and cucumber smelts etc	2
(b) Supports a residential fish population which is broadly consistent with the physical and hydrographical conditions	2
(c) Supports a benthic community which is broadly consistent with the physical and hydrographical conditions	2
(d) Absence of substantially elevated levels in the biota of persistent toxic or tainting substances from whatever source	4
Maximum number of points	10
Aesthetic Quality (Choose one description only)	
(a) Estuaries or zones of estuaries that either do not receive a significant polluting input or which receive inputs that do not cause significant aesthetic pollution	10
(b) Estuaries or zones of estuaries which receive inputs which cause a certain amount of aesthetic pollution but do not seriously interfere with estuary usage	6
(c) Estuaries or zones of estuaries which receive inputs which result in aesthetic pollution sufficiently serious to affect estuary usage(d) Estuaries or zones of estuaries which receive inputs which cause widespread public nuisance	3
Water Quality (Score according to quality)	
Dissolved oxygen exceeds the following saturation values:	
60%	10
40%	6
30%	5
20%	4
10%	3
below 10%	0

Classification of an estuary is summarised according to the length in each class. The length of an estuary should normally be measured along its centre line from the landward limit to the seaward limit of the survey. Where the classification is different from one side to the other, the length of estuary affected should be allocated proportionally between the different classes.

A5 The General Quality Assessment scheme

The General Quality Assessment (GQA) scheme is used by the Environment Agency to assess generally the quality of rivers and canals in the UK. Eventually, the scheme is intended to have chemical, biological, nutrients and aesthetic components which will be reported separately, but currently only the chemical component has been implemented.

A5.1 THE CHEMICAL COMPONENT

The chemical component of the scheme grades rivers class A to F (representing good to bad quality) on the basis of concentrations of BOD, ammonia and dissolved oxygen in the rivers (see Table A5.1). These three substances were selected as providing the best overall measure of basic chemical quality, because they are indicators of the extent to which waters are affected by wastewater discharges (such as sewage effluent) and rural land runoff containing organic matter. These discharges are the sources of the major polluting loads in UK rivers.

However, some types of pollution are not picked up by the chemical component of the scheme and this includes pollution from a huge range of other chemicals. The rcason why these other chemicals were not included in the scheme is two-fold. Firstly, the cost of analysing for all these chemicals everywhere would be prohibitively large and secondly, new chemicals are being developed all the time. The Environment Agency would need continually to update the chemical standards as new chemicals were developed. This would undermine the comparability of the grading from year to year. This way the chemical component of the GQA scheme provides a "yardstick" to measure the quality of rivers from year to year.

A5.2 THE BIOLOGICAL COMPONENT

Because the chemical component of the scheme does not address all types of possible pollution in rivers, the GQA scheme will also have a biological component.

The biological grading will be based on the monitoring of benthic macro-invertebrates (small creatures living on the river bed). Each particular species thrives best under a narrow range of environmental conditions and so the biological quality of the river will reflect the extent to which the river is affected by environmental stresses, including pollution from the whole range of chemicals.

The biological component of the scheme is expected to be introduced in time for the 1995 river quality survey.

A5.3 OTHER COMPONENTS

An aesthetic and nutrients component of the scheme is currently under development and, subject to a successful outcome of the research, these components may also be ready for use in the 1995 survey.

Table A5.1 *The chemistry component of the General Quality Assessment scheme*

Water Quality	Grade	Dissolved Oxygen % saturation 10-percentile	BOD (ATU) (mg/l) 90-percentile	Ammonia (mgN/l) 90-percentile
GOOD	A	80	2.5	0.25
	B	70	4	0.6
FAIR	C	60	6	1.3
	D	50	8	2.5
POOR	E	20	15	9
BAD	F	Where quality of river does not meet requirements of grade E in respect of one or more substances		

Note: BOD (ATU) refers to the Biological Demand measured when the sample is suppressed by adding allyl thio-urea.

A5.4 STANDARDS

BOD and ammonia standards require 90-percentile compliance - this means that the river should contain less than the specified levels of BOD and ammonia for at least 90 per cent of the time.

The dissolved oxygen standards require 10-percentile compliance - this means that levels of dissolved oxygen should not fall below the standard for more than 10 per cent of the time.

Compliance is assessed using data over three years.

A6 Chemical, fire and weathering resistance of protective systems

Protective System	Resistance to Acids and Alkalis				Resistance to Solvents					Resistance to fire	Resistance to weathering	Relevant British Standard
	Conc. Inorganic acids	Dilute Inorganic acids	Organic acids	Alkalis	Petrol	Paraffin, diesel and fuel oil	Aromatic and chlorinated solvents	Ethers, kerosene and esters	Alcohols			
Butyl Rubber	2	2	2	2	0	0	0			0	2	3227
Chlorosulphonated polyethylene					0 to 2	0 to 2				1	2	
Epoxide Resins	2	2	2	2	2	2	2			0	2	4994, 6374
Polychloroprene					1 to 2	2	0			1	2	2752
Polyester	2	2	2	0	2	2	2	0	2	0	1(UV-)	3532
Polyethylene (LD)	2	2	1	2	0	1	0	0	1	0	1(UV-)	1972, 1973
Polyethylene (HD)	2	2	1	2	0	1	0	0	2	0	1(UV-)	4646
Natural rubber	2	2	2	2	0	0	0			0	0	1711, 6716
Polypropylene	2	2	2	2	1	2	0	1	2	0	0 to 1	
Polysulphides	0	0	0	0	2	2	0-1			0	1 to 2	4254, 5215
Polytetraflouroethylene	2	2	2	2	2	2	2	2	2	2	2	6564
(PVC) rigid	0	2	2	2	2	2	0	0	2	1	1 to 2	
(PVC) plasticised	0	0 to 1	0	0 to 1	0 to 1	0 to 1	0	0	0 to 2	0 to 1	0 to 2	2571, 3869
Asphalt	1	2	2	1	0	0	0 to 1					
Brickwork	0 to 2	0 to 2	0 to 2	1 to 2	2	2	2	2	2	2	1 to 2	5628
Tiles	1 to 2	2	2	2	2	2	2	2	2	2	1 to 2	
Bituminous & polymer paints	1	1 to 2	1 to 2	1 to 2	0	0	0 to 1	0	0	0	1	

Key:

Resistance to Fire
0 Flammable
1 Combustible but self-extinguishing
2 Difficult to ignite

Resistance to Weathering
0 Nil
1 Poor
2 Good

Chemical Resistance
0 Unsuitable
1 Some attack
2 Resistant

A7 Undertank leak detection systems

(See Section 12.2.4)

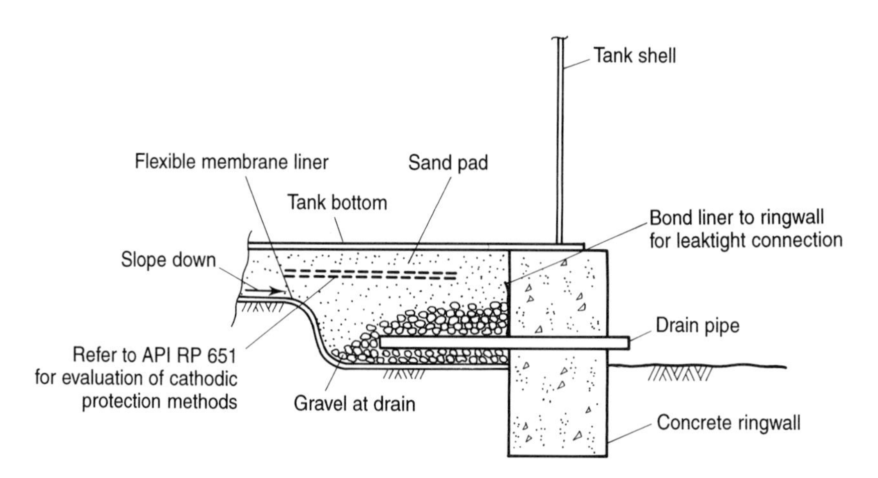

Figure A7.1 *Concrete ringwall with undertank leak* detection at the tank perimeter (Reproduced courtesy of the American Petroleum Institute from API, 1988)

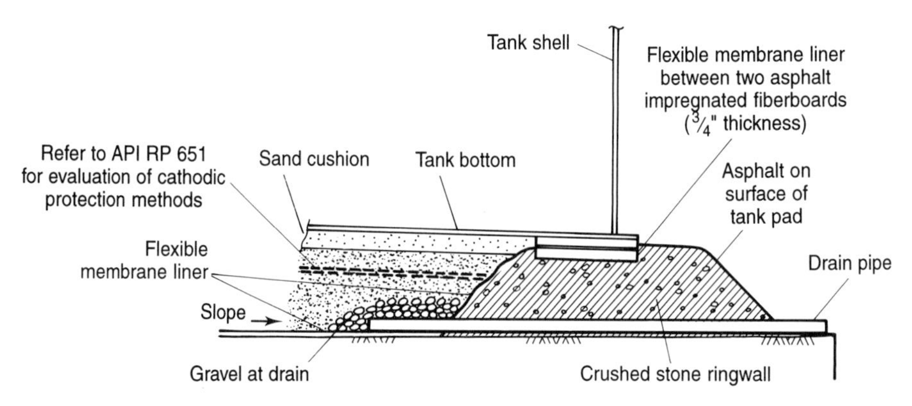

Figure A7.2 *Crushed stone ringwall with undertank leak detection at the tank perimeter* (Reproduced courtesy of the American Petroleum Institute from API, 1988)

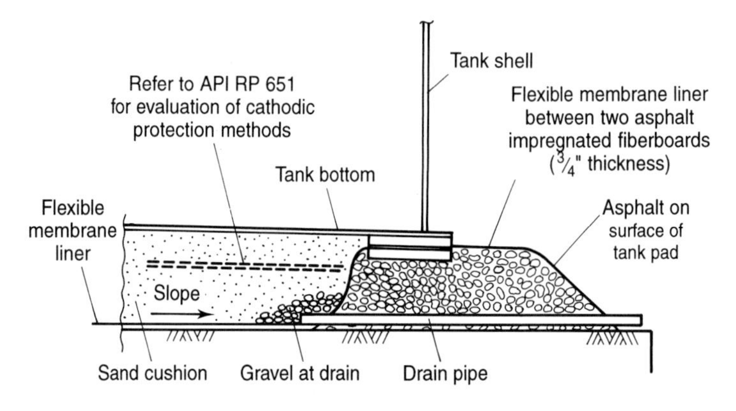

Figure A7.3 *Earthen foundation with undertank leak detection at the tank perimeter* (Reproduced courtesy of the American Petroleum Institute from API, 1988)

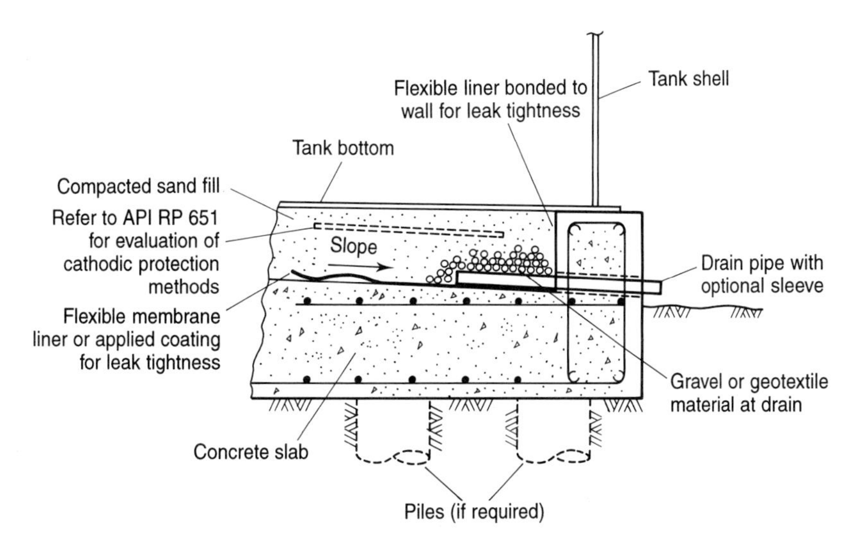

Figure A7.4 *Reinforced concrete slab with undertank leak detection at the perimeter* (Reproduced courtesy of the American Petroleum Institute from API, 1988)

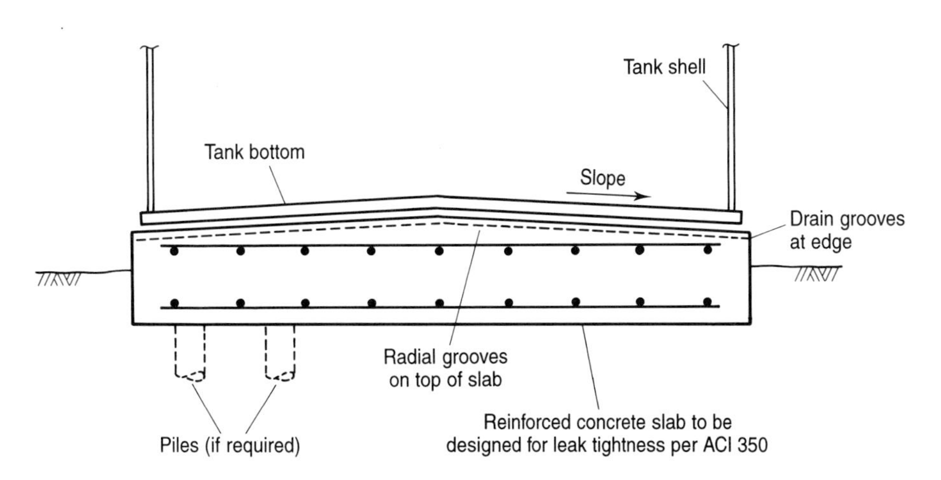

Figure A7.5 *Reinforced concrete slab with radial grooves for leak detection* (Reproduced courtesy of the American Petroleum Institute from API, 1988)

A8 Concrete repair techniques

Table A8.1 *Application of repair techniques for concrete* (Source: Shaw, 1981)

	Spalling over large area Reinforcement cover (mm)			Spalling over small area Reinforcement cover (mm)	
Repair material	**>25**	**12 - 25**	**6 - 12**	**12 - 25**	**6 - 12**
Concrete	✓				
Sprayed concrete	✓				
Cement/ sand mortars	✓				
Polymer modified cementitious mortars		✓		✓	✓*
Epoxy resin mortars			✓		✓
Polyester resin mortars					✓
Repair material	**Crack sealing**	**Crack repair**	**Bonding aid**	**Honeycomb concrete**	**Permeable concrete**
Moisture tolerant epoxy resins			✓		
SBR, acrylic and co-polymer emulsions	✓		✓		depends on permeability
Low-viscosity polyester and acrylic resins	✓				depends on permeability
Epoxy resin, low viscosity		✓		✓	depends on permeability
Penetrating polymer systems. 'In-surface' sealers					✓
Special coatings and penetrating 'in-surface' sealers					✓
Universal bonding aids, PVA, PVA modified mortars	usually not suitable for external work				

* May require additional anti-carbonation coating

Box A8.1 *Concrete repair using resin injection* (Source: Shaw, 1984)

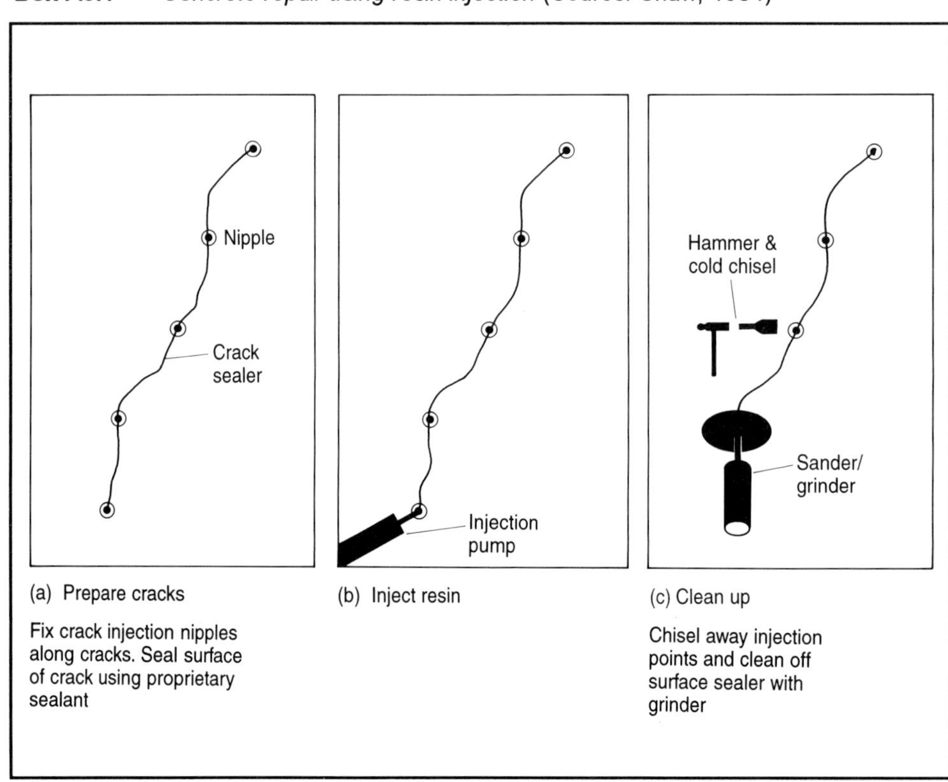

(a) Prepare cracks

Fix crack injection nipples along cracks. Seal surface of crack using proprietary sealant

(b) Inject resin

(c) Clean up

Chisel away injection points and clean off surface sealer with grinder

Box A8.2 *Patch repair to concrete* (Source: Shaw, 1984)

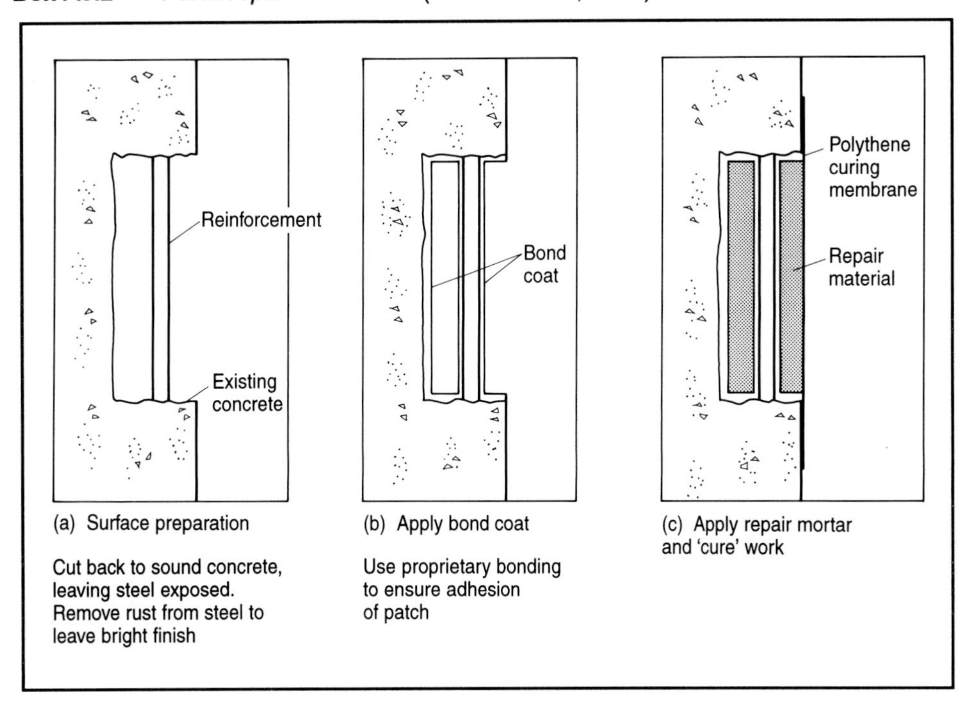

(a) Surface preparation

Cut back to sound concrete, leaving steel exposed. Remove rust from steel to leave bright finish

(b) Apply bond coat

Use proprietary bonding to ensure adhesion of patch

(c) Apply repair mortar and 'cure' work

Box A 8.3 *Thin surface bonded repair to concrete* (Source: Shaw, 1984)

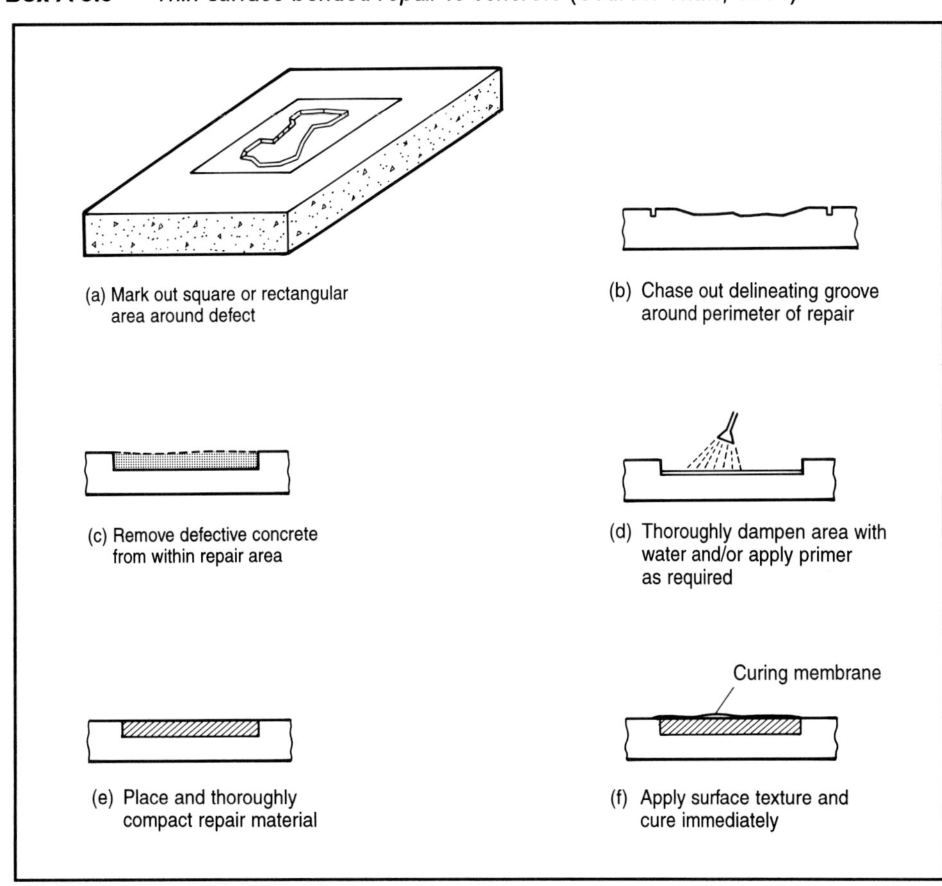

(a) Mark out square or rectangular area around defect

(b) Chase out delineating groove around perimeter of repair

(c) Remove defective concrete from within repair area

(d) Thoroughly dampen area with water and/or apply primer as required

(e) Place and thoroughly compact repair material

(f) Apply surface texture and cure immediately

Box A8.4 *Thin bonded repairs to joints in concrete* (Source: Shaw, 1984)

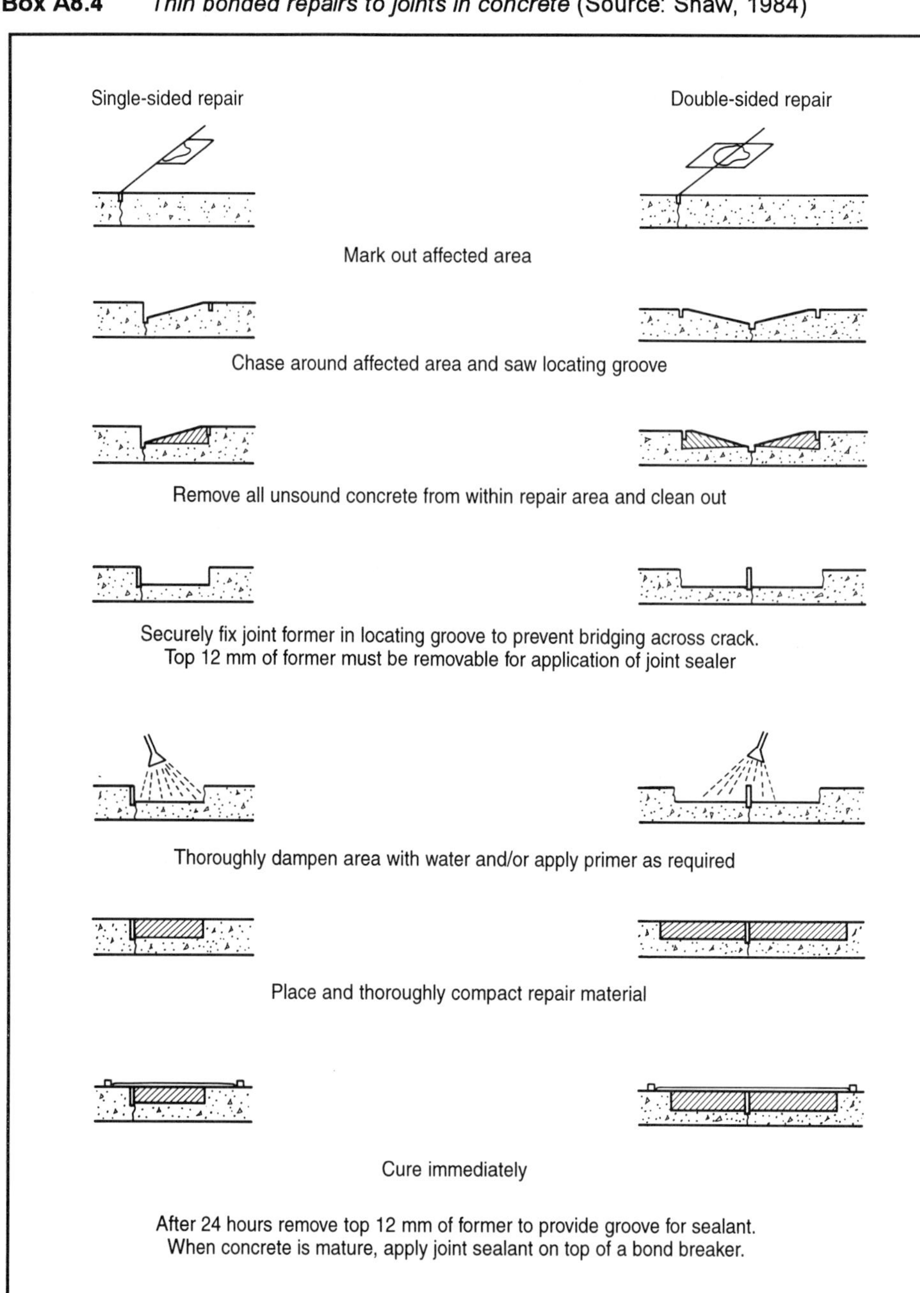

A9 Model design and description of works for small bunds

This Appendix comprises a series of detailed design drawings showing how to construct small bunds in reinforced concrete and masonry. For larger bunds an appropriately qualified and experienced structural engineer should be consulted.

The designs comply with the structural design recommendations set out in this report.

A9.1 DETAILED DESIGN DRAWINGS

This section comprises a series of detailed design drawings showing how to construct bunds in reinforced concrete and reinforced masonry. They cover a range of tank sizes, up to $25m^3$.

The designs comply with the structural design recommendations set out in this report. The drawings must be read in conjunction with the relevant parts of Section 6.2

Description of works. This description must be followed closely to ensure that the finished works are of the required quality.

In certain circumstances bunds may require Planning Permission. Further advice on this should be obtained from the Local Planning Authority.

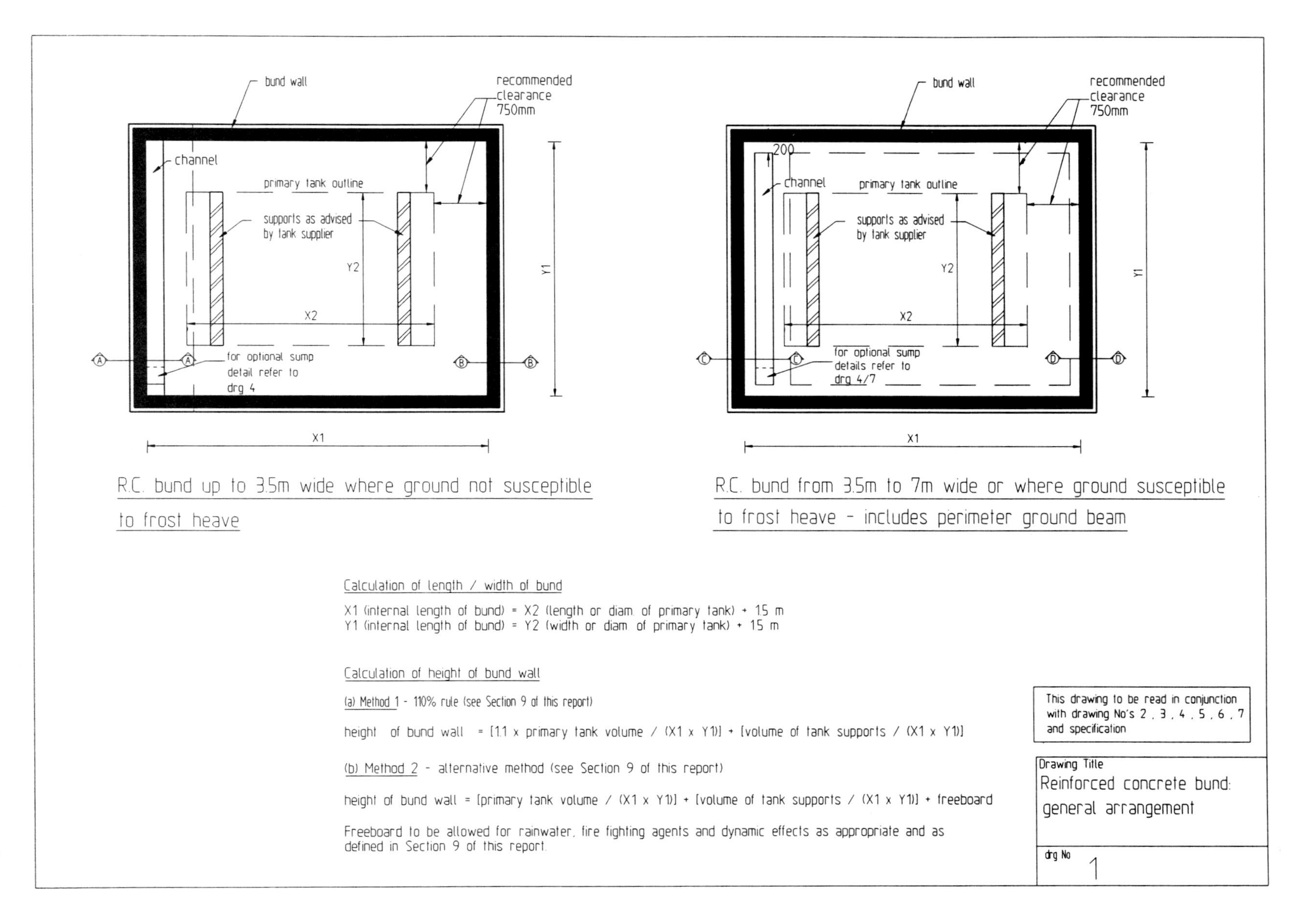

Drawing 1 *Reinforced concrete bund: general arrangement*

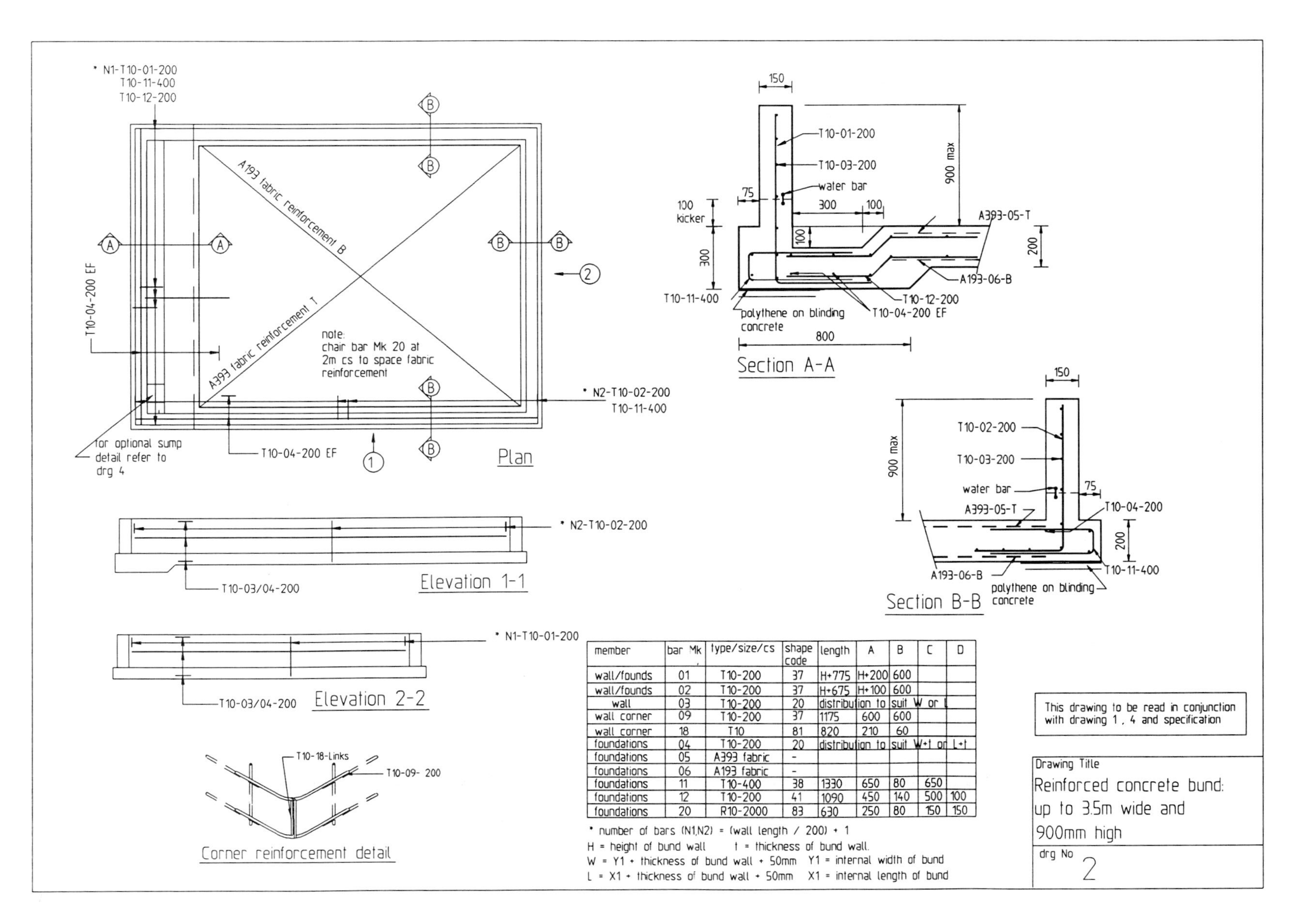

member	bar Mk	type/size/cs	shape code	length	A	B	C	D
wall/founds	01	T10-200	37	H+775	H+200	600		
wall/founds	02	T10-200	37	H+675	H+100	600		
wall	03	T10-200	20	distribution to suit W or L				
wall corner	09	T10-200	37	1175	600	600		
wall corner	18	T10	81	820	210	60		
foundations	04	T10-200	20	distribution to suit W+t or L+t				
foundations	05	A393 fabric	-					
foundations	06	A193 fabric	-					
foundations	11	T10-400	38	1330	650	80	650	
foundations	12	T10-200	41	1090	450	140	500	100
foundations	20	R10-2000	83	630	250	80	150	150

* number of bars (N1,N2) = (wall length / 200) + 1
H = height of bund wall   t = thickness of bund wall.
W = Y1 + thickness of bund wall + 50mm   Y1 = internal width of bund
L = X1 + thickness of bund wall + 50mm   X1 = internal length of bund

Drawing 2 *Reinforced concrete bund: up to 3.5m wide and 900mm high*

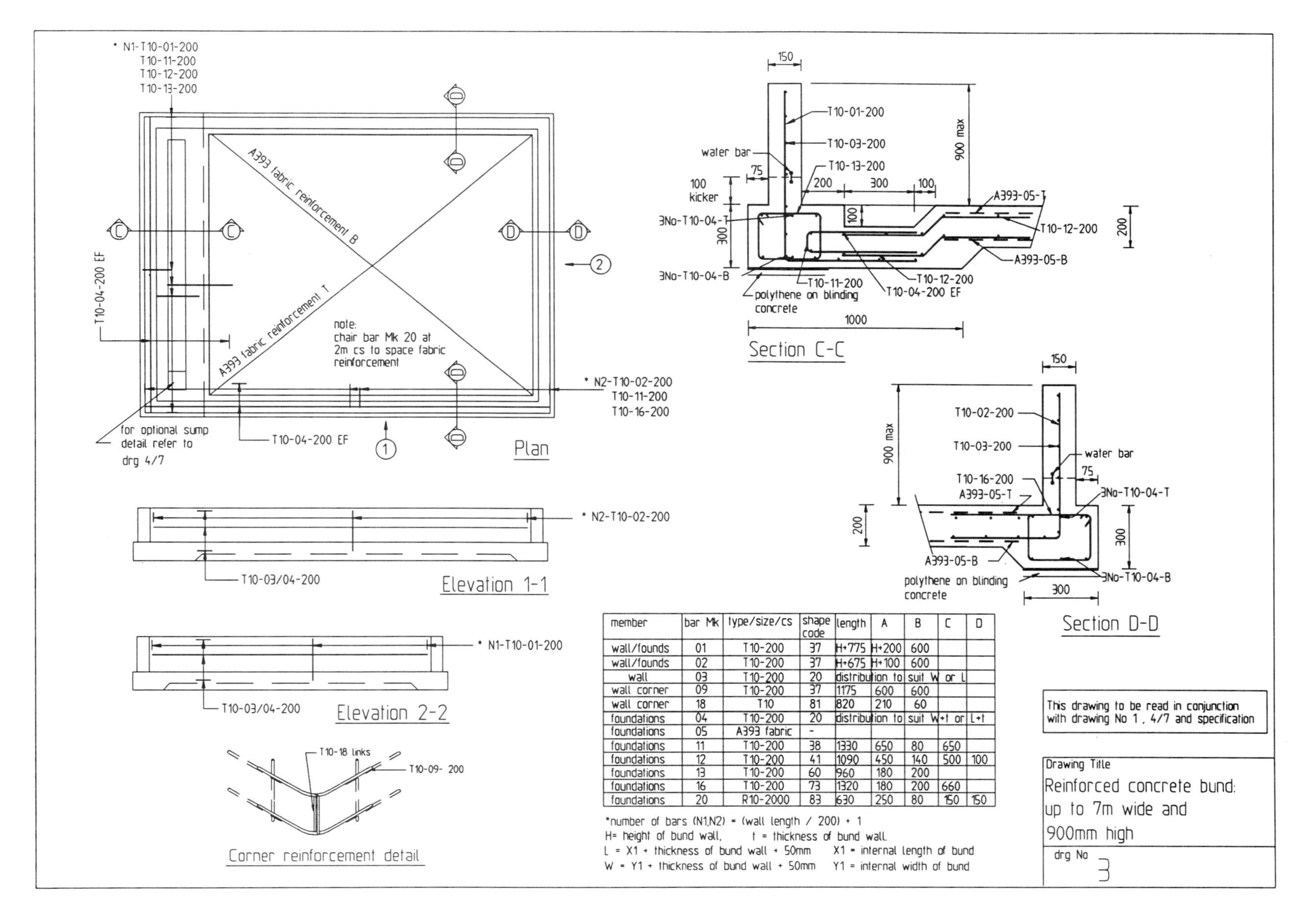

member	bar Mk	type/size/cs	shape code	length	A	B	C	D
wall/founds	01	T10-200	37	H+775	H+200	600		
wall/founds	02	T10-200	37	H+675	H+100	600		
wall	03	T10-200	20	distribution to suit W or L				
wall corner	09	T10-200	37	1175	600	600		
wall corner	18	T10	81	820	210	60		
foundations	04	T10-200	20	distribution to suit W+t or L+t				
foundations	05	A393 fabric	-					
foundations	11	T10-200	38	1330	650	80	650	
foundations	12	T10-200	41	1090	450	140	500	100
foundations	13	T10-200	60	960	180	200		
foundations	16	T10-200	73	1320	180	200	660	
foundations	20	R10-2000	83	630	250	80	150	150

*number of bars (N1,N2) = (wall length / 200) + 1
H= height of bund wall, t = thickness of bund wall.
L = X1 + thickness of bund wall + 50mm X1 = internal length of bund
W = Y1 + thickness of bund wall + 50mm Y1 = internal width of bund

Drawing 3 *Reinforced concrete bund: up to 7m wide and 900mm high*

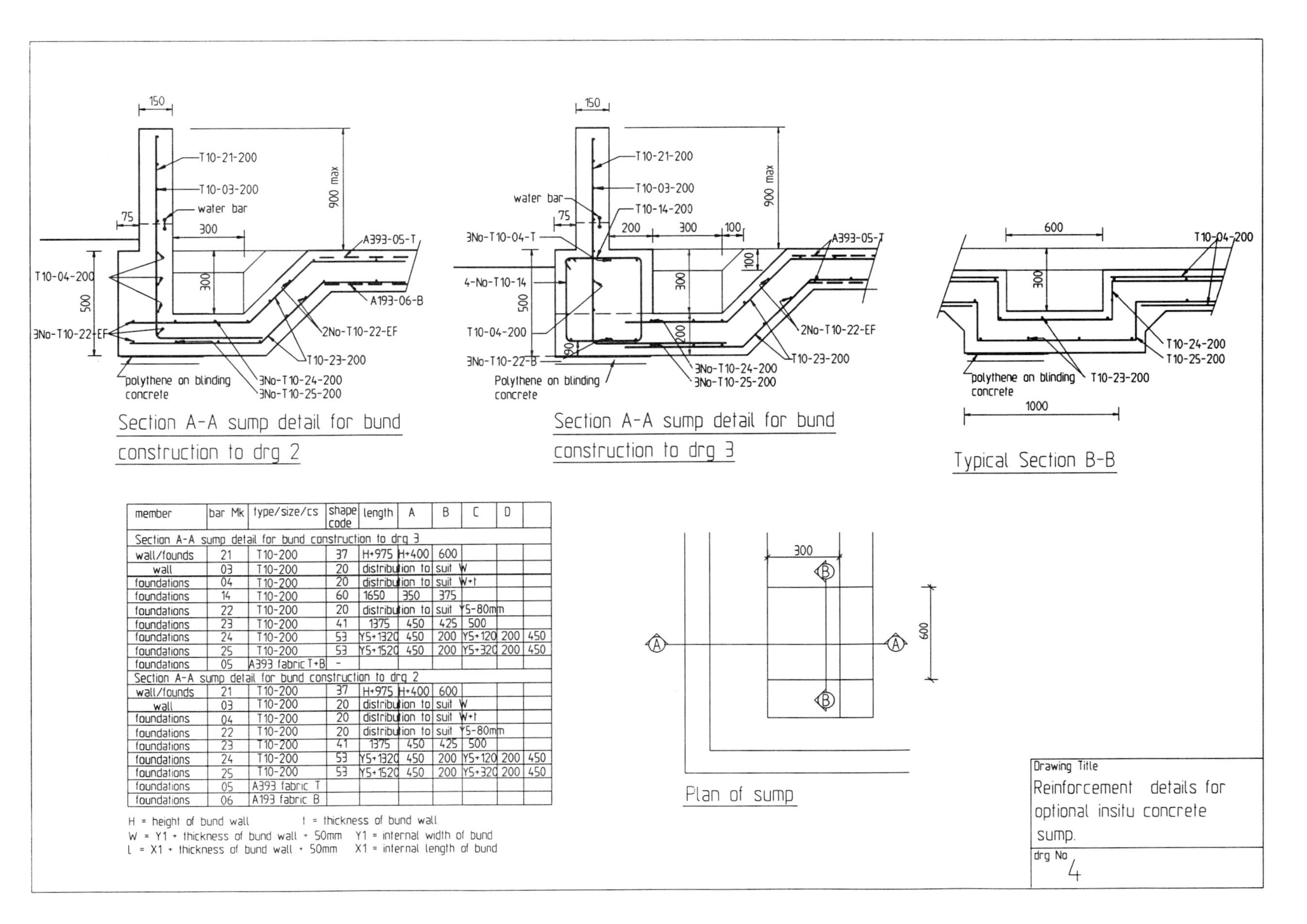

member	bar Mk	type/size/cs	shape code	length	A	B	C	D	
Section A-A sump detail for bund construction to drg 3									
wall/founds	21	T10-200	37	H+975	H+400	600			
wall	03	T10-200	20	distribution to suit W					
foundations	04	T10-200	20	distribution to suit W+t					
foundations	14	T10-200	60	1650	350	375			
foundations	22	T10-200	20	distribution to suit Y5-80mm					
foundations	23	T10-200	41	1375	450	425	500		
foundations	24	T10-200	53	Y5+1320	450	200	Y5+120	200	450
foundations	25	T10-200	53	Y5+1520	450	200	Y5+320	200	450
foundations	05	A393 fabric T+B	-						
Section A-A sump detail for bund construction to drg 2									
wall/founds	21	T10-200	37	H+975	H+400	600			
wall	03	T10-200	20	distribution to suit W					
foundations	04	T10-200	20	distribution to suit W+t					
foundations	22	T10-200	20	distribution to suit Y5-80mm					
foundations	23	T10-200	41	1375	450	425	500		
foundations	24	T10-200	53	Y5+1320	450	200	Y5+120	200	450
foundations	25	T10-200	53	Y5+1520	450	200	Y5+320	200	450
foundations	05	A393 fabric T							
foundations	06	A193 fabric B							

H = height of bund wall t = thickness of bund wall
W = Y1 + thickness of bund wall + 50mm Y1 = internal width of bund
L = X1 + thickness of bund wall + 50mm X1 = internal length of bund

Drawing 4 *Reinforcement details for optional in-situ concrete sump*

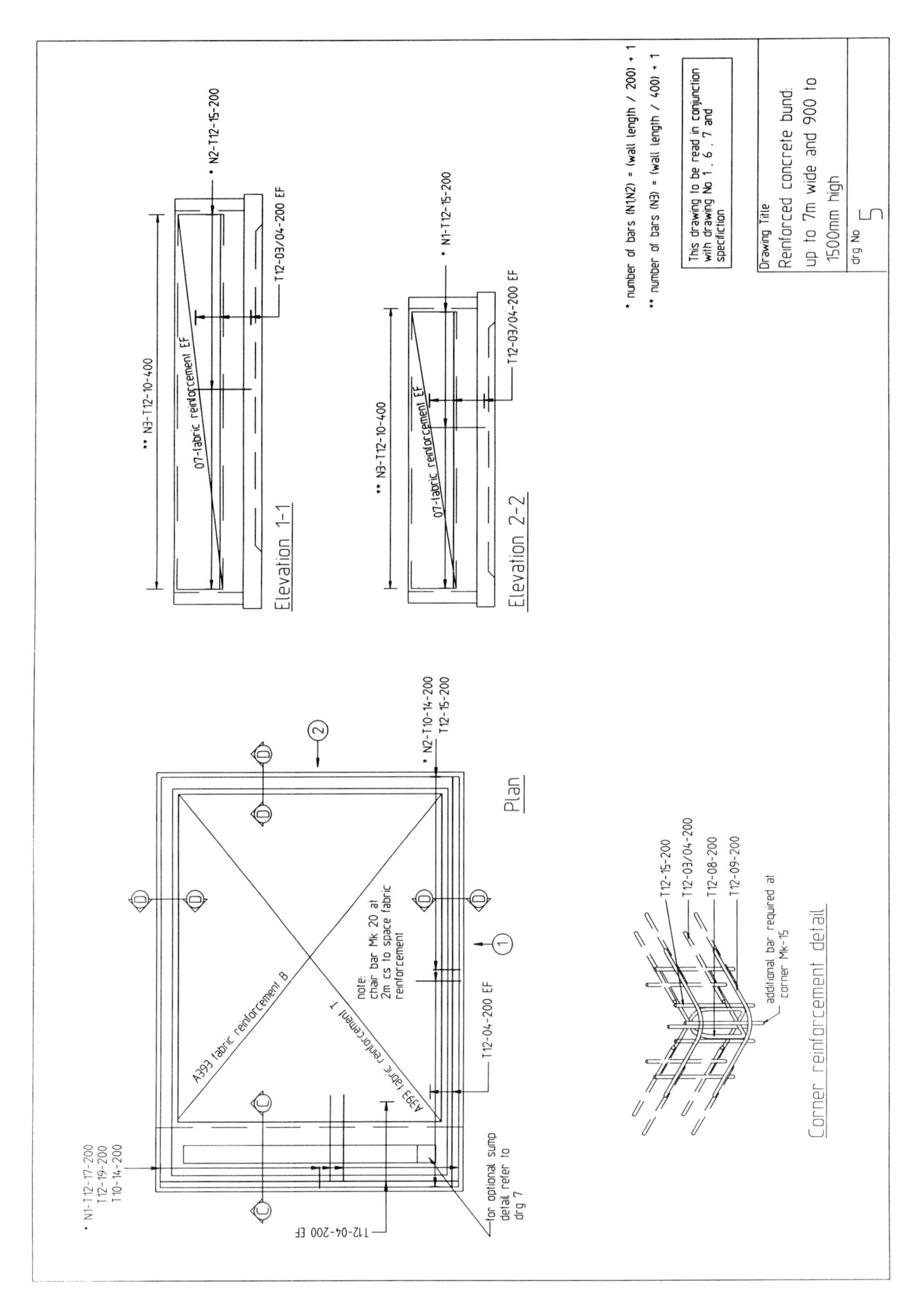

Drawing 5 *Reinforced concrete bund: up to 7m wide and 900 to 1 500mm high*

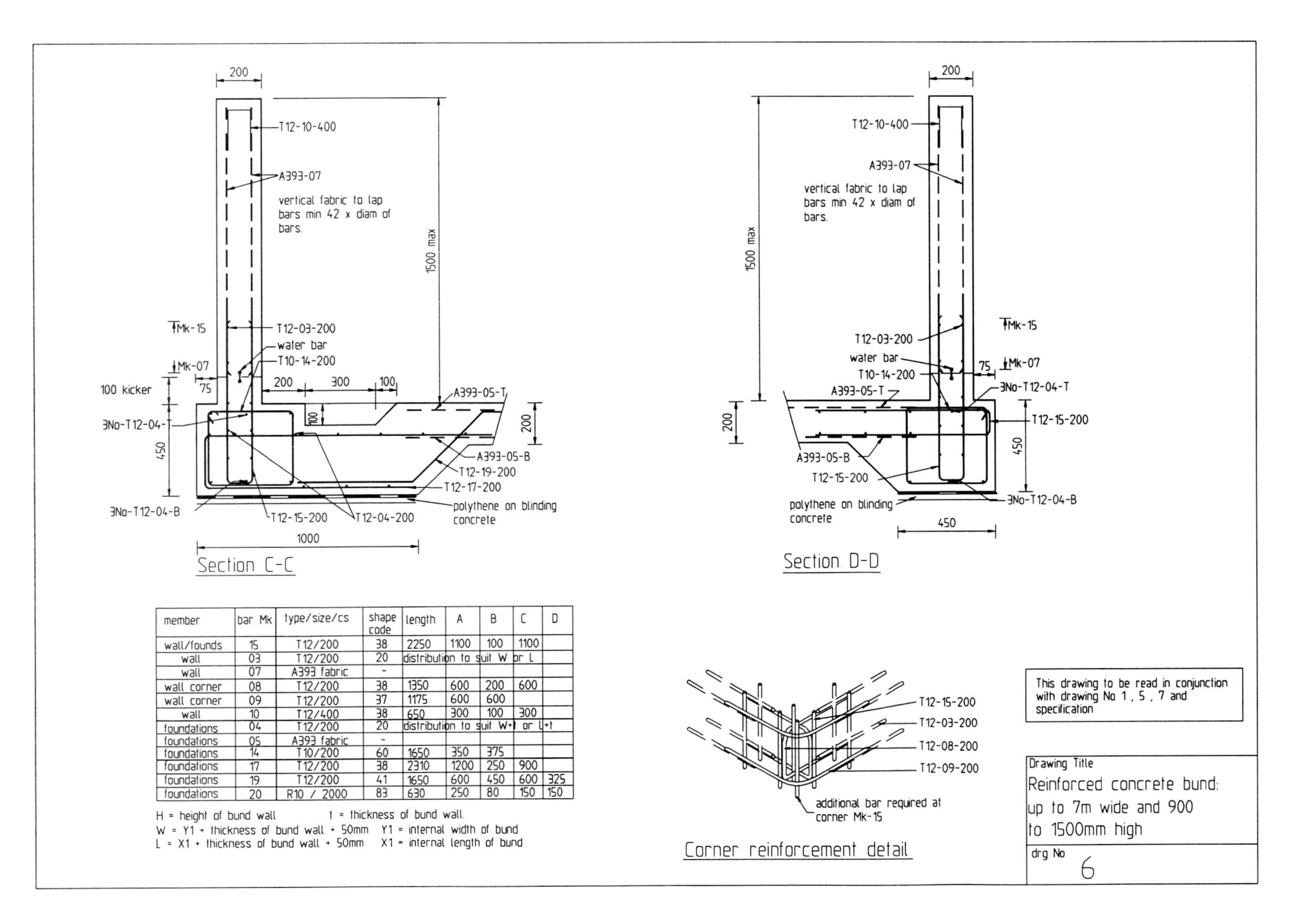

member	bar Mk	type/size/cs	shape code	length	A	B	C	D
wall/founds	15	T12/200	38	2250	1100	100	1100	
wall	03	T12/200	20	distribution to suit W or L				
wall	07	A393 fabric	-					
wall corner	08	T12/200	38	1350	600	200	600	
wall corner	09	T12/200	37	1175	600	600		
wall	10	T12/400	38	650	300	100	300	
foundations	04	T12/200	20	distribution to suit W+t or L+t				
foundations	05	A393 fabric	-					
foundations	14	T10/200	60	1650	350	375		
foundations	17	T12/200	38	2310	1200	250	900	
foundations	19	T12/200	41	1650	600	450	600	325
foundations	20	R10 / 2000	83	630	250	80	150	150

H = height of bund wall   t = thickness of bund wall.
W = Y1 + thickness of bund wall + 50mm   Y1 = internal width of bund
L = X1 + thickness of bund wall + 50mm   X1 = internal length of bund

Drawing 6 *Reinforced concrete bund: up to 7m wide and 900 to 1 500mm high*

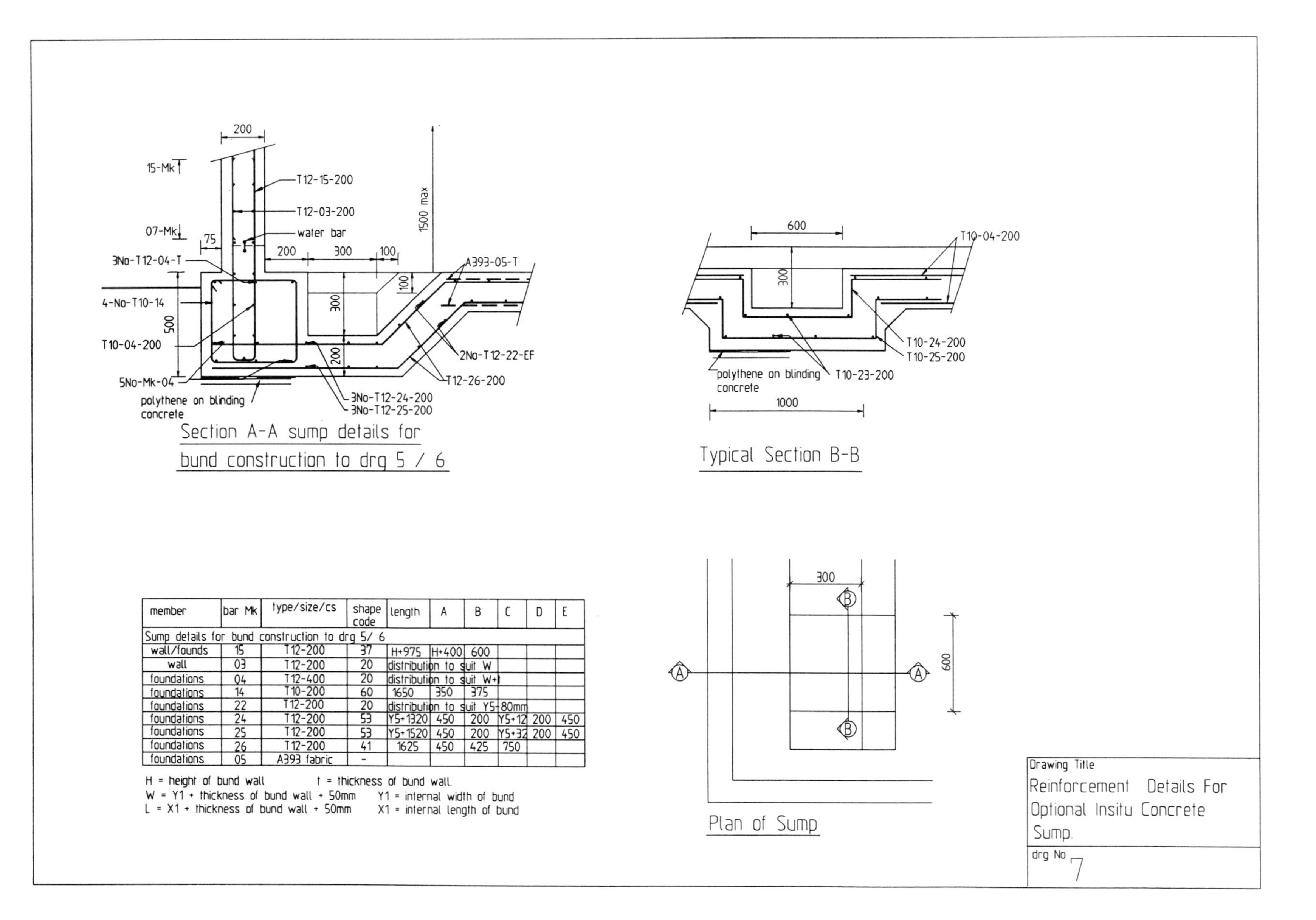

member	bar Mk	type/size/cs	shape code	length	A	B	C	D	E
Sump details for bund construction to drg 5/ 6									
wall/founds	15	T12-200	37	H+975	H+400	600			
wall	03	T12-200	20	distribution to suit W					
foundations	04	T12-400	20	distribution to suit W+t					
foundations	14	T10-200	60	1650	350	375			
foundations	22	T12-200	20	distribution to suit Y5+80mm					
foundations	24	T12-200	53	Y5+1320	450	200	Y5+12	200	450
foundations	25	T12-200	53	Y5+1520	450	200	Y5+32	200	450
foundations	26	T12-200	41	1625	450	425	750		
foundations	05	A393 fabric	-						

H = height of bund wall t = thickness of bund wall.
W = Y1 + thickness of bund wall + 50mm Y1 = internal width of bund
L = X1 + thickness of bund wall + 50mm X1 = internal length of bund

Drawing 7 *Reinforcement details for optional* in-situ *concrete sump*

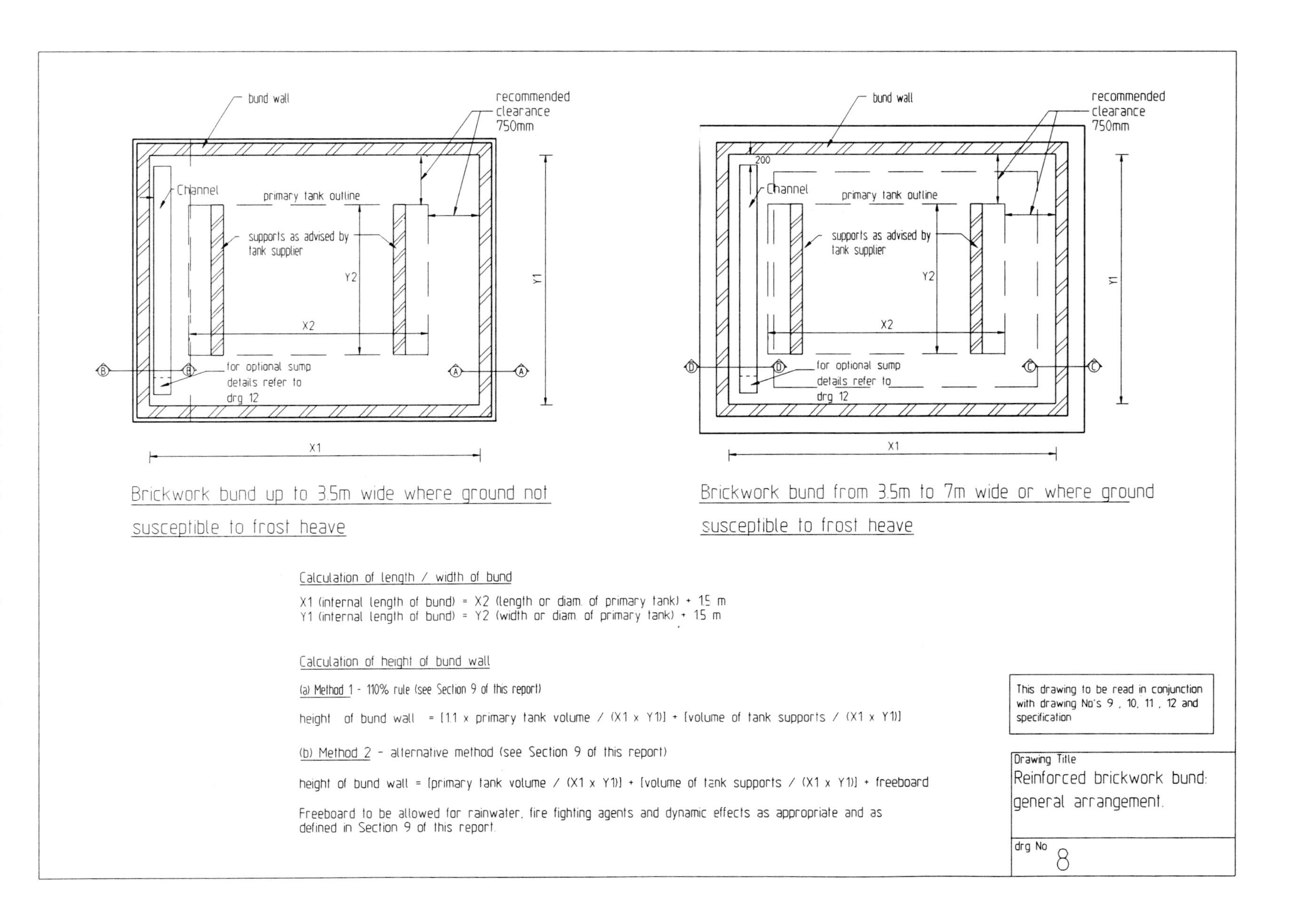

Drawing 8 *Reinforced brickwork bund: general arrangement*

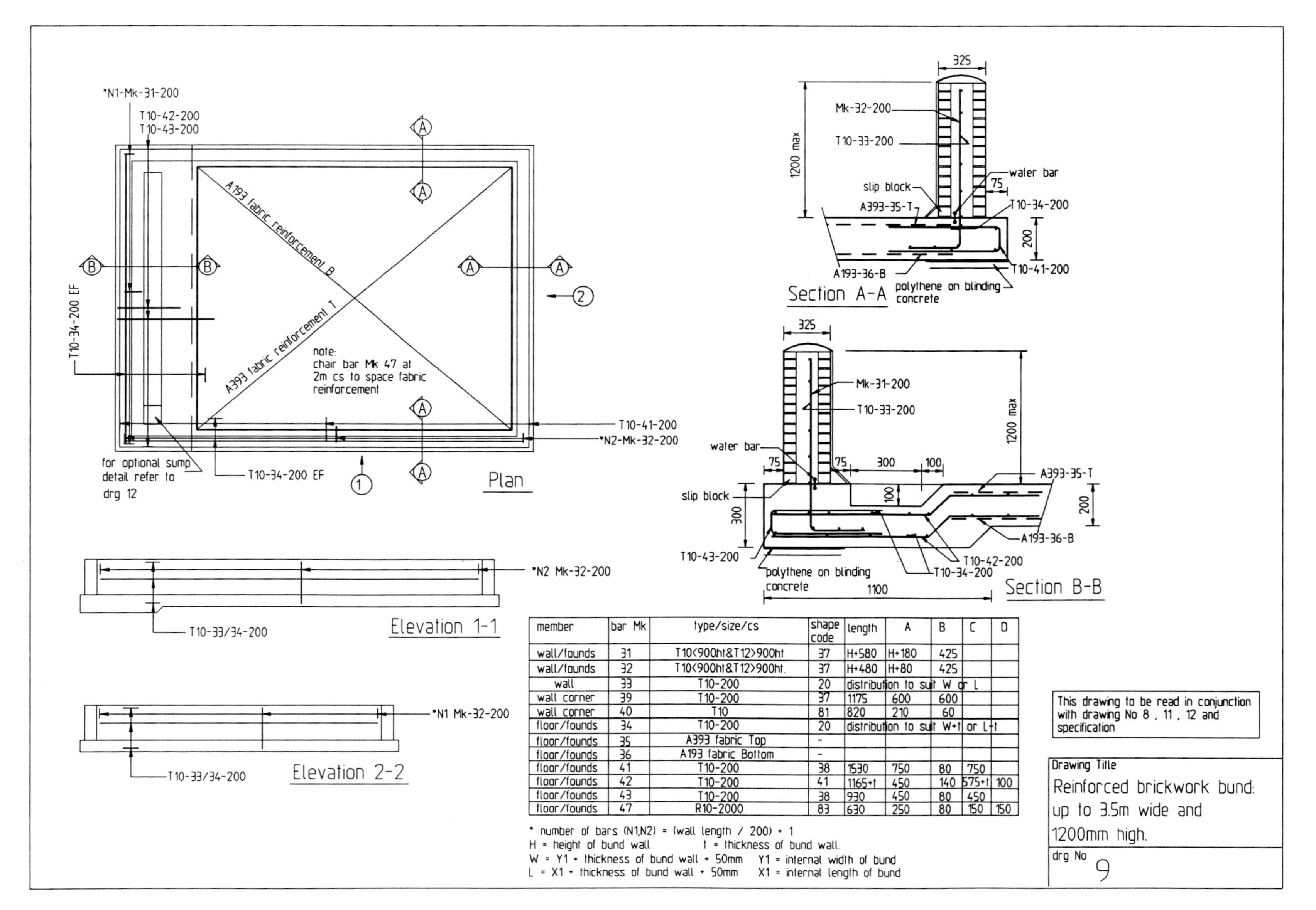

member	bar Mk	type/size/cs	shape code	length	A	B	C	D
wall/founds	31	T10<900ht&T12>900ht	37	H+580	H+180	425		
wall/founds	32	T10<900ht&T12>900ht.	37	H+480	H+80	425		
wall	33	T10-200	20	distribution to suit W or L				
wall corner	39	T10-200	37	1175	600	600		
wall corner	40	T10	81	820	210	60		
floor/founds	34	T10-200	20	distribution to suit W+t or L+t				
floor/founds	35	A393 fabric Top	-					
floor/founds	36	A193 fabric Bottom	-					
floor/founds	41	T10-200	38	1530	750	80	750	
floor/founds	42	T10-200	41	1165+t	450	140	575+t	100
floor/founds	43	T10-200	38	930	450	80	450	
floor/founds	47	R10-2000	83	630	250	80	150	150

* number of bars (N1,N2) = (wall length / 200) + 1
H = height of bund wall t = thickness of bund wall.
W = Y1 + thickness of bund wall + 50mm Y1 = internal width of bund
L = X1 + thickness of bund wall + 50mm X1 = internal length of bund

Drawing 9 *Reinforced brickwork bund: up to 3.5m wide and 1 200mm high*

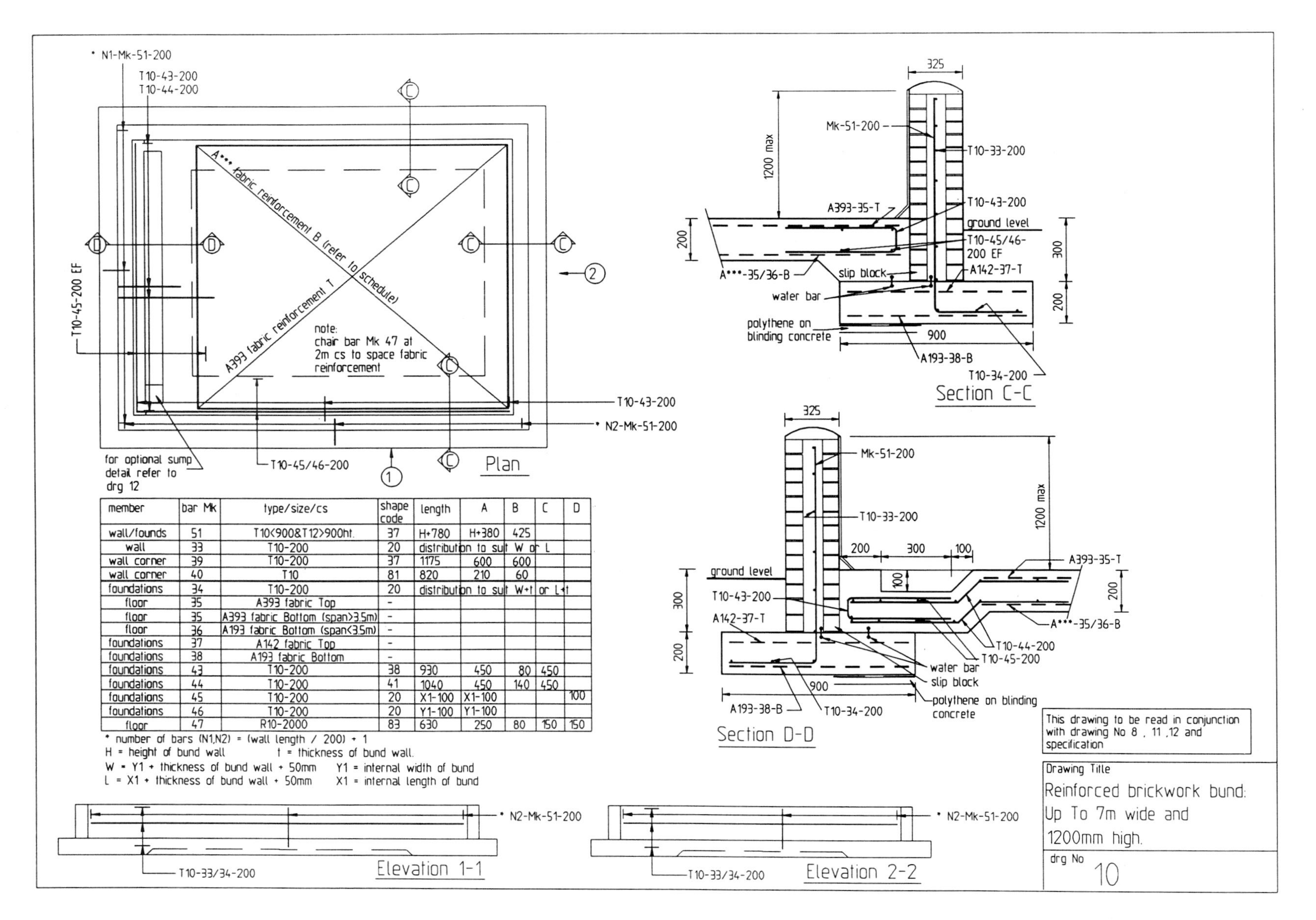

member	bar Mk	type/size/cs	shape code	length	A	B	C	D
wall/founds	51	T10<900&T12>900ht.	37	H+780	H+380	425		
wall	33	T10-200	20	distribution to suit W or L				
wall corner	39	T10-200	37	1175	600	600		
wall corner	40	T10	81	820	210	60		
foundations	34	T10-200	20	distribution to suit W+t or L+t				
floor	35	A393 fabric Top	-					
floor	35	A393 fabric Bottom (span>3.5m)	-					
floor	36	A193 fabric Bottom (span<3.5m)	-					
foundations	37	A142 fabric Top	-					
foundations	38	A193 fabric Bottom	-					
foundations	43	T10-200	38	930	450	80	450	
foundations	44	T10-200	41	1040	450	140	450	
foundations	45	T10-200	20	X1-100	X1-100			100
foundations	46	T10-200	20	Y1-100	Y1-100			
floor	47	R10-2000	83	630	250	80	150	150

* number of bars (N1,N2) = (wall length / 200) + 1
H = height of bund wall   t = thickness of bund wall.
W = Y1 + thickness of bund wall + 50mm   Y1 = internal width of bund
L = X1 + thickness of bund wall + 50mm   X1 = internal length of bund

Drawing 10 *Reinforced brickwork bund: up to 7m wide and 1 200mm high*

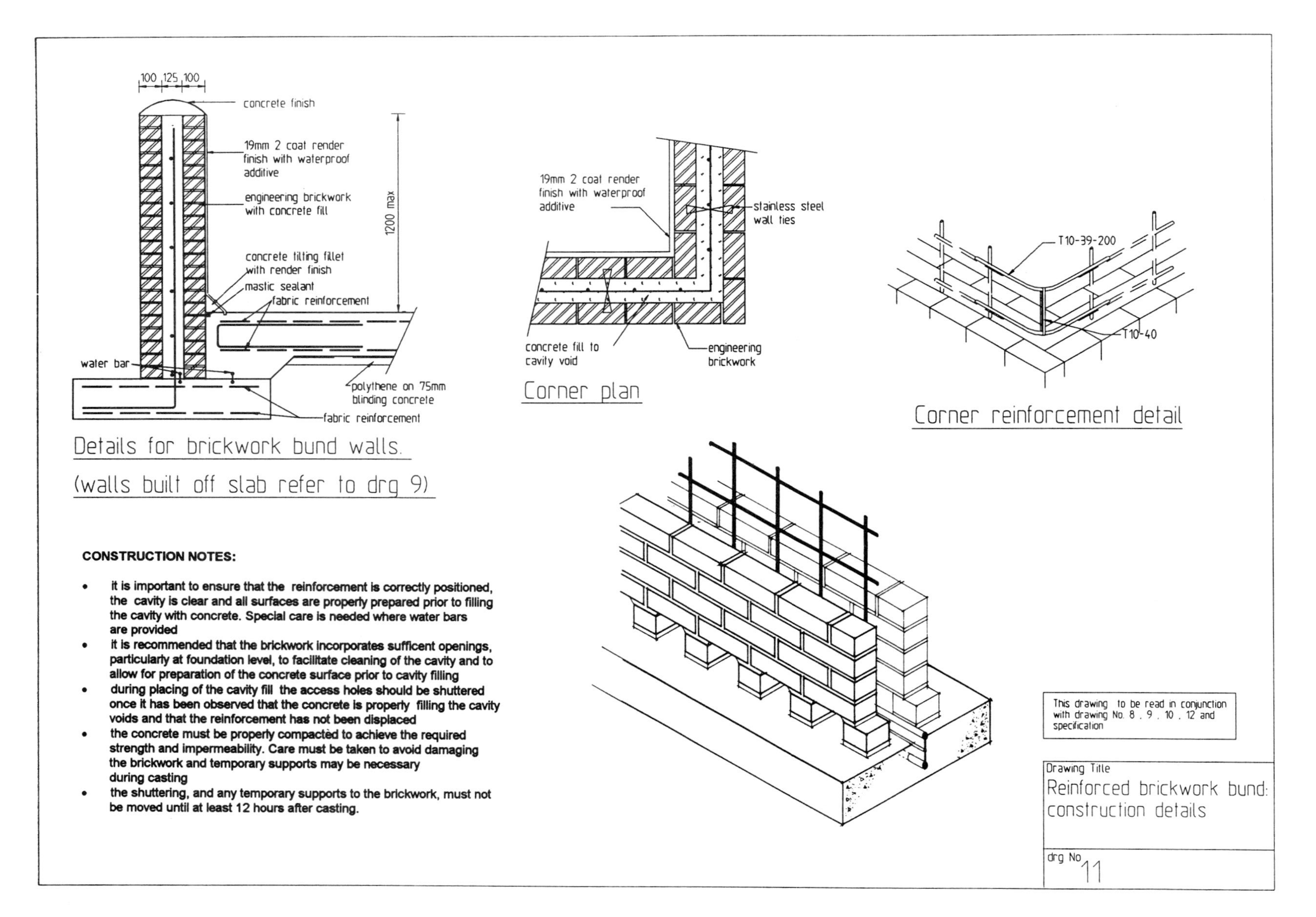

Drawing 11 *Reinforced brickwork bund: construction details*

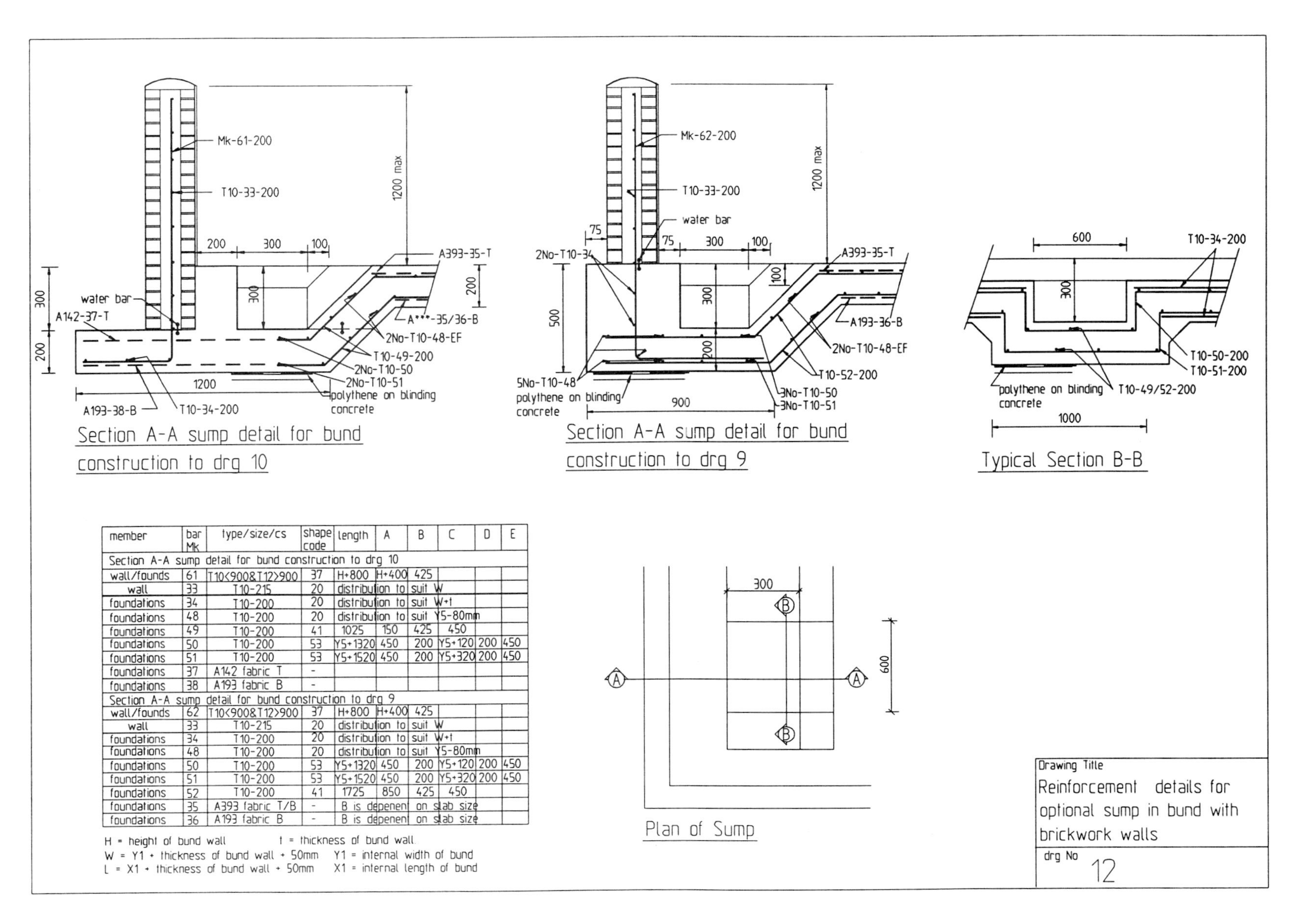

member	bar Mk	type/size/cs	shape code	length	A	B	C	D	E
Section A-A sump detail for bund construction to drg 10									
wall/founds	61	T10<900&T12>900	37	H+800	H+400	425			
wall	33	T10-215	20	distribution to suit W					
foundations	34	T10-200	20	distribution to suit W+t					
foundations	48	T10-200	20	distribution to suit Y5-80mm					
foundations	49	T10-200	41	1025	150	425	450		
foundations	50	T10-200	53	Y5+1320	450	200	Y5+120	200	450
foundations	51	T10-200	53	Y5+1520	450	200	Y5+320	200	450
foundations	37	A142 fabric T	-						
foundations	38	A193 fabric B	-						
Section A-A sump detail for bund construction to drg 9									
wall/founds	62	T10<900&T12>900	37	H+800	H+400	425			
wall	33	T10-215	20	distribution to suit W					
foundations	34	T10-200	20	distribution to suit W+t					
foundations	48	T10-200	20	distribution to suit Y5-80mm					
foundations	50	T10-200	53	Y5+1320	450	200	Y5+120	200	450
foundations	51	T10-200	53	Y5+1520	450	200	Y5+320	200	450
foundations	52	T10-200	41	1725	850	425	450		
foundations	35	A393 fabric T/B	-	B is depenent on slab size					
foundations	36	A193 fabric B	-	B is depenent on slab size					

H = height of bund wall   t = thickness of bund wall
W = Y1 + thickness of bund wall + 50mm   Y1 = internal width of bund
L = X1 + thickness of bund wall + 50mm   X1 = internal length of bund

Drawing 12 *Reinforcement details for optional sump in bund with brickwork walls*

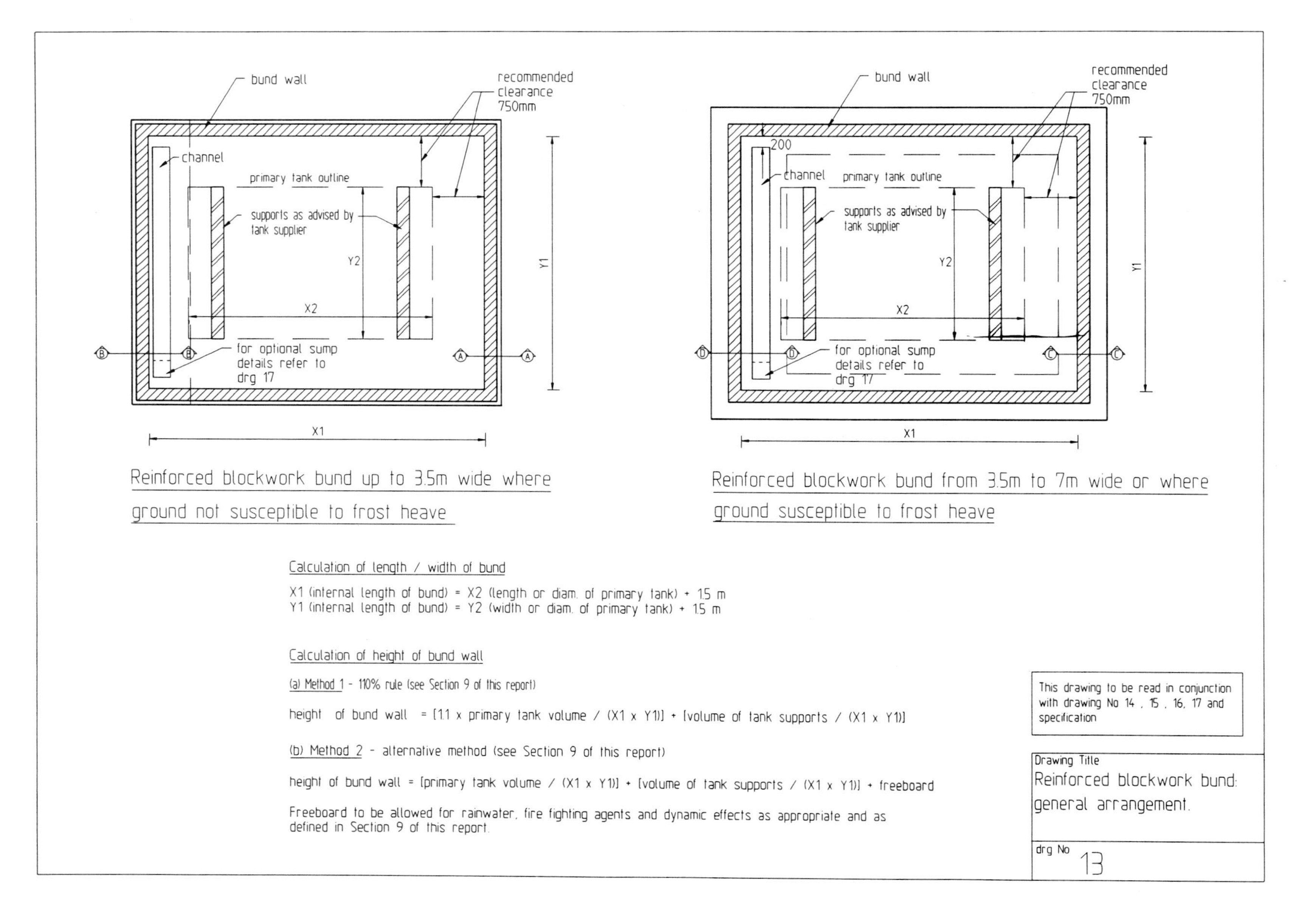

Drawing 13 *Reinforced blockwork bund: general arrangement*

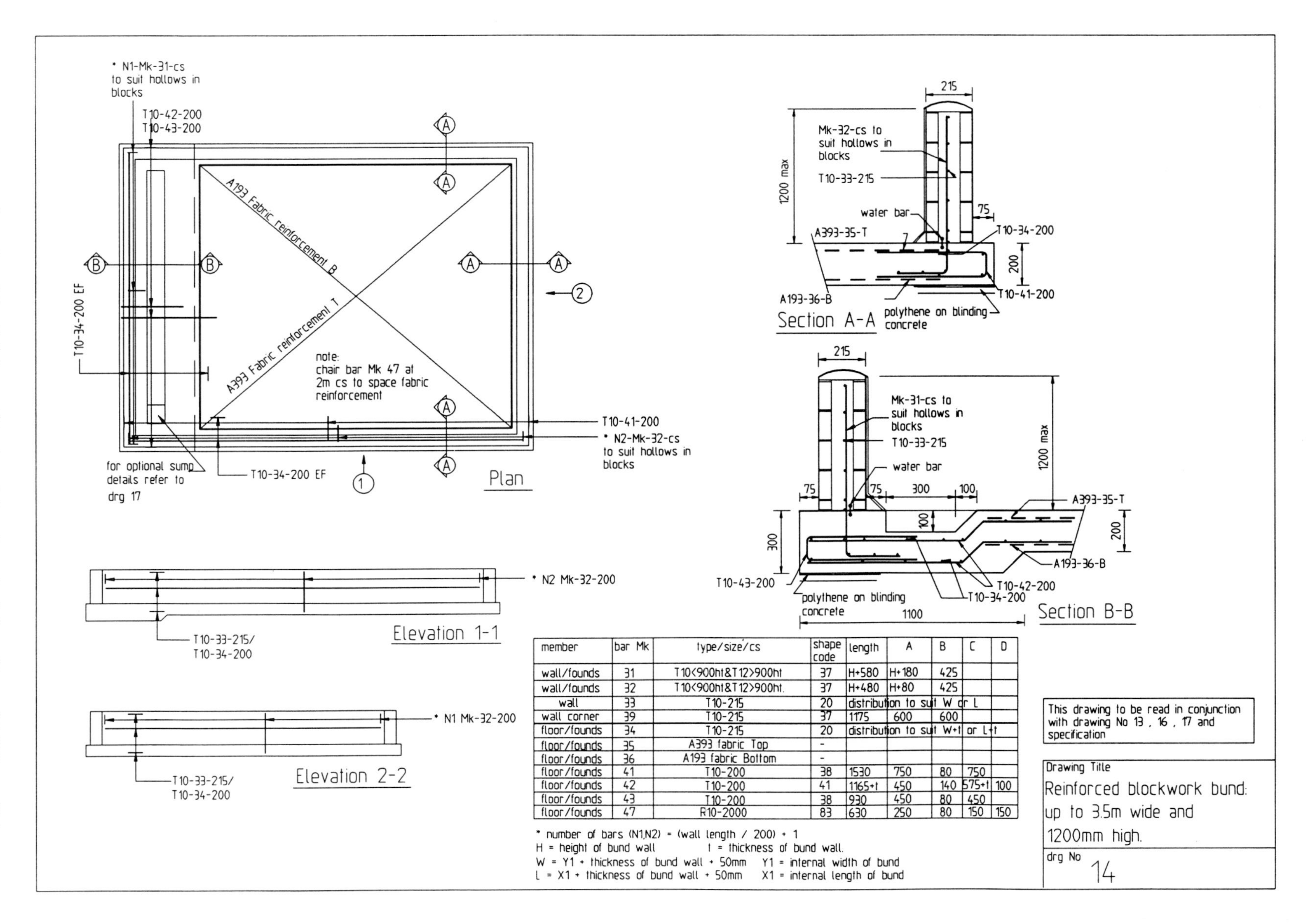

member	bar Mk	type/size/cs	shape code	length	A	B	C	D
wall/founds	31	T10<900ht&T12>900ht	37	H+580	H+180	425		
wall/founds	32	T10<900ht&T12>900ht.	37	H+480	H+80	425		
wall	33	T10-215	20	distribution to suit W or L				
wall corner	39	T10-215	37	1175	600	600		
floor/founds	34	T10-215	20	distribution to suit W+t or L+t				
floor/founds	35	A393 fabric Top	-					
floor/founds	36	A193 fabric Bottom	-					
floor/founds	41	T10-200	38	1530	750	80	750	
floor/founds	42	T10-200	41	1165+t	450	140	575+t	100
floor/founds	43	T10-200	38	930	450	80	450	
floor/founds	47	R10-2000	83	630	250	80	150	150

* number of bars (N1,N2) = (wall length / 200) + 1
H = height of bund wall   t = thickness of bund wall.
W = Y1 + thickness of bund wall + 50mm   Y1 = internal width of bund
L = X1 + thickness of bund wall + 50mm   X1 = internal length of bund

Drawing 14 *Reinforced blockwork bund: up to 3.5m wide and 1 200mm high*

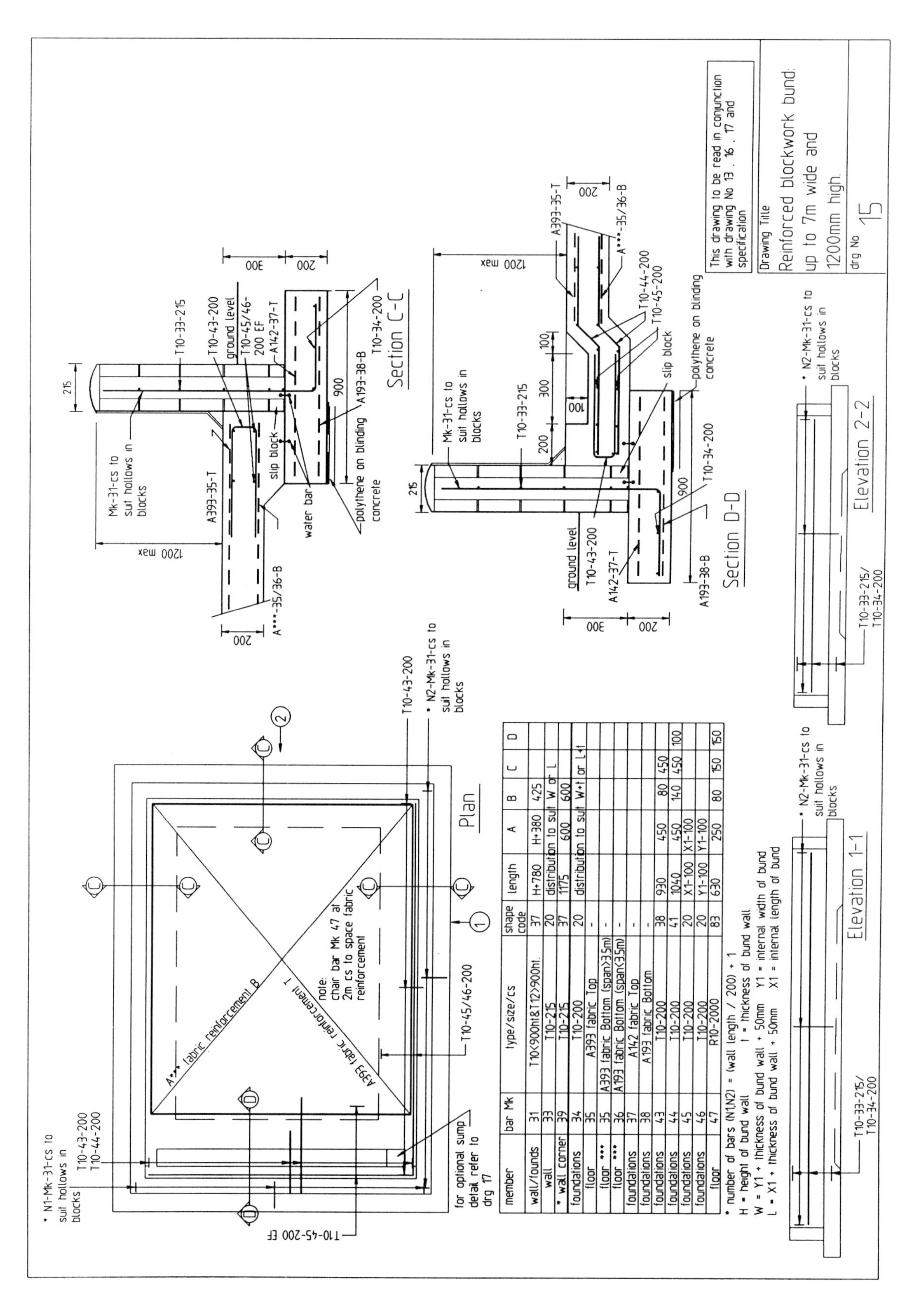

member	bar Mk	type/size/cs	shape code	length	A	B	C	D
wall/founds	31	T10<900ht & T12>900ht.	37	H+780	H+380	425		
wall	33	T10-215	20	distribution to suit W or L				
* wall corner	39	T10-215	37	1175	600	600		
foundations	34	T10-200	20	distribution to suit W+t or L+t				
floor	35	A393 fabric Top	-					
floor ***	35	A393 fabric Bottom (span>3.5m)	-					
floor ***	36	A193 fabric Bottom (span<3.5m)	-					
foundations	37	A142 fabric Top	-					
foundations	38	A193 fabric Bottom	-					
foundations	43	T10-200	38	930	450	80	450	
foundations	44	T10-200	41	1040	450	140	450	100
foundations	45	T10-200	20	X1-100	X1-100			
foundations	46	T10-200	20	Y1-100	Y1-100			
floor	47	R10-2000	83	630	250	80	150	150

* number of bars (N1,N2) = (wall length / 200) + 1
H = height of bund wall   t = thickness of bund wall
W = Y1 + thickness of bund wall + 50mm   Y1 = internal width of bund
L = X1 + thickness of bund wall + 50mm   X1 = internal length of bund

Drawing 15 *Reinforced blockwork bund: up to 7m wide and 1 200mm high*

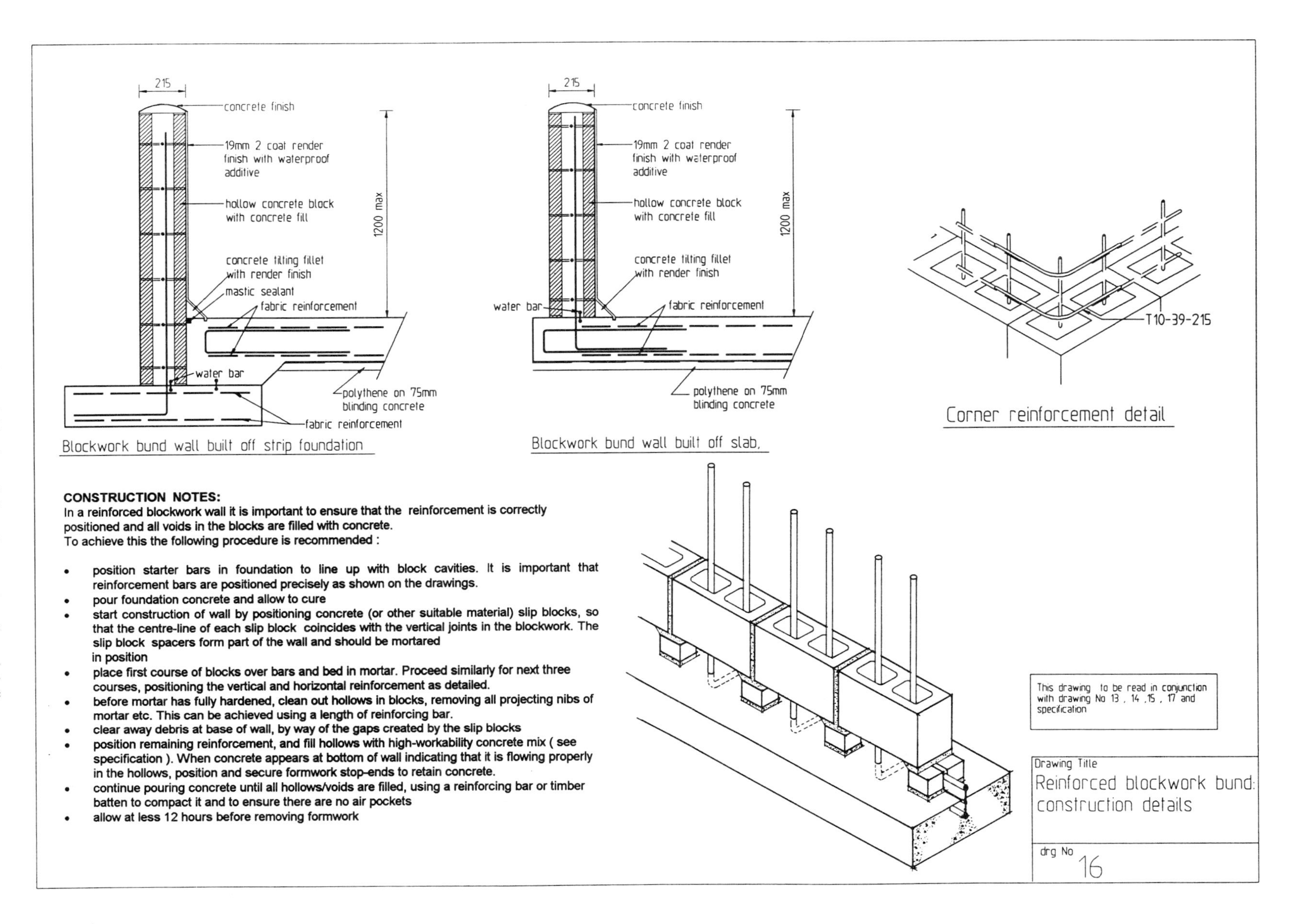

Drawing 16 *Reinforced blockwork bund: construction details*

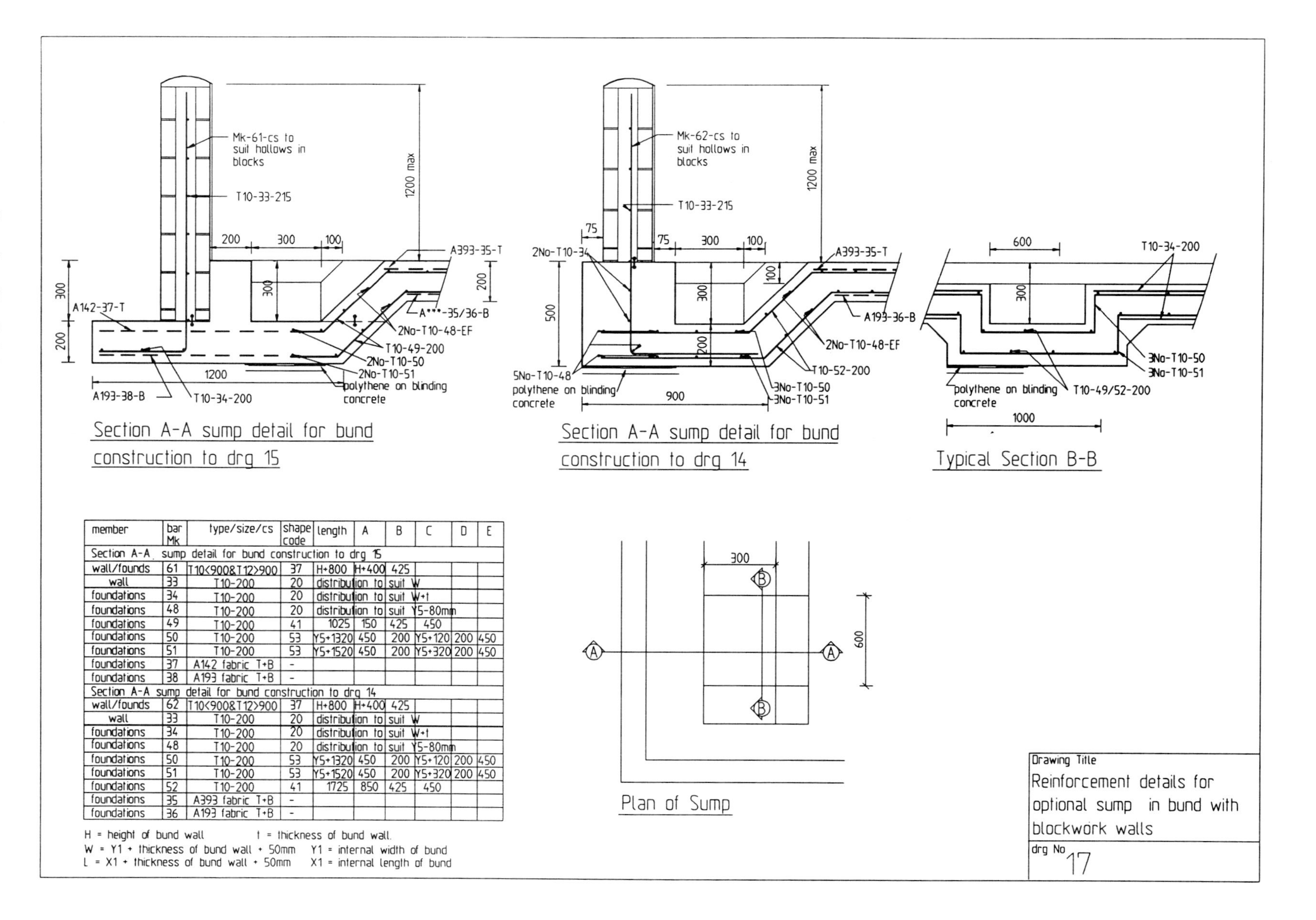

member	bar Mk	type/size/cs	shape code	length	A	B	C	D	E
Section A-A sump detail for bund construction to drg 15									
wall/founds	61	T10<900&T12>900	37	H+800	H+400	425			
wall	33	T10-200	20	distribution to suit W					
foundations	34	T10-200	20	distribution to suit W+t					
foundations	48	T10-200	20	distribution to suit Y5-80mm					
foundations	49	T10-200	41	1025	150	425	450		
foundations	50	T10-200	53	Y5+1320	450	200	Y5+120	200	450
foundations	51	T10-200	53	Y5+1520	450	200	Y5+320	200	450
foundations	37	A142 fabric T+B	-						
foundations	38	A193 fabric T+B	-						
Section A-A sump detail for bund construction to drg 14									
wall/founds	62	T10<900&T12>900	37	H+800	H+400	425			
wall	33	T10-200	20	distribution to suit W					
foundations	34	T10-200	20	distribution to suit W+t					
foundations	48	T10-200	20	distribution to suit Y5-80mm					
foundations	50	T10-200	53	Y5+1320	450	200	Y5+120	200	450
foundations	51	T10-200	53	Y5+1520	450	200	Y5+320	200	450
foundations	52	T10-200	41	1725	850	425	450		
foundations	35	A393 fabric T+B	-						
foundations	36	A193 fabric T+B	-						

H = height of bund wall t = thickness of bund wall

W = Y1 + thickness of bund wall + 50mm Y1 = internal width of bund

L = X1 + thickness of bund wall + 50mm X1 = internal length of bund

Drawing 17 *Reinforcement details for optional sump in bund with blockwork walls*

A9.2 DESCRIPTION OF WORKS

The description of works must be read in conjunction with detailed design drawings (Section 6.1). The British Standards referred to in this section are listed on page 59.

Part 1 Siting and site preparation

1.00 Site investigation

Avoid sites with:-

- variations in substrata which may give rise to differential settlement
- unstable slopes that may cause slip or other movement
- geological faults, below ground voids or fissures
- deleterious matter present, which may have an adverse effect on construction materials
- a site history which may cause structural problems, e.g. previous mining, made-up ground, underground services, etc.
- low ground bearing pressure (the model designs are based on a permissible ground bearing pressure of not less than 200kN/m^2).

Note: the siting of a bund is dictated by the location of the primary tank, which may in turn be dictated by the general layout of the facility of which it is part. It may not be possible, therefore, to avoid sites with the characteristics listed above, in which case suitable precautions must be taken. The model drawings assume that the proposed site does not have any of the undesirable features listed above. Where sites with any of the above undesirable features cannot be avoided, specialist advice should be sought.

2.00 Site preparation

2.01 Remove all vegetation and organic top soil from the site to expose the subsoil. Suitable subsoils or substrata include:

- firm or stiff clay
- firm or stiff sandy clay
- boulder clay
- shale clay (non sulphurous)
- compact sand or sandy gravel
- chalk
- rock.

2.02 Excavate and trim the surface of the excavations to the level surfaces necessary for the formation of wall foundations and the base slab.

2.03 Soft ground should be excavated and removed, and the formation levelled down to firm ground with a permissible ground bearing pressure of not less than 200kN/m^2.

2.04 Compact the excavated surfaces using a vibrating roller or vibrating plate compactor.

2.05 Protect the prepared formation from the elements prior to construction of base.

3.00 **Hardcore and blinding concrete**

Note: **Blinding concrete.** The British Standard for water retaining structures, BS 8007, requires that at least 75mm of C20 blinding concrete (see Section 3.00) be placed directly over the prepared formation. The structural concrete is then poured onto a polythene slip membrane placed on top of the blinding concrete.

In accordance with BS 8007, this Description of works requires ground level reinforced concrete slabs and wall foundations to be built on blinding concrete rather than hardcore. Blinding concrete is also required for making up discrepancies in level between the formation and the underside of the structure.

Hardcore should only be used for make up beneath the blinding concrete where the method of placing and compaction gives the hardcore sufficient strength to support the structure without any long-term adverse effect.

3.01 Where hardcore is to be used for making up levels, it should conform with Granular Sub Base Type 2, Table 8/3 of the Department of Transport Specification. The following gradation should be used:

BS sieve size	Percentage by mass passing
75mm	100
37.5mm	85 -100
10mm	45 - 100
5mm	25 - 85
600μm	8 - 45
75μm	0 - 10

Suitable materials include natural sands, gravels, crushed rock and concrete. The soluble sulphate content should be within the DoT specified limits.

3.02 The hardcore should be placed in uniform horizontal layers not exceeding 150mm in depth. Each layer must be compacted prior to the placing of the next.

3.03 The compaction method should be sufficient to achieve the required permissible bearing pressure of not less than 200kN/m^2.

3.04 Suitable compaction plant includes vibrating rollers having a mass per metre equivalent to 1,300kg with a total weight of 1,000kg, and vibrating plate compactors having an equivalent mass of 1,000kg.

3.05 Granular fill in combination with geotextiles may be used to stabilise soft ground. Construction and design techniques are site-specific and expert advice should be obtained before using geotextiles.

Part 2 Reinforced concrete construction

4.00 **Formwork**

4.01 Formwork should be properly designed and accurately constructed to produce a finished concrete to the lines, levels and dimensions indicated on the drawings and to acceptable tolerances (vertical and horizontal tolerances bund walls and floor: plus or minus 3mm).

4.02 The formwork should be of an appropriate type to provide the required surface appearance. External surfaces should normally be smooth unless an alternative finish is specified for aesthetic reasons. Internal surfaces should be smooth except where required to be roughened to receive surface rendering, or at temporary construction joints where bonding between adjacent pours of concrete is required.

4.03 A suitable mould oil should be applied to the surface prior to casting to prevent damage during stripping.

4.04 Temporary shutter ties should be of an approved type, preferably of a type which do not leave holes through walls or other parts of the structure which, if improperly filled, could cause the finished structure to leak.

4.05 The formwork should be designed for ease of striping, i.e. without re-entrant angles which might cause damage to the cast concrete during removal.

4.06 The formation of *in situ* kickers between 50mm and 150mm high is recommended at the junction between wall and slab. These provide an upstand at the base of the wall against which the formwork of the wall may be clamped, so minimising leakage of concrete fines at base level during the casting of the remaining wall sections.

4.07 The formwork must be designed so that proper compaction of the concrete can be achieved and so that the reinforcement is not displaced during concrete pouring or compaction.

5.00 Reinforcement

5.01 Reinforcement sizes, lengths and shapes must comply with the bar bending schedules.

5.02 Hot rolled bars to conform to BS 4449
Cold worked bars to conform to BS 4449
Steel fabric reinforcement to conform to BS 4483

5.03 High tensile deformed bars to be grade 460
Mild steel plain bars to be grade 250

5.04 Cutting and bending to comply with BS 4466 and BS 8110

5.05 Prior to use, reinforcement should be brushed to remove loose deposits of rust, mill scale, grease and other deleterious matter. Surface rust is acceptable provided that the reinforcement is not deeply pitted or the section reduced in size.

5.06 The reinforcement should be properly fixed at the positions and centres indicated on the drawings in such a way that it will not be displaced from the indicated position during or after concrete casting.

5.07 The nominal cover of concrete for all steel reinforcement, including links, stirrups, fabric and spacers, should be not less than 40mm (in compliance with BS 8110 Part 1 Section 3.3.4 - 'severe' exposure category).

5.08 Laps in fabric reinforcement should be not less than one mesh. Bar laps should be not less than 42 times the bar diameter unless otherwise indicated on the drawings.

6.00	**Concrete Mixes**
6.01	Concrete mixes should conform to BS 5328, BS 8110 and BS 8007. Ready mixed concrete should be obtained from a supplier currently registered with an approved quality assurance scheme, e.g. QSRMC (Quality Scheme for Ready Mixed Concrete).
6.02	Site mixing is not recommended unless a weight batching cement mixer conforming to the appropriate British Standards is used and aggregates, cements and mix designs can be demonstrated to conform to BS 5328 for a mix equivalent to those specified in 6.03.
6.03	Recommended Concrete mixes :

Reinforced concrete to ground slabs	To conform with BS 8007 specification: C35A mix designation Minimum cement content 325kg/m^3 Water cement ratio not to exceed 0.55 for OPC mixes and 0.50 for blended cement mixes.
Reinforced concrete to walls and foundations	To conform with either: C40 mix to BS 8110 Minimum cement content 325kg/m^3 Water cement ratio should not exceed 0.55 or RC 40 mix to BS 5328 1991 or C35A to BS 8007
Blinding concrete under slabs and foundations	GEN 2 to BS 5328 1991 or C20 to BS 8007
Concrete fill to reinforced blockwork	GEN 2 to BS 5328 1991 or as otherwise specified. Maximum size coarse aggregate 10mm

6.04 For blended cements - *OPC/GGBS (Ground Granulated Blast Furnace Slag) or OPC/PFA (Pulverised Fuel Ash)* - the mean proportion should not exceed 50% for GGBS and 35% for PFA as stated in BS 8110 Part 1, 6.1.2. The water cement ratio for BS 8007 C35A mixes should not exceed 0.50 when blended cements are used. Sulphate resisting cement to be used in ground slabs where sulphates are present in the ground.

6.05 Aggregates should conform to BS 882 or BS 1047 and have an absorption of not greater than 3% when tested in accordance with BS 812.

6.06 The recommended sizes of coarse aggregate are as follows:

Structural and blinding concrete	20mm nominal
Concrete fill to reinforced blockwork	10mm nominal

6.07 Concrete workability should be specified so that the concrete can be placed without the risk of segregation, and so that it can be properly compacted around all reinforcement without leaving voids.

6.08 A water cement ratio of less than 0.5 will produce an impermeable concrete but the workability will be low. Increasing the water cement ratio from 0.5 to 0.6 will result in a threefold increase in permeability making the concrete unsuitable for bund construction. The optimum water cement ratio for producing impermeable concrete is 0.35, but a plasticiser will have to be incorporated to give the mix sufficient workability. The approved supplier should be consulted on workability requirements. GGBFS and PFA may be used as replacements for plasticisers to increase workability.

7.00 Concrete Placing and Workmanship

7.01 A polythene membrane (1,000g/m^2 gauge) should be placed on top of the blinding concrete to prevent leaching of water and fines out of the concrete mix during placing. Membrane joints should be sealed or lapped for added protection against leakage.

7.02 Concrete must not be placed during adverse weather conditions, during freezing conditions or against frozen or frost covered surfaces.

7.03 The concrete must be transported by appropriate means and placed in its final position in such a way that no segregation occurs.

7.04 Concrete should be placed in one continuous operation up to construction joints and formwork surfaces. The concrete should be placed in layers no thicker than can be compacted effectively with the equipment available.

7.05 The placed concrete should be properly compacted using mechanical vibrators such that no segregation, voids, or other defects that might affect the final condition of the cast concrete can occur.

7.06 After casting and formwork removal, the exposed surfaces of the cast concrete should be cured in accordance with the requirements of BS 8110. Proper curing is essential to help prevent loss of moisture during hydration of the cement and to minimise cracking. It also helps the concrete reach its design strength more rapidly.

7.07 Curing may be achieved by covering the work and wetting the surface of the concrete from time to time as necessary. Alternatively, a proprietary curing agent may be sprayed on to the surface of the concrete in accordance with the manufacturers instructions.

7.08 Other precautions should be taken as necessary to protect the finished concrete against damage or adverse weather during curing.

7.09 Concrete cube test results or test certificates should be provided by the contractor or concrete supplier to demonstrate compliance of the concrete mix with BS 8110 or BS 8007 as appropriate.

8.00 Joints

8.01 The model drawings assume continuous construction of the base slab with no requirement for expansion, contraction or construction joints. Where vertical construction joints are required in walls, 'water stops' should be provided. Water stops should be provided at the junction between the walls and base slab.

8.02 Contractors who are sufficiently experienced in constructing water-tight reinforced concrete structures may bc able to dispense with the water stop at the junction of the wall and base slab. The technique requires the top surface of the kicker to be prepared by washing away the cement and fine sand before the concrete hardens completely. The clean aggregate which is exposed forms a surface to which the next layer of poured concrete in the walls can bond. To be successful, this method requires a skilled contractor and careful attention to detail.

8.03 Water stops should conform to BS 6213. The types that are likely to be used in bund construction are as follows:

- centre bulb type (flexible polymer) - for expansion, contraction and partial contraction joints, set centrally in walls. The distance of the waterstop from the nearest exposed concrete face should be not less than half the waterstop width
- dumbell type (flexible polymer) - similar to above but without centre bulb. Used only for contraction, partial contraction or construction joints
- surface types (flexible polymer) - used on the underside of concrete slabs and occasionally on the outer face of below ground walls
- rigid type (metal, e.g. copper or steel strip) - used for construction joints where no movement is anticipated.

8.04 Water stops should be joined using proprietary joining pieces and heat welded, vulcanised or bonded together in accordance with the manufacturers' instructions.

8.05 Joint sealing compounds should be resistant to the contained liquid, in this case oil, and the work detailed and constructed to the manufacturer's recommendations. BS 6213 specifies the various sealants and their application.

Part 3 Reinforced masonry walls

9.00 **Reinforced Blockwork to BS 5628**

9.01 Reinforced blockwork walls may be built off independent reinforced concrete strip foundations or constructed integrally with the reinforced concrete bund floor.

9.02 In either case the L-shaped reinforcement bars, which give the blockwork its tensile strength, must be set in the correct positions before the foundation or slab is cast. Longitudinal lacer reinforcement bars are required to maintain the alignment of the L-shaped bars during concrete casting and, where appropriate, to tie the bottom leg of the bar to the mesh reinforcement of the foundations or slab.

9.03 Hollow concrete blocks must comply with BS 6073. Block thickness must be not less than 215mm and net block strength should be not less than 10N/mm^2.

9.04 The mortar mix for bedding and jointing the blocks shall conform to BS 5628 class (i) and comprise the following proportions by volume:

1 part of ordinary Portland cement, 1/4 part of hydrated lime, 3 parts sand.

9.05 The concrete for infilling the block cores shall be GEN 2 or a mix comprising the following proportions measured by *weight* :

1 part OPC, 3 parts sand to BS 882, 2 parts coarse aggregate to BS 882.

The maximum size of the coarse aggregate shall be 10mm. The mix should have a high workability with a concrete slump of 150mm to 200mm. A superplasticiser may be used to help provide the necessary workability.

9.06 The blocks must be laid and bonded so that the cores align vertically and that the completed work is in horizontal and vertical alignment and to the required dimensions. U-shaped flat metal ties (one per core) may be used to bond the block cavities at straight-bonded pier positions.

9.07 In order to facilitate proper placing of the infill concrete, the cores of the blocks and the reinforcement must be kept clean of any extraneous or adhered mortar. This can be achieved by placing the joints of the first row of blocks on small concrete block or brick spacers (see drawing no. 16).

9.08 The vertical reinforcement should be fixed prior to laying the blocks.

9.09 Horizontal reinforcement must be placed as the work proceeds.

9.10 On completion of the blockwork, the hollow cores should be cleaned out and, when the mortar has hardened sufficiently, filled with concrete. When concrete begins to flow out of the bottom of the wall, the voids between the spacer blocks should be shuttered. The infill concrete should be poured and tamped continuously until all core voids are filled. The cores should be slightly overfilled and the surplus concrete should be trowelled over to form a rounded coping along the top of the wall.

9.11 External surfaces and joints should be finished or prepared as necessary to be compatible with any subsequent surface coatings.

9.12 The wall should not be fully loaded until it has achieved the specified 28-day design strength.

9.13 It is extremely difficult to make watertight movement and construction joints in blockwork walls. Joints should therefore be avoided wherever possible, although vertical movement joints are essential to help minimise shrinkage cracking in long walls. Joint component and joint sealant manufacturers should be consulted about suitable products and techniques for forming joints. Hydrostatic pressure limitations on wall joints may dictate the maximum wall height and therefore the bund capacity.

Where a masonry wall would be so long (e.g. in multi-tank installations) that a vertical contraction joint would be needed, it is recommended that either more than one bund is constructed (thus negating the need for joints), or that the bund is constructed from reinforced concrete.

10.00 Reinforced Brickwork to BS 5628

10.01 Reinforced brickwork, either concrete filled cavity or concrete pocket construction, may be founded on individual reinforced concrete strip footings or constructed directly off the reinforced concrete bund floor slab.

10.02 In either case the specified wall reinforcement bars must be set into the concrete foundations and located at the correct positions to provide the reinforcement for the walls. Lacer bars are necessary to maintain the position of the reinforcement bars during concrete casting (see 9.02). All wall reinforcement (including horizontal distribution bars in cavity fill construction) must be fixed prior to construction of the wall.

10.03 Clay, solid class B engineering bricks to BS 3921 Table 1 and Table 6, (Class B) shall be used for reinforced brickwork.

10.04 Cavity brickwork should be of bonded stretchers, with the two skins tied together with stainless steel wall ties.

10.05 110mm thick walls and single skin blockwork shall be in stretcher bond: other walls shall be in English bond.

10.06 The mortar mix must conform to BS 5628 class (i) and comprise the following proportions by volume:

1 part of OPC, 1/4 part of hydrated lime, 3 parts sand.

10.07 The concrete for infilling cavity brickwork should be high workability grade RC30.

10.08 The mix for infilling blockwork cores shall be GEN2 or a mix comprising the following proportions measured by *weight*:

1 part OPC, 3 parts sand to BS 882, 2 parts coarse aggregate to BS 882.

The maximum size of the coarse aggregate shall be 10mm. The mix should have a high workability with a concrete slump of 150mm to 200mm. A superplasticiser may be used to help achieve the necessary workability.

10.09 Beds shall be level with joints of uniform thickness and perpends plumb. All beds and joints should be flush and plumb. All bricks should be well wetted prior to laying.

10.10 Brickwork should be true, square and properly bonded and each course levelled round in a uniform manner.

10.11 Reinforcement must be fixed in the correct position and all cavities and pockets kept free from all deleterious matter and extraneous mortar prior to concrete infilling.

10.12 Concrete infill should be uniformly placed and properly compacted in a continuous operation taking care not to displace or damage the brickwork during this operation.

10.13 Internal surfaces of the walls should be prepared ready to receive any subsequent surface coatings.

Part 4 Coatings and surface treatments

11.01 *In situ* reinforced concrete work. Providing that the concrete is in compliance with the specified mixes and the method of placing, compaction and all other workmanship conforms to BS 8110 or 8007 as appropriate, surface coatings should not be necessary. Any minor defects in the finished surfaces should be treated, if possible, while the concrete is still green. Pin holes and small cracks may be filled by bag rubbing the surface with a grout mix comprising 1 part OPC, and 1 part of the fine sand aggregate as used in the concrete mix. The floor slab should be trowelled or finished to a smooth finish and to the required falls leading to the sump.

Where other finishes are required for aesthetic reasons, they should be applied in accordance with the manufacturers specification. Care must be taken to ensure that any paints or other coatings that are specified are not be affected by oil.

11.02 The sump has to withstand significantly more abrasion than other parts of the bund. It is therefore recommended that the surface of the sump is coated with an epoxy mortar render. The concrete surface should be properly prepared ready to receive the coating and the mortar applied strictly in accordance with the manufacturer's instructions.

11.03 Masonry is inherently more permeable than *in situ* concrete. It is therefore recommended that all internal masonry surfaces are rendered with at least 19mm of sharp sand/OPC 3:1 mix applied in two coats. Surfaces should be properly prepared and wetted prior to application. Special attention should be given to the junction of the wall and floor slab. A mastic sealant is recommended at the junction. Grooves should be cut in the concrete ready to receive the finished coatings. The second render coat should be applied so that joints in work areas do not coincide with joints in the first coat.

List of British Standards referred to in this section

BS 8007:	1987	*Design of concrete structures for retaining aqueous liquids*
BS 4449:	1988	*Specification for carbon steel bars for the reinforcement of concrete*
BS 4483:	1985	*Specification of steel fabric for the reinforcement of concrete*
BS 4466:	1989	*Specification for scheduling, dimensioning, bending and cutting of steel reinforcement for concrete*
BS 8110:	1985	*Structural use of concrete*
BS 5328 Part 2:	1991	*Methods for specifying concrete mixes*
BS 882:	1992	*Specification for aggregates from natural sources for concrete*
BS 1047:	1983	*Specification for air-cooled blastfurnace slag aggregate for use in construction*
BS 812:		*Testing aggregates* (comprising 24 Parts)
BS 6213:	1992	*Guide to selection of constructional sealants*
BS 5628 Part 3:	1985	*Use of masonry*
BS 6073:	1981	*Precast concrete masonry units*
BS 3921:	1985	*Clay bricks*

A9.3 ADDITIONAL GUIDANCE ON DRAWINGS AND DESCRIPTION OF WORKS

The guidance provided in this section supplements the information already given Section 6. It is intended particularly for non-specialist engineers and contractors.

A9.3.1 Site factors

The site topography and the nature of the ground are important aspects to consider when deciding on the suitability of a location. Since the bund designs included in this report are restricted to a maximum individual site area of approximately 50 m^2, it should be relatively easy to investigate a site and determine its suitability for bund location. A methodical and careful site reconnaissance should be undertaken, involving both a site survey and subsoil investigation.

Site survey

Where the ground conditions are well known, the survey need involve only a simple visual inspection to identify any topographical or other features that may cause construction problems. Slopes and all other relevant features, including natural features such as trees, water-courses and ponds, and artificial features such as made up ground, drains, overhead power lines and service pipes, any of which could affect site suitability, should be noted. Examination of old records and drawings of previous site development may provide useful information.

Subsoil investigation

It is important that the nature and properties of the ground lying below the topsoil are identified, since the substratum formation will provide the foundation support for the bund. Bunds should not be built on organic topsoils or any other deleterious material which may impair structural integrity or affect construction materials. It may be necessary to dig trial holes to determine the characteristics of the site subsoils; such holes must be properly backfilled before construction commences. The investigation should include an examination of the water table fluctuations on the site.

A9.3.2 Points to note when assessing a site

1. Sloping sites (above 10%) on poorly drained sand or clay soils can lead to instability problems. Where a potentially unstable soil condition is suspected, expert advice must be sought.

2. Where it is necessary to build a bund on a sloping site, the site should be excavated to form a horizontal shelf sufficiently large to accommodate the total area of the bund. Bunds should not be built on fill on sloping ground since this may lead to instability problems.

3. Sites which are liable to flooding and sites with a high water table may lead to instability problems. Where sites are liable to flooding, additional pollution prevention measures may be necessary and the possibility of floatation should be taken into account.

4. The proximity of nearby buildings and structures, and particularly their foundations, may affect the construction of the bund.

5. Tree and other large roots must be removed before construction. On clay soils large trees immediately adjacent to the site may cause long-term foundation problems and expert advice should be sought.

6. Where there is existing under-drainage to the site of a bund it should either be re-routed or stopped (both upstream and downstream of the site). Under-drains running beneath a bund (or of course a primary storage tank) are a potential pollution pathway.

7. Where there are variations in the subsoil across the proposed construction site these may give rise to differential settlement. Expert advice should be sought.

8. Where deleterious matter is present in the subsoil, special precautions may be required to protect the construction materials. Sulphurous shale, for example, adversely affects Ordinary Portland Cement. Where deleterious matter is suspected, expert advice should be sought.

9. On sites where the required minimum permissible bearing pressure of 200kN/m^2 cannot be achieved, it will be necessary either to improve the permissible ground-bearing pressure or to modify the foundation designs presented in the model drawings. In either case, expert advice should be sought.

10. The permissible ground-bearing pressure may be assessed approximately by driving a 50mm square pointed wooden peg vertically into the ground using a 5kg hammer. If the penetration is less than 50mm for ten firm hammer blows, it is likely that the permissible ground-bearing pressure exceeds 200kN/m^2.

A9.3.3 Forms of construction

The model drawings detail three forms of bund construction:

Type 1. *In situ* reinforced concrete construction.

The details shown for the construction of bund floors and walls are in accordance with the recommendations of BS 8007.

The walls are tied into the floor slab using starter bars.

Type 2. Reinforced brickwork wall construction.

The drawings detail reinforced concrete cavity-filled brickwork walls to BS 5628, and an *in situ* reinforced concrete floor to BS 8007.

Type 3. Reinforced concrete blockwork wall construction.

The drawings detail hollow concrete blockwork walls to BS 5628, with cores filled with reinforced concrete, and an *in situ* concrete floor to BS 8007.

The above list ranks the forms of construction in order of superiority in terms of structural robustness and degree of impermeability. While each of the forms of construction detailed in the model drawings is capable of providing an effective bund, Type 1 walls provide the highest level of security and, although they are the most costly, may be the preferred option for sensitive sites.

The choice of type of construction will be influenced by the site situation and the resources and skills available. The impermeability of Type 2 and Type 3 walls may be improved by specifying high integrity watertight coatings. In all cases special attention should be given to ensure watertight joints in the construction.

Unreinforced masonry walls are susceptible to cracking and vulnerable to impact damage and are not therefore recommended for bund wall construction.

Unreinforced masonry may be suitable for the internal subdivision of bunds.

Bund wall height

This report recommends that, where possible, bund wall height is limited to 1.5m for safety reasons, and this is the maximum height shown on the model drawings. For structural and buildability reasons the maximum height shown on the model drawings for reinforced masonry bunds is limited to 1.2m.

Floor construction

Where the walls are to be constructed using reinforced concrete, the drawings detail two forms of floor construction (see drawing no. 1). For bunds up to 3.5m wide *and* where the ground is not susceptible to frost heave (e.g. free draining ground) a flat slab form of construction is shown. For bunds exceeding 3.5m wide *or* where the ground is susceptible to frost heave (e.g. clay soil), the floor slab construction includes an integral perimeter ground beam.

Where the walls are to be constructed using reinforced masonry, the floor slab arrangement for bunds up to 3.5m wide *and* where the ground is not susceptible to frost heave are as described above. Where bunds exceed 3.5m wide or where the ground is susceptible to frost heave, the walls are shown constructed off separate strip footings (see drawing no. 8).

Each of the floor slab designs incorporates a 100mm deep channel running along the inside face of one of the walls. Where the floor slab incorporates a ground beam, as described above, the channel is constructed 200mm in from the wall face to allow for the ground beam reinforcement. The purpose of the channel is to facilitate emptying and cleaning of the bund. A channel is recommended in preference to a sump since it is generally easier to clean and easier to construct. There may, however, be some circumstances where a sump is preferred and the drawings provide for this as an option. It should be noted that the construction of a sump adds considerably to the complexity.

A9.3.4 Bund size and capacity

The range of drawings provides for a maximum bund footprint area of $49m^2$ (i.e. 7m square). Taking the maximum bund heights of 1.5m for concrete and 1.2m high for masonry, this gives maximum capacities (net of bund wall thickness and primary tank supports) of approximately $60m^3$ and $40m^3$ respectively.

The methods for calculating the bund area and height are described on drawings no. 1 and 8 which assume a horizontal clearance of 750mm between the primary tank and the inside face of the bund (see Section 4.2.2). In practice, on existing sites, the plan shape of a bund may be constrained by other factors such as the proximity of neighbouring buildings. Two methods are described for calculating the necessary height of the bund to give the required total capacity (in relation to the primary tank) and/or freeboard (see Section 3).

A9.3.5 Tank supports

The details shown on the drawings are suitable for use with most forms of primary tank up to $25m^3$ capacity. It is assumed that tanks will be supported across their full width on walls or cradles and that the weight of the tank and its contents will therefore be distributed as a uniform line load across the width of the bund floor.

It may be necessary to allow additional thickenings and reinforcement in the concrete floor slab to allow for holding down bolts, or other fixings and baseplates, particularly for tanks supported on legs and for tall slender tanks or other abnormal tank types and tank supports. This information, and all other detailed instructions for supporting the primary tank, should be provided by the tank supplier.

A9.3.6 Example of use of model drawings

A bund is to be provided around an $8m^3$ primary oil storage tank which is 3m long, 2m wide and 1.33m high. The site is level with 250mm depth of topsoil overlying a firm sandy clay. The location is at an industrial site in East Nottingham and its south boundary is the River Trent at approximately 80m from the tank location. The site is well managed. It is not intended to cover the bund to keep out rainwater.

Capacity and freeboard calculation (see Section 3)

Method (a) - 110% rule

According to the 110% rule the required net capacity of the bund is $8.8m^3$. Although there is no specific requirement for freeboard within the 110% rule, the 10% additional capacity of the bund is effectively a freeboard. The depth of this freeboard will depend on the plan area of the bund (see Section B6.3).

Method (b) - alternative method

Capacity requirement for bund = capacity of primary oil tank, i.e. $8m^3$.

Refer to Figure 3.2 to define rainfall Region for site. Since Nottingham is on the boundary of Regions 1 and 2, select Region 2.

Refer to Table 3.3 for freeboard. For rainfall Region 2, require 282mm freeboard (including dynamic effects) or 132mm (excluding dynamic effects).

Form of construction

Inspection of the site reveals no problems with the topography or subsoil characteristics and it is concluded that the model drawings are applicable. It is decided to use reinforced concrete construction for the walls.

Bund dimensions

Calculate required plan dimensions of bund using formula in drawing no. 1:

Internal length of bund (*X1*)
= length or diameter of primary tank (*X2*) + 1.5 = 3 + 1.5 = 4.5m

Internal width of bund (*Y1*)
= width or diameter of primary tank (Y2) + 1.5 = 2 + 1.5 = 3.5m

Calculate required height of bund using formulae in drawing no. 1:

Method (a) - 110% rule

Height of bund
= (1.1 x 8) / (4.5 x 3.5) + 0.86 / (4.5 x 3.5) = 0.613m

(The effective freeboard is 0.613m minus the bund height calculated on the basis of 100% primary tank capacity, which is 0.563m, i.e. 50mm.)

Method (b) - alternative method

Height of bund (including dynamic effects)
= 8 / (4.5 x 3.5) + 0.86 / (4.5 x 3.5) + 0.282 = 0.845m

Height of bund (excluding dynamic effects)
= 8 / (4.5 x 3.5) + 0.86 / (4.5 x 3.5) + 0.132 = 0.694m

It is clear that on this site the alternative method results in a higher bund wall with greater freeboard and, consequently, safety margin against overtopping. Bearing in mind the sensitivity of the site, and following consultation with the Environment Agency, it is decided to adopt Method (b), including dynamic effects. The bund height will therefore be 0.845m.

In consultation with the regulators, it is concluded that it is not necessary in this case to make specific provision for 'jetting'.

Construction details

Drawing no. 3 provides details of construction for *in situ* reinforced concrete bund walls up to 900mm high and for reinforced concrete ground slabs up to 7m plan dimension. The main features of the drawing are described below.

Floor slab

The floor slab is 200mm thick, reinforced with two layers (one each, top and bottom) of steel reinforcement fabric mesh, designation A393. This is a square reinforcing mesh with wires spaced and welded together at 200mm centres. The 393 designation indicates the cross-sectional area of the reinforcing steel in mm^2 per metre width.

Walls and perimeter ground beam

Sections C-C and D-D show that the wall is 150mm thick. The wall foundation comprises a reinforced concrete perimeter edge beam cast integrally with the floor slab. (The perimeter beam is to provide frost protection beneath the floor slab. Freezing of moist soils causes expansion and the resulting forces can fracture unprotected ground slabs.)

The perimeter ground beam is reinforced with a reinforcement cage consisting of a series of square link bars and straight lacing bars. There are three lacing bars in both the top and bottom of the ground beam (six in all) and the square link bars are spaced equally along them. The link bars are tied to the lacing bars with proprietary ties or with soft mild steel wire.

Section D-D shows detail of three sides of the slab: Section C-C shows detail of the fourth side which has a channel and an optional sump allowed for in the design. The link bars in Section C-C are designated as T10 -13 - 200, and in Section D-D as T10 - 16 - 200. 'T' signifies a high tensile bar, '10' indicates the bar diameter in millimetres, '13' (or '16' in the latter case) denotes the bar mark number and '200' indicates the spacing in millimetres between each link bar. The bar mark relates each bar to the bending and cutting instructions which are detailed in the bending schedule set out on the drawing. The lacing bars are designated as T10 - 04, i.e. 10mm diameter high tensile bars, bar mark '04'.

Section D-D shows walls reinforced with L-shaped bars T10 - 02 - 200 (i.e. high tensile, 10mm diameter at 200mm centres, bar mark 02). Section C-C shows that the wall next to the channel is reinforced with L-shaped bars T10 - 01 - 200 (i.e. high tensile, 10mm diameter at 200mm centres, bar mark 01). The L-shaped bars are held in position using horizontal steel lacing bars (T10 - 03 - 200) which also serve to distribute stresses within the concrete.

Reinforcement bending schedule

The method for specifying the bending dimensions of steel reinforcement is standardised in BS 4466. This standard tabulates reinforcing bar details, assigning the various shapes of bent bars to a range of standardised shape codes. Free copies of these tables may be obtained from most major reinforcement bar suppliers. Drawing no. 3 shows a typical reinforcement bending schedule with shape codes and relevant dimensions.

The following shape codes are used in drawing no. 3:

Shape code	Shape description
20	Straight bar
37	L-shaped bar
38	U-shaped bar
41	Cranked bar with parallel outstanding legs
60	Link bar
73	Variation on link bar: one side of link protruding
81	Variation on link bar: 'sausage' shaped
83	'Chair' bar to support top and bottom steel reinforcement in the floor slab. (This bar is shown in the bending schedule as R10-23 but not shown in the drawing. Note that 'R' denotes a round mild steel bar.)

When ordering reinforcement bars it is necessary to give the supplier the following information for each bar mark on the drawing:

- type of bar - e.g. round mild steel (R) or deformed round high tensile steel (T)
- bar diameter in millimetres
- the shape code
- the bending dimensions, including the overall length
- the number of bars required.

For example, for bar mark 02 shown on drawing no. 3:

Type of bar: high tensile deformed bar (T) grade 460 (see Section 5.03 of the Description of works)

Bar diameter: 10mm

Shape code: 37 (i.e. L-shaped)

Dimensions: the dimensions required and given in the bending schedule are:

upstand length, A = 1000mm
horizontal leg length, B = 60mm, and
length, L = 1,575mm.

Note: these dimensions relate to a wall 900mm high. If the wall is to be constructed less than 900mm high, the upstand length (A) and the overall length (L) must be reduced accordingly.

Number of bars in walls:

Bars are at 200mm centres. and the total number ($N2$) required in each wall is calculated using the equation:

$N2$ = (wall length / 200) + 1

In this example the total number of bars mark 02 is therefore:

N2

$$= 2\ (4500\ /\ 200 + 1) + 1\ (3500\ /\ 200 + 1) = 66$$ (rounded up to nearest whole number)

This process must be repeated for each of the 12 bar marks identified on drawing no. 3.

A10 Bund wall designs

The model designs in this Appendix are reproduced by courtesy of the Institute of Petroleum.

The first three designs are for free standing perimeter bund walls with heights of 1200 mm, 1500 mm and 2000 mm respectively. The second three are for intermediate bund walls (i.e. can be loaded from either side) in the same height range.

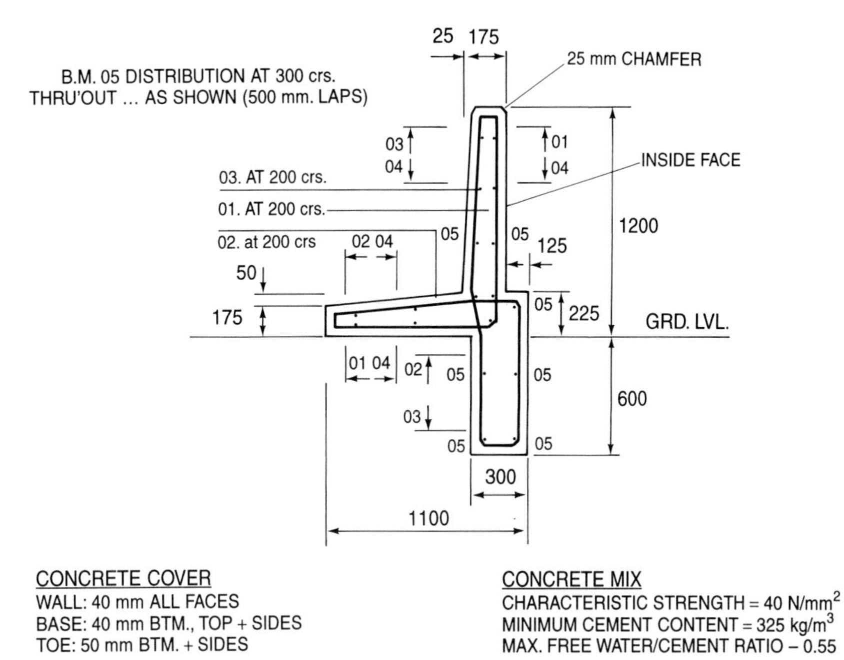

REINF'T. QUANTITIES FOR 4.0 M LENGTH OF WALL										
Bar Mark	Type Size	C/C	No. off	Length	Total Length	Shape Code	Bending Dim's.			
							A	B	C	D
01	T.10	200	21	1950	40.95m	37	875			
02	T.10	200	21	2350	49.35m	99	1000, 87°, 725, 200			
03	T.10	200	21	1675	31.175m	41	550	250		75
04	T.8	200	42	750	31.5m	38	350	95		
05	T.8	300	18	4200	75.6m	20	STR			

Figure A10.1 *Bund wall 1200 mm high* (Source: IP)

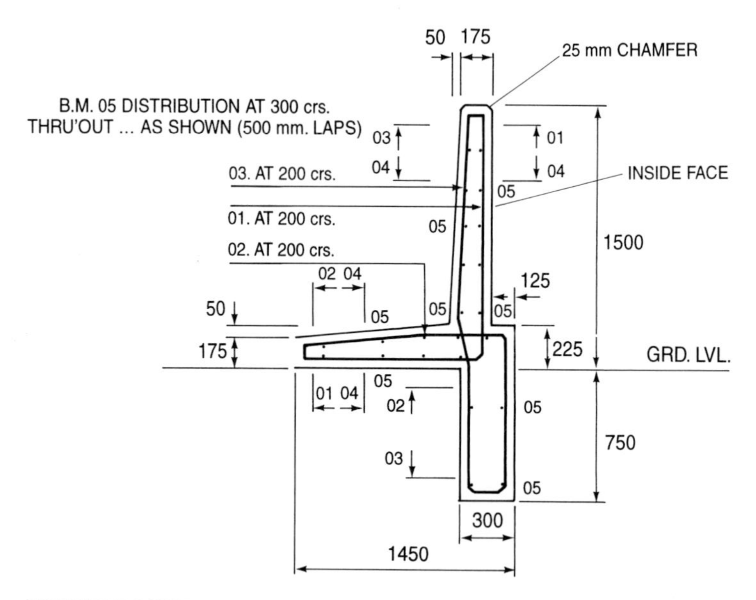

<u>CONCRETE COVER</u>
WALL: 40 mm ALL FACES
BASE: 40 mm BTM., TOP + SIDES
TOE: 50 mm BTM. + SIDES

<u>CONCRETE MIX</u>
CHARACTERISTIC STRENGTH = 40 N/mm^2
MINIMUM CEMENT CONTENT = 325 kg/m^3
MAX. FREE WATER/CEMENT RATIO – 0.55

REINF'T. QUANTITIES FOR 4.0 M LENGTH OF WALL										
Bar Mark	Type Size	C/C	No. off	Length	Total Length	Shape Code	Bending Dim's.			
							A	B	C	D
01	T.10	200	21	2575	54.075m	37	1200			
02	T.10	200	21	3000	63.0m	99	1325, 88°, 825, 200			
03	T.10	200	21	2100	44.1m	41	650	250		75
04	T.8	200	42	750	31.5m	38	350	95		
05	T.8	300	23	4200	96.6m	20	STR			

Figure A10.2 *Bund wall 1500 mm high* (Source: IP)

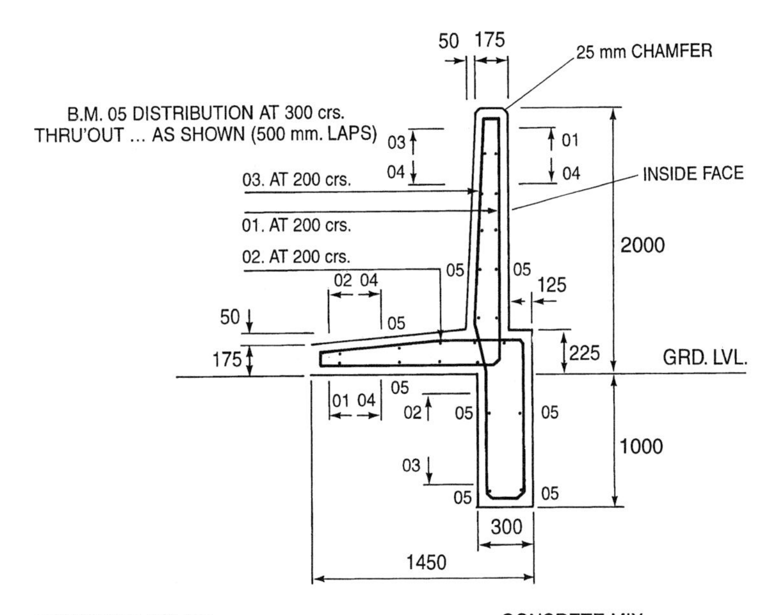

CONCRETE COVER
WALL: 40 mm ALL FACES
BASE: 40 mm BTM., TOP + SIDES
TOE: 50 mm BTM. + SIDES

CONCRETE MIX
CHARACTERISTIC STRENGTH = 40 N/mm^2
MINIMUM CEMENT CONTENT = 325 kg/m^3
MAX. FREE WATER/CEMENT RATIO – 0.55

REINF'T. QUANTITIES FOR 4.0 M LENGTH OF WALL										
Bar Mark	Type Size	C/C	No. off	Length	Total Length	Shape Code	Bending Dim's.			
							A	B	C	D
01	T.10	200	21	3200	67.2m	37	1450			
02	T.10	200	21	3450	72.45m	99	1675, 87°, 1125, 200			
03	T.10	200	21	2850	59.85m	41	900	250		75
04	T.8	200	42	750	31.5m	38	350	95		
05	T.8	300	18	4200	121.8m	20	STR			

Figure A10.3 *Bund wall 2000 mm high* (Source: IP)

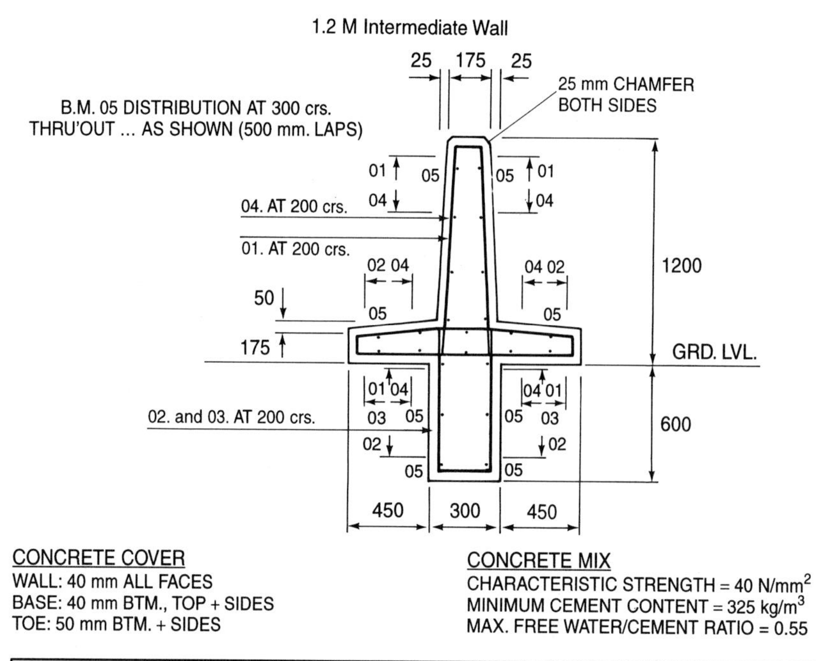

REINF'T. QUANTITIES FOR 4.0 M LENGTH OF WALL

Bar Mark	Type Size	C/C	No. off	Length	Total Length	Shape Code	Bending Dim's.			
							A	B	C	D
01	T.10	200	42	1700	71.4m	37	625			
02	T.10	200	42	1325	55.65m	37	650			
03	T.10	200	21	1250	26.25m	38	550	200		
04	T.8	200	63	750	47.25m	38	350	95		
05	T.8	300	24	4500	108.0m	20	STR			

Figure A10.4 *Intermediate bund wall 1200 mm high* (Source: IP)

1.5 M Intermediate Wall

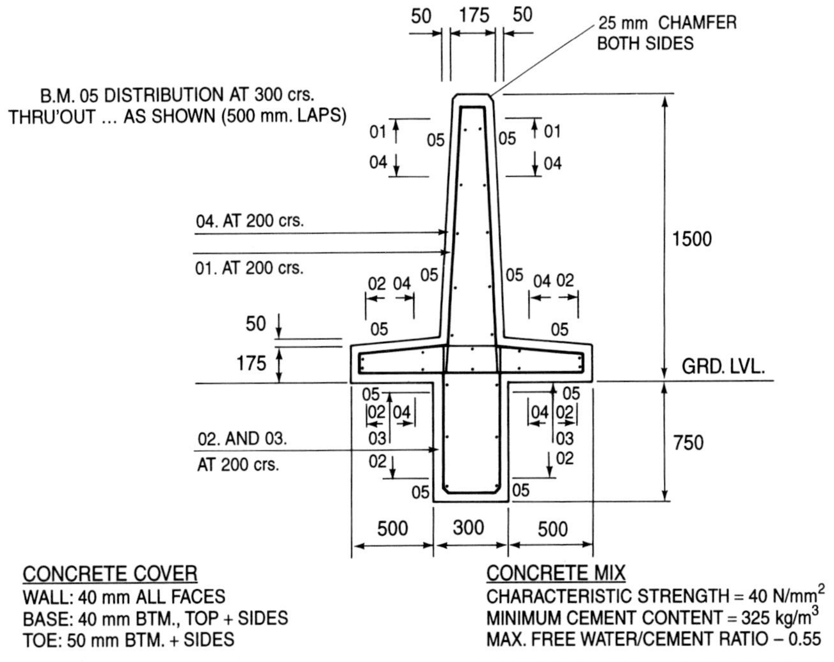

REINF'T. QUANTITIES FOR 4.0 M LENGTH OF WALL

Bar Mark	Type Size	C/C	No. off	Length	Total Length	Shape Code	Bending Dim's.			
							A	B	C	D
01	T.10	200	42	2050	86.10m	37	625			
02	T.10	200	42	1550	65.10m	37	700			
03	T.10	200	21	1250	26.25m	38	550	200		
04	T.8	200	63	750	47.25m	38	350	95		
05	T.8	300	26	4500	117.0m	20	STR			

Figure A10.5 *Intermediate bund wall 1500 mm high* (Source: IP)

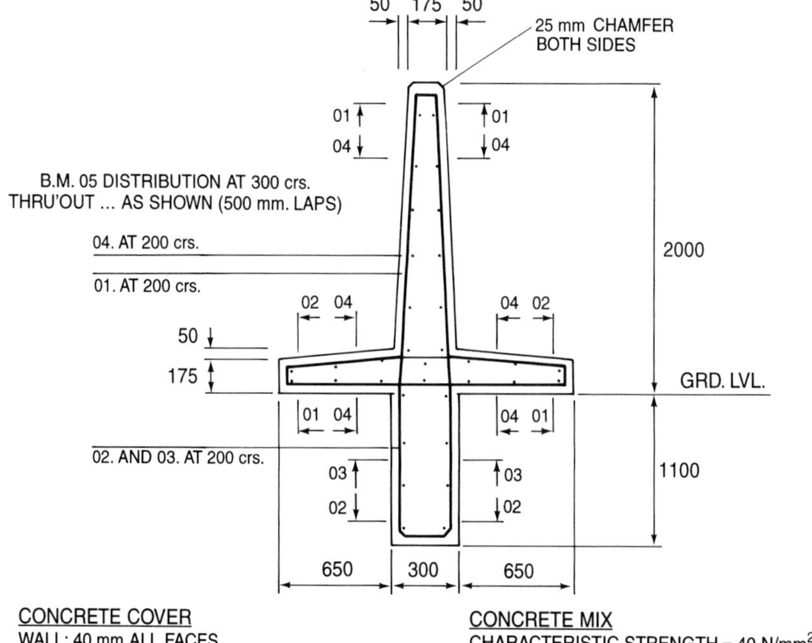

CONCRETE COVER
WALL: 40 mm ALL FACES
BASE: 40 mm BTM., TOP + SIDES
TOE: 50 mm BTM. + SIDES

CONCRETE MIX
CHARACTERISTIC STRENGTH = 40 N/mm^2
MINIMUM CEMENT CONTENT = 325 kg/m^3
MAX. FREE WATER/CEMENT RATIO – 0.55

REINF'T. QUANTITIES FOR 4.0 M LENGTH OF WALL

Bar Mark	Type Size	C/C	No. off	Length	Total Length	Shape Code	Bending Dim's.			
							A	B	C	D
01	T.10	200	42	2700	113.40m	37	825			
02	T.10	200	42	2050	86.10m	37	850			
03	T.10	200	21	1250	26.25m	38	550	200		
04	T.8	200	63	750	47.25m	38	350	95		
05	T.10	300	34	4500	153.0m	20	STR			

REVISION MEMO

3/95 Rev. 2 Intermediate walls added, all heights consolidated on one Standard and general update.

Figure A10.6 *Intermediate bund wall 2000 mm high (Source: IP)*

A11 Permeability assessment: soil settlement test

The *soil settlement* test described in Box A11.1 is a simple and quick method for assessing the approximate clay content of a soil and, therefore, whether the soil is likely to be sufficiently impermeable for the construction of lagoons or earth bunds. If soils are to be used *in situ* it is recommended that samples are taken from just below the soil surface and at 1 m below the proposed lagoon base level, using a narrow-bore soil auger.

There are two main drawbacks to this test:

- it is often difficult to distinguish where the clay and silt layer finishes and the fine sand starts
- it is diffficult to achieve proper dispersion of soil particles, even when using sodium silicate as a dispersant, and this can lead to an underestimation of the clay and silt fraction.

Box A11.1 *Soil settlement test* (Source: MAFF Bulletin 202)

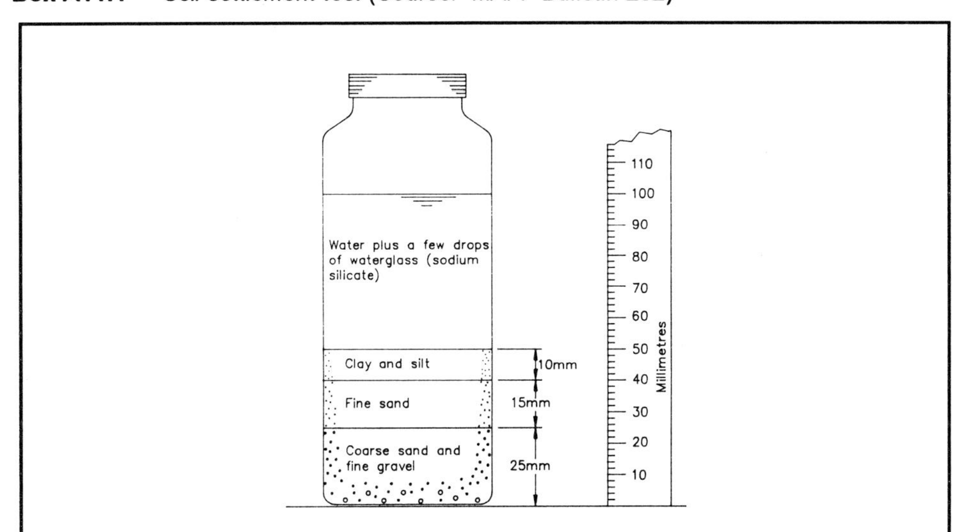

The proportions of the various component materials that form a particular soil may be estimate with reasonable accuracy by making the following test. Half fill a narrow parallel-sided glass bottle and shake well to mix the soil and water thoroughly. If waterglass (sodium silicate) is available add 2 or 3 drops to water and stir vigorously. Stand the bottle on a firm surface and allow the soil to settle for 24 hours. Coarse sand will settle out immediately, fine sand within a few minutes and the silt and clay last. These will stratify into clearly visible layers from which the approximate solid proportions can be estimated by measuring the depth of each layer.

For example:

Total depth of soil = 50mm

Proportion of clay and silt = $\frac{10mm}{50mm} \times 100 = 20\%$

Proportion of fine sand = $\frac{15mm}{50mm} \times 100 = 30\%$

Proportion of coarse sand and fine gravel = $\frac{25mm}{50mm} \times 100 = 50\%$

A12 Permeability assessment: hand texturing

The 'hand texturing' test, Box A12.1, involves assessing the soil according to a range of basic physical characteristics, each of which is distinguishable by moulding the soil sample in the hand. No tools are required. It is another simple and quick method for assessing the approximate clay content of a soil. With experience, this method can give accurate and consistent results.

Box A12.1 *Assessment of soil clay content by hand texturing (Source: ADAS and Soil Survey and Land Research Centre)*

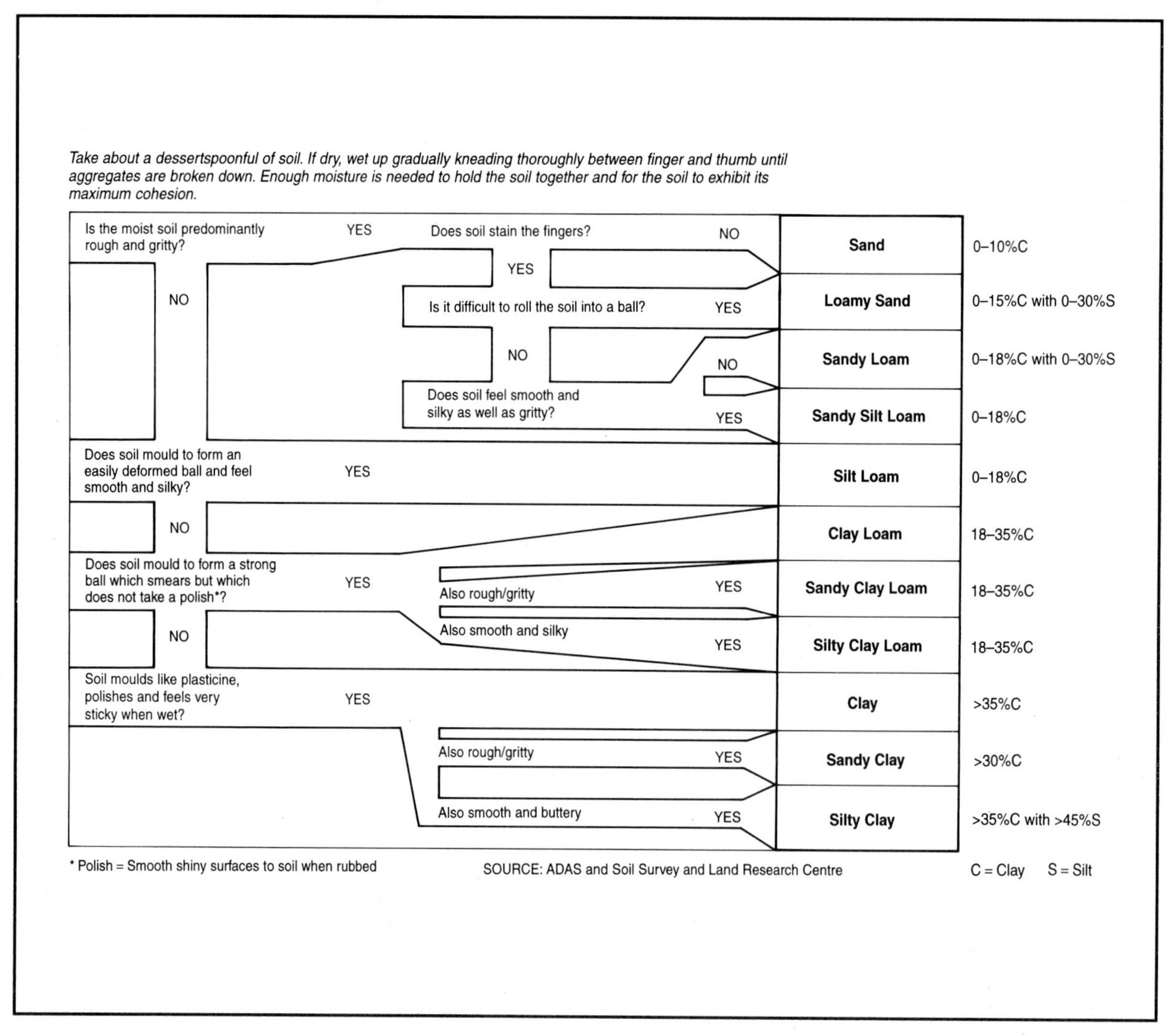

A13 Permeability assessment: soil classification

The permeability of a clay soil and its susceptibility to problems such as drying shrinkage are related principally to the type and quantity of clay present. The type and quantity of clay can be established through standard testing procedures including *soil particle analysis* and *liquid* and *plastic limits.* Empirical links have been established between soil classification according to the British Soil Classification System for Engineering Purposes (BSCS) (and the very similar American Unified Soil Classification System (USCS)) and soil permeability, and consequently their suitability for impermeable embankment construction.

The BSCS is described fully in BS 5930 (BSI, 1981a). Box A13.1 reproduces Table 8 of BS 5930 which summarises the BSCS and defines the 'Group symbols' by which soils are classified. The section of the Table headed 'FINE SOILS' is relevant to soils suitable for lagoon and earth bund construction. The tests which have to be carried out in order to classify soils according to the BSCS are described in detail in BS 1377 (BSI, 1990b).

Although the BSCS is based on closely defined laboratory tests, BS 6031 (BSI, 1981b), Table 1 (part of which is reproduced below in Box A13.2), provides some useful additional field identification notes against each of the soil Groups.

Box A13.3 reproduces Figure 31 of BS 5930 which positions soil Groups in relation to plasticity index and liquid limit. The 'A-line' may be taken to represent the boundary between clays (above the line) and silts (below). It is recommended that in the absence of direct measurements of soil permeability, only those soils in Groups CH, CI and MH are considered suitable for lagoons and earth bund construction.

This recommendation is consistent with the typical coefficients of permeability reported by Sharma and Lewis (1994) and reproduced in Box A13.4, in relation to soils classified under the American USCS.

Box A13.1 *British Soil Classification System* (Source: BS 5930)

Soil groups (see note 1)			Subgroups and laboratory identification				
GRAVEL and SAND may be qualified Sandy GRAVEL and Gravelly SAND, etc. where appropriate			Group symbol (see notes 2 & 3)	Subgroup symbol (see note 2)	Fines (% less than 0.06mm)	Liquid limit %	Name
COARSE SOILS less than 35% of the material is finer than 0.06mm	GRAVELS more than 50% of the material is of gravel size (coarser than 2mm)	Slightly silty or clayey GRAVEL	G GW GP	GW GPg GPu	0 to 5		Well graded GRAVEL Poorly graded/ Uniform/Gap graded GRAVEL
		Silty GRAVEL Clayey GRAVEL	G-F G-M G-C	GWM GPM GWC GPC	5 to 15		Well graded/Poorly graded silty GRAVEL Well graded/Poorly graded clayey GRAVEL
		Very silty GRAVEL Very clayey GRAVEL	GF GM GC	GML, etc. GCL GCI GCH GCV GCE	15 to 35		Very silty GRAVEL: subdivided as for GC Very clayey GRAVEL (clay of low, intermediate, high, very high, extremely high plasticity)
	SANDS more than 50% of the coarse material is of sand size (finer than 2mm)	Slightly silty or clayey SAND	S SW SP	SW SPg SPu	0 to 5		Well graded SAND Poorly graded/Uniform/Gap graded SAND
		Silty SAND Clayey SAND	S-F S-M S-C	SWM SPM SWC SPC	5 to 15		Well graded/Poorly graded silty SAND Well graded/Poorly graded clayey SAND
FINE SOILS more than 35% of the material is finer than 0.06mm		Very silty SAND Very clayey SAND	SF S-M S-C	SML, etc. SCL SCI SCH SCV SCE	15 to 35		Very silty SAND: subdivided as for SC Very clayey SAND (clay of low, intermediate, high, very high, extremely high plasticity)
	Gravelly or sandy SILTS and CLAYS 35% to 65% fine	Gravelly SILT Gravelly CLAY(see note 4)	FG MG CG	MLG, etc. CLG CIG CHG CVG CEG		<35 35 to 50 50 to 70 70 to 90 >90	Gravelly SILT: subdivide as for CG Gravelly CLAY of low plasticity
		Sandy SILT Sandy CLAY (see note 4)	FS MS CS	MLS, etc. CLS, etc.			Sandy SILT: subdivide as for CG Sandy CLAY: subdivide as for CG
	SILTS and CLAYS 65% to 100% fine	SILT (M-SOIL) CLAY (see notes 5 & 6)	F M C	ML, etc. CL CI CH CV CE		<35 35 to 50 50 to 70 70 to 90 >90	SILT: subdivide as for C CLAY of low, intermediate, high, very high, extremely high plasticity
ORGANIC SOILS		Descriptive letter 'O' suffixed to any group or sub-group symbol			Organic matter suspected to be a significant constituent. Example MHO: Organic SILT of high plasticity.		
PEAT		Pt Peat soils consist predominantly of plant remains which may be fibrous or amorphous.					

NOTE 1. The name of the soil group should always be given when describing soils, supplemented, if required, by the group symbol, although for some additonal applications (e.g. longitudinal sections) it may be convenient to use the group symbol alone

NOTE 2. The group symbol or sub-group symbol should be placed in brackets if laboratory methods have not been used for identification, e.g. (GC).

NOTE 3. The designation FINE SOIL of FINES, F, may be used in place of SILT, M, or CLAY, C, when it is not possible or not required to distinguish between them.

NOTE 4. GRAVELY if more than 50% of coarse material is of gravel size. SANDY if more than 50% of coarse material is of sand size.

NOTE 5 SILT (M-SOIL), M, is material plotting below the A-line, and has a restricted plastic range in relation to its liquid limit, and relatively low cohesion. Fine soils of this type include clean silt-sized materials and rock flour, micaceous and distomaceous. Organic soils also usually plot below the A-line on the plasticity chart, when they are designated ORGANIC SILT, MO.

NOTE 6 CLAY, C, is material plotting above the A-line, and is fully plastic in relation to its liquid limit.

Box A13.2 *BSCS field identification of soils* (Source: BS 6031)

Soil groups*		Subgroups and laboratory identification						Field identification
		Description	Group symbol	Subgroup symbol	Liquid limit %	Subgroup name	Casagrande group symbol	
FINE SOILS more than 35% finer than 0.06 mm	Gravelly or sandy SILTS and CLAYS 35% - 65% fines	Gravelly SILT	MG (FG)	MG		Gravelly SILT (sub-divide like CG)	–	Coarse particles visible to naked eye. Silt fraction dries moderately quickly and can be dusted off the fingers. Clay fraction can be rolled into threads when, moist, smooth to touch and plastic, sticks to fingers and dries slowly.
		Gravelly CLAY	CG	CLG	35	Gravelly CLAY of low plasticity		
				CIG	35-50	Gravelly CLAY of intermediate plasticity		
				CHG	50-70	Gravelly CLAY of high plasticity		
				CVG	70-90	Gravelly CLAY of very high plasticity		
		Sandy SILT	MS (FS)	MS		Sandy SILT	ML	Sandy silts and sandy clays feel gritty when rubbed between the fingers. Silts and sandy silts dry quickly and can be dusted off the fingers, exhibit marked dilatancy. Dry lumps have some cohesion but can be powdered easily in the fingers.
		Sandy CLAY	CS	CLS etc.		Sandy CLAY (sub-divide like CG)	MI	
	SILTS and CLAYS 65% - 100% fines	SILT (M-SOIL)	M (F)	M		SILT (sub-divide like C)	ML/MI	Clays, silty clays and sandy clays are plastic and can be readily rolled into threads when moist. Dry lumps can be broken but not powdered, but they disintegrate under water. They stick to fingers and dry very slowly. Clays feel smooth to the touch.
		CLAY	C	CL	35	CLAY of low plasticity	CL	
				CI	35-50	CLAY of intermediate plasticity	CI	
				CH	50-70	CLAY of high plasticity	CH	
				CV	70-90	CLAY of very high plasticity	–	
ORGANIC SOILS		Descriptive letter 'O' suffixed to any group or sub-group symbol Organic matter suspected to be a significant constituent. Example MHO: Organic SILT of high plasticity						Usually dark in colour, plant remains may be visible, often with distinctive smell.
PEAT		Peat soils consist predominantly of plant remains which may be fibrous or amorphous					Pt	Usually black or brown in colour. Very compressible. Easily identifiable visually.

* Gravel and sand my be qualified as Sandy GRAVEL and Gravelly SAND where appropriate

Box A13.3 *Plasticity chart* (Source: BS 5930)

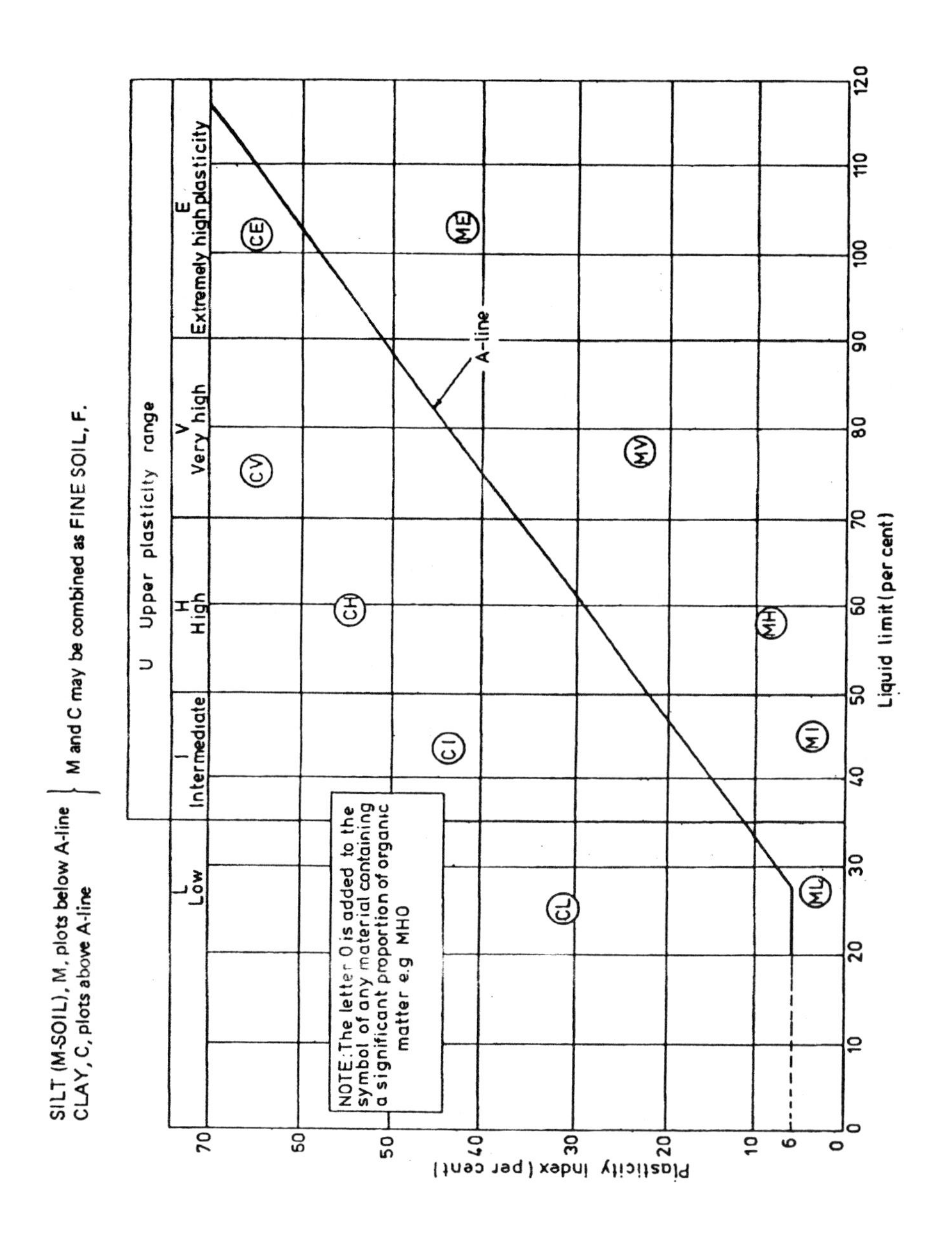

Box A13.4 *Typical permeabilities according to USCS soil Groups* (Source: Sharma and Lewis)

Group Symbol USCS	SOIL TYPE	Typical Coefficient of Permeability (cm/sec)
GW	Well graded, clean gravels, gravel-sand mixture	2 x 10^{-2}
GP	Poorly graded, clean gravels, gravel-sand mixture	5 x 10^{-2}
GM	Silty gravels, poorly graded, gravel-sand-silt	5 x > 10^{-7}
GC	Clayey gravels, poorly graded, gravel -sand-silt	5 x >10^{-8}
SW	Well graded, clean sands, gravelly sands	5 x >10^{-4}
SP	Poorly graded, clean sands, sand-gravel mixture	5 x >10^{-4}
SM	Slty sands, poorly graded, sand-silt mixture	2 x >10^{-5}
SM - SC	Sand-silt clay mixed with slight plastic fines	1 x >10^{-6}
SC	Clayey sands, poorly graded, sand-clay mixture	2 x >10^{-7}
ML	Inorganic silts and clayey silts	5 x >10^{-6}
ML - CL	Mixture of inorganic silt and clay	2 x >10^{-7}
CL	Inorganic clays of low to medium plasticity	5 x >10^{-8}
OL	Organic silts and silt-clays, low plasticity	--
MH	Inorganic clayey silts, elastic silts	2 x >10^{-7}
CH	Inorganic clays of high plasticity	5 x >10^{-8}
OH	Organic clays and silty clays	--

A14 Permeability assessment: falling head cell method and field-testing

A14.1 FALLING HEAD CELL METHOD

The falling head permeameter is a laboratory test apparatus designed for testing low-permeability soils.

In summary, a small diameter glass capillary tube (cross-sectional area a) is fixed above a larger tube containing the soil sample (cross-sectional area A, length L). The tubes are filled with water and the period of time (t) is noted for a fall in head (h_1 to h_2) to occur in the capillary tube as the liquid flows through the soil sample. The fine bore of the capillary tube relative to the diameter of the soil sample under test permits very small flows to be measured.

Applying Darcy's law:

k (the sample permeability) $= (aL / At)\log_n(h_1 / h_2)$

A14.2 FIELD TESTING

BS 5930 describes two methods for determining soil permeability in the field by measuring flow in boreholes. Each method relies upon measurement of the flow induced by the application of a hydraulic pressure in the borehole different from that in the surrounding ground. Where the pressure is increased by introducing water into the borehole the method is commonly known as the falling head, or inflow, test. Where the pressure in the borehole is decreased by pumping, the method is known as the rising head, or outflow, test.

The technique is best suited to measuring the permeability of soils below the natural groundwater level. Tests should be carried out on a number of boreholes over a site to establish the range of permeabilities.

A15 Technical properties of geotextiles

A15.1 PHYSICAL AND MECHANICAL PROPERTIES

Mass/unit area

May also be used as an indicator for cost and strength comparisons between fabrics of similar type.

Woven tape, thin needle punched and heat bonded fabrics; typically 100 to 250g/m^2. Thick needle punched, woven tape yarns and heavyweight woven polyesters, meshes and grids; up to 1000g/m^2.

Strength

Usually quoted in units of kilonewtons per metre width kN/m.

Puncture resistance

Usually measured as CBR values (Californian Bearing Ratio).

Pore size

Required for the assessment of filtration characteristics, usually given as O_{90} or O_{95}; O_{90} = 500 means that 90% of the pores have an equivalent size of 500 microns. Typically the O_{90} for non-wovens is in the range 50 μm to 350 μm, and the O_{90} range for wovens is 150 μm to 600 μm.

Surface friction

The bond between the soil and the geotextile that prevents the soil from sliding or the geotextile from being pulled out of a soil mass. This is usually expressed as the ratio of the angle of bond stress between the geotextile and the soil (δ) to the internal angle of shearing resistance of the soil (ϕ). The surface friction ratio (δ)/(ϕ) is normally about 0.75. For clays this is sometimes expressed as an adhesion factor, α, which defines a proportion of the undrained shear strength.

Short-term load deformation

This property gives an indication of the tensile strength of a geotextile and the axial strain at rupture. It should generally only be used in the design of temporary structures in low risk situations since the test does not give an indication of long-term behaviour, e.g. creep.

Long-term load deformation (creep)

Creep performance under long-term load, and ultimate strength, are fundamental considerations in the selection of a suitable fabric in the context of structural durability. After an initial strain, a stressed fabric continues to deform at a lower rate for several months. The phenomenon causes stretching and physical rearrangement of the polymer structure. Failure as a result may occur at loadings significantly below the predicted ultimate strength. Long-term creep varies according to the load condition, although all fabrics will fail in the long term at less than 60% of the ultimate load. This period may vary from about one month to as much as one year according to the properties and characteristics of the geotextile. Stress strain curves allow comparisons of performance.

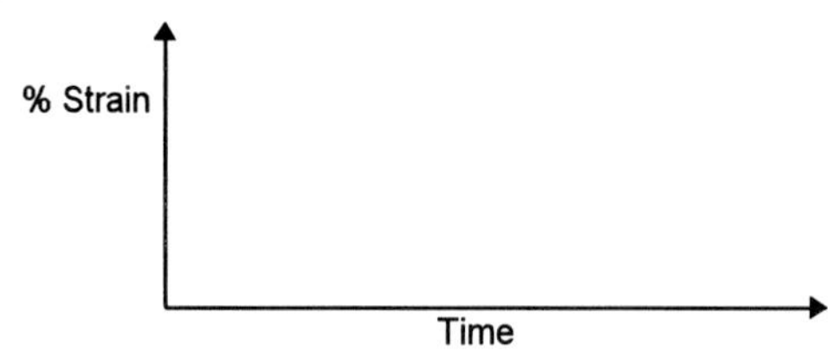

Strain increase of a geosynthetic under constant load (creep)

Modulus

Geosynthetics do not demonstrate a perfectly linear relationship between stress and strain (i.e. Young's Modulus). The modulus of a geosynthetic is defined as the stiffness of the material when subject to stress. The higher the modulus, the less strain is generated by any given stress. The modulus is calculated by dividing the increase in dimension by the original dimension for any unit of stress. If 1 metre length is stressed by 1 kN and the length increases to 1.1 metres (i.e. increased by 0.1m), then the modulus is 1/0.1, i.e. 10 kNm/m. The modulus for a particular strain is used as the basis for assessing allowable unit loads. The modulus at 10% strain would be referred to as E10.

Permittivity, permeability and transmissivity

A measure of the permeability gives an indication of the ability to allow water to flow through the geotextile, an essential characteristic for filtration and drainage. Water permeability is to some extent related to pore size although flows are not laminar and permeability may vary according to the stress on the geotextile or the static head. Units are expressed as l/s per m^2 for a head of either 5 or 10 cm. Typically, non-wovens exhibit a maximum water permeability of around
200 l/s per m^2 at 10 cm, reducing according to head. Woven monofilaments range from 100 -1000 l/s per m^2 at 10 cm. Permittivity is sometimes used to describe the flow rate per unit thickness in a plane normal to the fabric of the geotextile. The transmissivity of a geotextile is used to describe a measure of the flow rate per unit thickness in the plane of the fabric.

A15.2 DURABILITY

The extremely varied nature and composition of the many different forms of geosynthetics makes it impossible to give specific advice. The durability will be influenced by many factors. For example, a polymer may exhibit good UV resistance and also be resistant to a certain chemical, but might degrade significantly when attacked simultaneously by these elements. Chemical attack includes biodegradation which is essentially a chemical process initiated through enzymes. The possibility of mechanical damage, e.g. by rodents or construction equipment, should be allowed for in the design.

A16 Structural design of liners: geomembranes and geotextiles

A16.1 TENSILE FORCE IN A GEOSYNTHETIC WITH A UNIFORMLY THICK COHESIONLESS OVERBURDEN

(Basis: 2D wedge analysis with no allowance for construction plant or seepage)

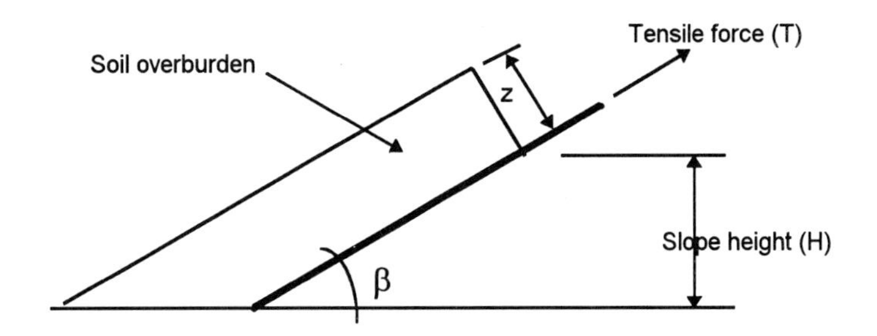

$$T = (\Upsilon z^2/\sin 2\beta)\{[2H\cos\beta/z - 1]\sin(\beta - \phi_1)/\cos\phi_1 - \sin\phi_2/\cos(\beta + \phi_2)\}$$

where

T	=	geosynthetic tensile force per m width
z	=	thickness of soil overburden
H	=	slope height
β	=	slope angle
Υ	=	density of overburden
ϕ_1	=	minimum soil geotextile interface friction angle
ϕ_2	=	overburden friction angle

A16.2 HORIZONTAL ANCHORAGE LENGTH OF A GEOSYNTHETIC WITH A UNIFORMLY THICK COHESIONLESS OVERBURDEN

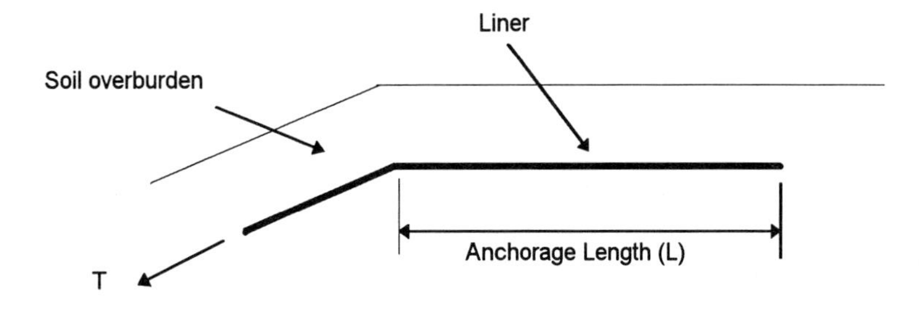

$$L = T(\cos\beta - \sin\beta\tan\phi_1) / \gamma\tan\phi_1$$

The above equation ignores the frictional resistance developed between the overburden and the geotextile since the overburden is likely to move with the geotextile/geomembrane.

A16.3 STRESS DUE TO SELF WEIGHT OF MEMBRANE (SIGNIFICANT FOR STEEP OR LONG SLOPES ONLY)

$f = (W\sin\beta - F)/A$

where

f	=	stress on membrane
W	=	weight of total slope length of geosynthetic (per m width)
β	=	slope angle
F	=	frictional restraint between geosynthetic and underlying soil ($= W\cos\beta\tan\phi_i$)
A	=	cross-sectional area of membrane (per m width)

A16.4 OTHER FACTORS

A16.4.1 Wind

Estimates of wind effects may be obtained using BS CP3, Chapter V, Part 2 (BSI, 1972). Membranes may need to be weighted to prevent uplift during and after construction.

A16.4.2 Basis of tensile strength analysis

Load limited allowable unit loading is the basis of design for the tensile strength of a geotextile. It is necessary to work in terms of unit load instead of stresses, i.e. force per unit width rather than force per unit area.

Most geosynthetic manufacturers state the ultimate tensile unit strength of a material in units of kN per metre width. This is usually given as the short-term ultimate tensile load. These loads should be factored to give the safe design load, taking account of the characteristics of the material which may result in long-term creep failure at less than the stated short-term ultimate load.

A16.4.3 Selection of allowable tensile unit load

There is currently no common approach to assessing allowable unit load. Stress strain modulus graphs are useful when designing permanent *Class 2* and *3* works. The characteristic tensile load per unit width (modulus) is selected for an acceptable level of strain which will give stable conditions either (a) without long-term creep or (b) creep within acceptable limits for the particular structure. This load is then factored to allow for durability and mechanical damage to give the allowable unit load.

Allowable tensile unit load per metre width = *K1* x *K2* x *K3* x *K4* x modulus for the value of strain that is judged appropriate, where *K1*, *K2*, *K3* and *K4* are reduction factors to take account of operating conditions as described below.

Suggested reduction factors applicable to allowable unit tensile load are:

K1 - Chemical

This cannot currently be quantified. It will depend upon the contact time and aggressiveness of the particular chemicals or substances to be contained.

K2 -Abrasion from contact with soils and strata

Table A16.1

Material underlying geomembrane	Reduction factor *K2*
Sand	0.9
Rock	0.7
Sharp rock	0.6

K3 - Construction

Suggested multiplying reduction factor to allow for 10% breakage of fibres and strands of geosynthetic during installation:

$K3 \approx 0.9$

K4 - Deterioration due to Age

Depends upon properties of the geosynthetic. Limited data are available but it is estimated that reduction in strength may be in the order of 50% for many geotextiles over a period of 25 years.

K4 (25 year) $\approx$ 0.5 (50% strength reduction due to 25 year ageing)

A17 Site zoning pro forma

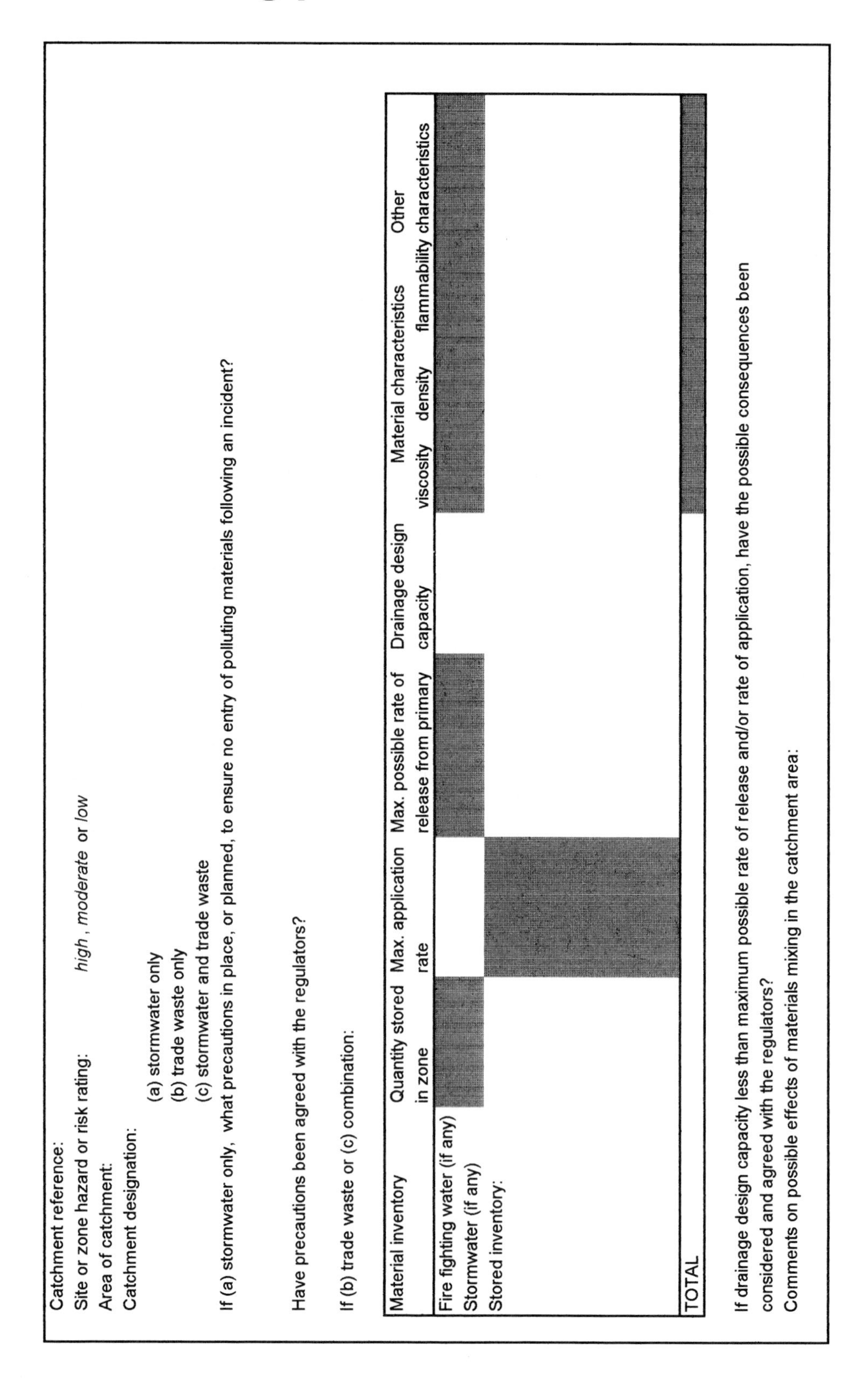

Catchment reference:

Site or zone hazard or risk rating: *high* , *moderate* or *low*

Area of catchment:

Catchment designation:

(a) stormwater only
(b) trade waste only
(c) stormwater and trade waste

If (a) stormwater only, what precautions in place, or planned, to ensure no entry of polluting materials following an incident?

Have precautions been agreed with the regulators?

If (b) trade waste or (c) combination:

Material inventory	Quantity stored in zone	Max. application rate	Max. possible rate of release from primary	Drainage design capacity	Material characteristics viscosity	density	flammability	Other characteristics
Fire fighting water (if any)								
Stormwater (if any)								
Stored inventory:								
TOTAL								

If drainage design capacity less than maximum possible rate of release and/or rate of application, have the possible consequences been considered and agreed with the regulators?

Comments on possible effects of materials mixing in the catchment area:

A18 Design of pipes and pipe networks

The two main design considerations relate to hydraulic and structural performance.

A18.1 HYDRAULIC DESIGN

Networks are designed by sizing the pipes according to the previously assessed design capacity requirement. Systems operate either under gravity or by pumping, or a combination of gravity and pumping.

A18.1.2 Gravity systems

The Colebrook White formula is commonly used in the determination of pipe flow capacities.

Nomograms based on the above formula may be used to determine pipe capacities. Box A18.1 shows discharge capacities for pipes in good condition with diameters in the range 75 mm to 300 mm.

A18.1.3 Pumped systems

Pipe flows and pumping requirements are calculated using Bernoulli's theory. BS 8005 provides information on pumping mains and pumps. The assessment of energy losses throughout a pipe network is critical to the design. Given the variations throughout most systems it is sufficient in most cases to assess these empirically. The Hazen Williams formula is commonly used to obtain estimates of pipe friction losses. Losses from constrictions, bends and junctions should also be considered. These may be estimated empirically using the expression:

$$H = \frac{KV2}{2g}$$

where

H	=	head loss
V	=	flow velocity
k	=	constant (0.8 for medium 90° bend, 1.50 for sharp 90° bend, 1.50 for valve, 1.00 for flush entrance/exit, etc.)

On simple single pipe systems losses are sometimes allowed for by assuming an increase in pipe length to allow for pipe friction and bends. This is often taken as 10% and for each end valve an additional 15 to 45 m according to type. To avoid deposition of sediment, flow velocities should exceed 1m/sec. Pipe design strength must allow for surge resulting from increased pressure due to the closure or switching of valves or pumps, and the effects of pressure waves.

Box A18.1 *Discharge capacities of drains running full, 3/4, 2/3 proportional depth: used pipes in good condition (hydraulic roughness, k= 0.6 mm)* (Source: BSI)

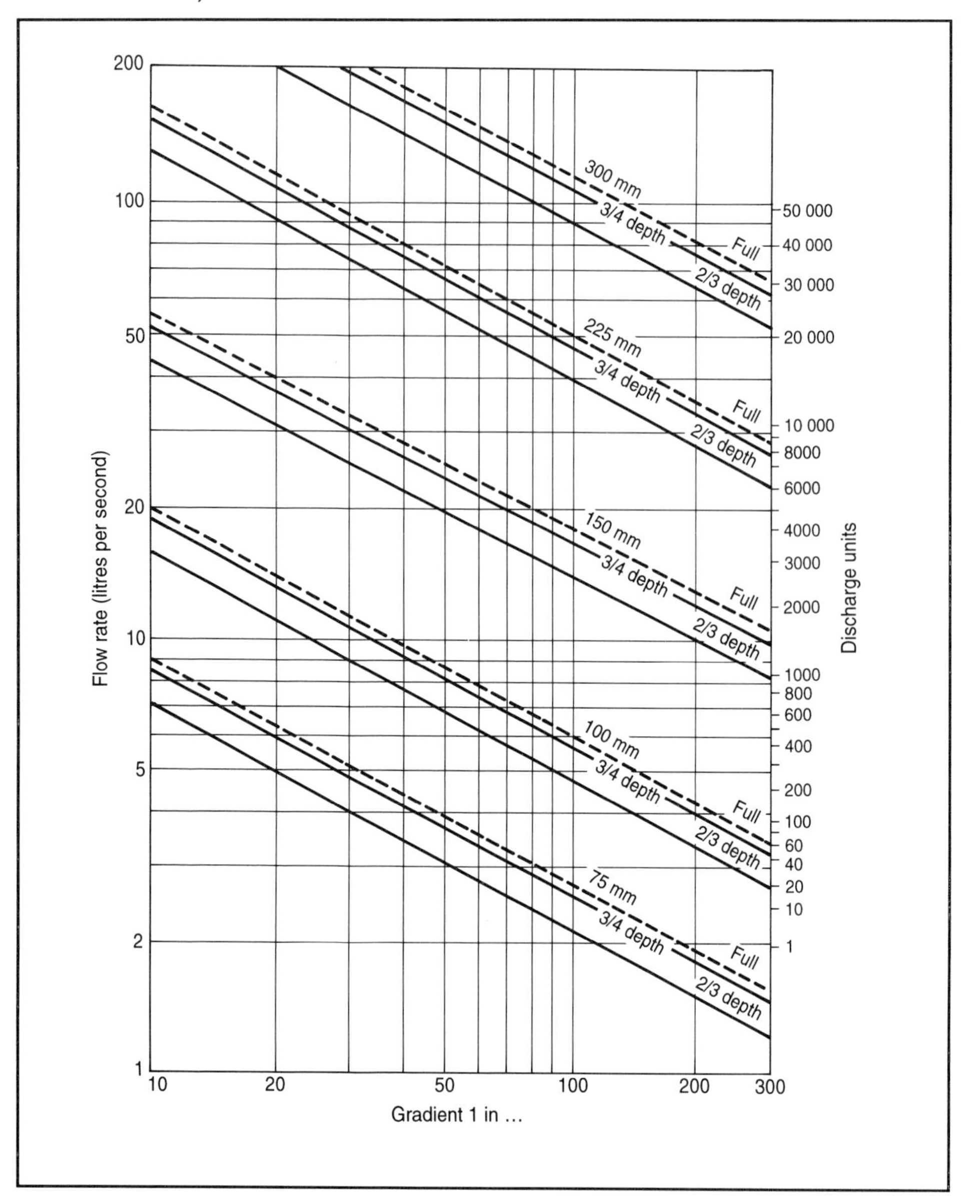

A18.2 STRUCTURAL DESIGN

Depending on ground conditions and pipe supports, above- and below-ground pressure pipelines require concrete thrust blocks at each change of pipe direction or gradient and at penstocks. Anchorage is also required at pipe ends.

Below-ground pipes may be considered in three categories for structural requirement: (a) rigid, (b) semi-rigid and (c) flexible.

Protection and embedment requirements are dependent on the method of laying, traffic loads, soil loads and compaction, the ground conditions, the type of pipe and the permissible diametric deflection. Tabulated bedding factors are available from many sources including manufacturing associations.

A19 Separators and interceptors

Although separators, also described as interceptors, are not in the true sense containment systems, they are an essential part of the surface water drainage of bunded areas and vehicular movement areas such as roads, parking areas, hardstandings etc. where spills of oil, petroleum and similar substances may occur. Separators may also be used as part of trade effluent drainage systems. The main functions and characteristics of separators are summarised as follows:

- separation by flotation of immiscible liquids having a specific gravity < 1
- efficiency dependent on retention time and quiescent flow condition
- detention period allows some sedimentation of detritus and solids
- not suitable for processing industrial oil/water effluents where degreasing or detergents are used
- unable to deal with large and sudden flows or high concentrations
- discharge consents usually require additional facilities such as filtration chambers containing absorbents, proprietary systems (e.g. moving floating belts) to mop up oil, floating absorbents, physical separation by vortex or similar systems and final polishing through special aeration chambers
- turbulent flow may result in emulsification, and pumps should therefore be sited downstream of the unit
- flow through several compartments can cause remixing and a reduction in efficiency. Tests have shown that for an influent concentration of 1000pmm and oil/water flowing at 75 l/second, an efficiency of 98% can be achieved for a two-compartment chamber, compared with 90 to 93% for a three-compartment chamber (see Figure A19.1)
- an effluent standard of less than 20 ppm can be achieved although 20 to 50 ppm is more normal.

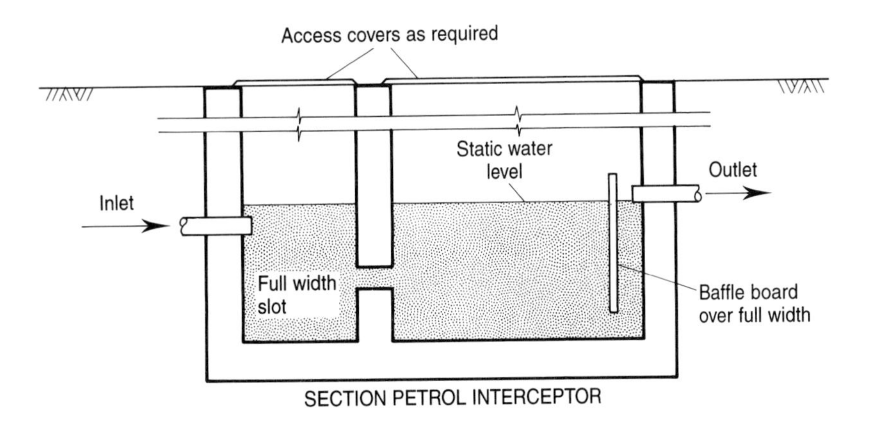

Figure A19.1 *Two compartment interceptor* (Source: BS 8301)

Separators cannot deal with the large spills that might result from a major incident. Cut-off valves should be provided and the means to divert flows to a bund or other containment facility in the event of a major spill. A reception tank fitted with diversion pipes to containment, and pumps where required, should be built into the separator. Alternatively an adjacent interception chamber may provide this facility.

A19.1 DESIGN AND CAPACITY REQUIREMENTS OF SEPARATORS

Environment Agency minimum requirements - recommended design and hydraulic loading criteria:

- six minutes minimum retention calculated in accordance with the drainage criteria - usually based on 50 mm rainfall per hour
- single chamber preferred with no integral bypass
- multiple chambers discouraged but when used six minute retention applies to each chamber or the largest chamber
- minimum capacity equal to maximum contents of compartment of a tanker but not less than six minutes catchment rainfall
- no integral bypass or storm overflow to controlled waters.

The American Petroleum Institute (API) separator

Figure A19.2 illustrates the type of separator recommended by the American Petroleum Institute.

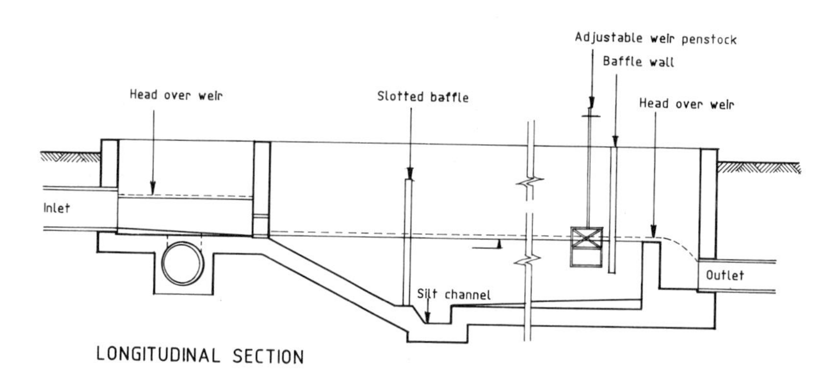

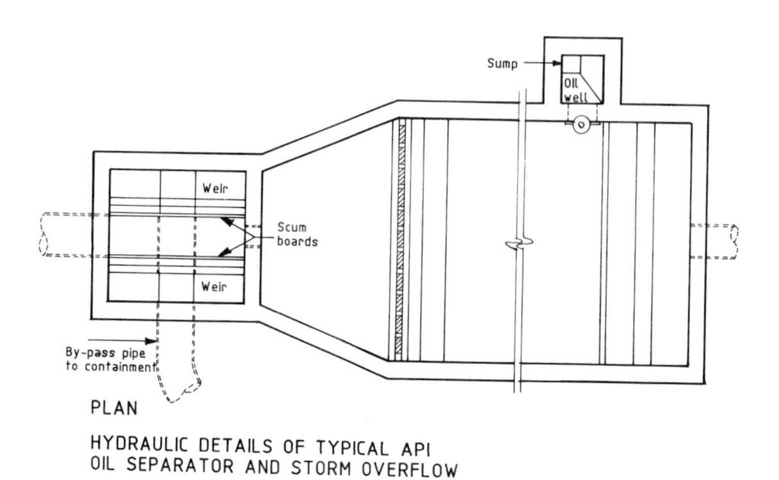

Figure A19.2 *Gravity type separator as recommended by the API* (Reproduced courtesy American Petroleum Insitute (API, 1988))

References

AEA TECHNOLOGY (1994) 1[1]
The PRAIRIE model
Developed for DoE, HSE and NRA.
AEA Technology, Risley, Warrington

AMERICAN PETROLEUM INDUSTRY (API) (1988) 2
Welded steel tanks for oil storage
API Standard 650

AUBREY (1992) 3
Performance of sealants in concrete joints in wet conditions
CIRIA Technical Note 144, London

BARNES, D S (1990) 4
The design of bunds
Report by UKAE Safety and Reliability Directorate
HSE, London

BOND, R P M (1991) 5
Major spills and incidents
The Centre for Exploitation of Science and Technology, London

BRITISH AGROCHEMICAL STANDARDS INSPECTION SCHEME (BASIS) (1992) 6
Inspection and approval of agrochemical stores by river and fire authorities for BASIS registration
BASIS, Ashbourne

BRITISH STANDARDS INSTITUTION (1969-1975) 7
The use of structural steel in building
BS449:BSI, London

BRITISH STANDARDS INSTITUTION (1972a) 8
Code of basic data for the design of buildings
CP3, Chapter V, Pt 2: BSI, London

BRITISH STANDARDS INSTITUTION (1972b) 9
Pre-stressed concrete pressure pipes (including fittings)
BS 4625: BSI, London

BRITISH STANDARDS INSTITUTION (1973) 10
Aggregates from natural resources for concrete (including granolithic)
BS 882: BSI, London

[1] See Appendix A1

BRITISH STANDARDS INSTITUTION (1975) 11
Testing aggregates
BS812: BSI, London

BRITISH STANDARDS INSTITUTION (1978) 12
Specification for cast iron spigot and socket drain pipes and fittings
BS 437:BSI, London

BRITISH STANDARDS INSTITUTION (1981a) 13
Code of Practice for site investigations
BS 5930: BSI, London

BRITISH STANDARDS INSTITUTION (1981b) 14
Code of Practice for earthworks
BS 6031: BSI, London

BRITISH STANDARDS INSTITUTION (1982a) 15
Concrete admixtures
BS 5075: Parts 1, 2 and 3: BSI, London

BRITISH STANDARDS INSTITUTION (1982b) 16
Code of practice for falsework
BS 5975: BSI, London

BRITISH STANDARDS INSTITUTION (1982-1994) 17
Precast concrete pipes, fittings and ancillary products
BS 5911: in nine Parts, BSI, London, 1982-1994

BRITISH STANDARDS INSTITUTION (1983a) 18
Specification for pressed steel sectional rectangular tanks
BS 1564: BSI, London

BRITISH STANDARDS INSTITUTION (1983b) 19
Specification for air-cooled blast furnace slag aggregate for use in construction
BS 1047: BSI, London

BRITISH STANDARDS INSTITUTION (1983c) 20
Steel plate, sheet and strip
BS 1449:BSI, London

BRITISH STANDARDS INSTITUTION (1985a) 21
Structural use of concrete
BS 8110: BSI, London

BRITISH STANDARDS INSTITUTION (1985b) 22
Code of Practice for building drainage
BS 8301: BSI, London

BRITISH STANDARDS INSTITUTION (1985c) 23
Clay bricks
BS 3921: BSI, London

BRITISH STANDARDS INSTITUTION (1986a) 24
Specification for weldable structural steels
BS 4360:BSI, London

BRITISH STANDARDS INSTITUTION (1986b) 25
Code of Practice for foundations
BS 8004: BSI, London

BRITISH STANDARDS INSTITUTION (1987a) 26
Sewerage
BS 8005: Parts 0 and 1, BSI, London

BRITISH STANDARDS INSTITUTION (1987b) 27
Code of Practice for design of concrete structures for retaining aqueous liquids
BS 8007: BSI, London

BRITISH STANDARDS INSTITUTION (1987c) 28
Specification for design and construction of vessels and tanks in reinforced plastics
BS 4994: BSI, London

BRITISH STANDARDS INSTITUTION (1987d) 29
Specification for oil burning equipment: Pt 5 Oil storage tanks
BS 799: BSI, London

BRITISH STANDARDS INSTITUTION (1987-1991) 30
Methods of test for geotextiles
BS 6906: in eight Parts, BSI, London, 1987-1991

BRITISH STANDARDS INSTITUTION (1987-1993) 31
Code of Practice for pipelines
BS 8010: various Parts, BSI, London

BRITISH STANDARDS INSTITUTION (1987-1994) 32
Structural use of steelwork in buildings
BS 5950: in nine Parts, BSI, London, 1987-1994

BRITISH STANDARDS INSTITUTION (1988a) 33
Report on strengthened/ reinforced soils and other fills
PD 6517: BSI, London

BRITISH STANDARDS INSTITUTION (1988b) 34
Specification for ductile iron pipe and fittings
BS 4772: BSI, London

BRITISH STANDARDS INSTITUTION (1989a) 35
Specification for unplasticized PVC pipe and fittings for gravity sewers
BS 5481: BSI, London

BRITISH STANDARDS INSTITUTION (1989b) 36
Specification for manufacture of vertical steel welded non-refrigerated storage tanks with butt-welded shells for the petroleum industry
BS 2654: BSI, London

BRITISH STANDARDS INSTITUTION (1989, 1990) 37
Workmanship on building sites
BS 8000: Part 1, 1989: Part 2, 1990: Part 4, 1989: BSI, London

BRITISH STANDARDS INSTITUTION (1990a) 38
Fire extinguishing installations and equipment on premises - specification for sprinkler systems
BS 5306: Part 2, BSI, London

BRITISH STANDARDS INSTITUTION (1990b) 39
Methods of test for soils for civil engineering purposes
BS 1377: Parts 1 to 9, BSI, London

BRITISH STANDARDS INSTITUTION (1990c) 40
Specification for glass reinforced plastics (GRP) pipes and fittings for use for water supply or sewerage
BS 5480: BSI, London

BRITISH STANDARDS INSTITUTION (1990, 1991) 41
Concrete specifications and procedures
BS 5328: in four Parts, BSI, London, 1990 and 1991

BRITISH STANDARDS INSTITUTION (1991) 42
Specification for vitrified clay pipes, fittings and ducts, also flexible mechanical joints for use solely with surface water pipes and fittings
BS 65: BSI, London

BRITISH STANDARDS INSTITUTION (1992a) 43
Hot rolled asphalt for roads and other paved areas
BS 594: BSI, London

BRITISH STANDARDS INSTITUTION (1992b) 44
Guide to the selection of constructional sealants
BS 6213: BSI, London

BRITISH STANDARDS INSTITUTION (1992, 1985) 45
Code of Practice for use of masonry
BS 5628: Parts 1, 2 and 3, BSI, London, 1992 and 1985

BRITISH STANDARDS INSTITUTION (1992, 1993) 46
Hot applied joint sealant systems for concrete pavements
BS 2499: Parts 1 and 2, BSI, London, 1992 and 1993

BRITISH STANDARDS INSTITUTION (1993a) 47
Coated Macadam for roads and other paved areas
BS 4987: BSI, London

BRITISH STANDARDS INSTITUTION (1993b) 48
Flat bottomed, vertical, cylindrical storage tanks for low temperature service
BS 7777: in four Parts, BSI, London

BRITISH STANDARDS INSTITUTION (1993c) 49
Code of Practice for design, construction and use of storage tanks and reception pits for livestock slurry
BS 5502: Part 50, BSI, London

BRITISH STANDARDS INSTITUTION (1994a) 50
Specification for environmental management systems
BS 7750: BSI, London

BRITISH STANDARDS INSTITUTION (1994b) 51
Code of Practice for earth retaining structures
BS 8002: BSI, London

BRITISH STANDARDS INSTITUTION (1995) 52
Ductile iron pipes, fittings, accessories and their joints for sewerage applications - requirements and test methods
BS EN 598: BSI, London

BRITISH STANDARDS INSTITUTION (1996) 53
Unfired fusion welded pressure vessels
BS 5500:BSI,London

BUGLER, J et al (1993) 54
Guidance on bunding of chemical storage vessels
Health and Safety Executive, Specialist Inspectors Report No. 39
HMSO, London

BUILDING RESEARCH ESTABLISHMENT (BRE) (1987) 55
Concrete
BRE Digest No. 325

BUILDING RESEARCH ESTABLISHMENT (BRE) (1994) 56
Slurry trench cut-off walls to contain contamination
BRE Digest No. 395

BUILDING RESEARCH ESTABLISHMENT (BRE) (1995) 57
National foreword and national application document for Eurocode 7: geotechnical data. Part 1: general rules
UK draft prepared for BSI by BRE under contract to DoE

CIRIA (1991) 58
Manual on good sealant application
CIRIA Special Publication 80, London

CLAY PIPE DEVELOPMENT ASSOCIATION (1983) 59
Design tables for determining the bedding construction of vitrified clay pipelines
Fourth Ed., CPDA

CONCRETE PIPE ASSOCIATION (1981) 60
Laying precast concrete pipelines in trenches
Technical bulletin No. 6, CPA

CONCAWE (1990) 61
European soil and groundwater legislation: implications for the oil refining industry
CONCAWE Report No 4/90, Brussels

CURTIN *et al* (1983) 62
Designing in reinforced brickwork
Brick Development Association

DEPARTMENT OF TRANSPORT (1991) 63
Manual of contract documents for highway works Volume 1- specification for highway works. Series 600 - earthworks
HMSO, London

DoE (1988) 64
Input of dangerous substances to water: proposals for a unified system of control
HMSO, London

DoE (1989) 65
Environmental Assessment - a guide to the procedures
HMSO, London, 1989

DoE (1991a) 66
Interpretation of major accidents to the environment for the purposes of the CIMAH Regulations
DoE, London

DoE (1991b) 67
The Control of Pollution (Silage, Slurry and Agricultural Fuel Oil) Regulations 1991 - guidance note for farmers
HMSO, London

DoE (1993) 68
Risk assessment of existing substances
Guidance produced by UK Government/ Industry working group, DoE

DoE/ WELSH OFFICE (1988) 69
Environmental assessment
DoE Circular 15/88 & Welsh Office Circular 23/88, July

DoE/ WELSH OFFICE (1991) 70
Integrated pollution control - a practical guide
HMSO, London

EDWARDS, P R (1992) 71
Dangerous substances in water: a practical guide
Environmental Data Services (ENDS), London

ENDS (1991, 1993, 1994) 72
Report No 226, p43, 1993
Report No 231, p5, 1994
Report No 201, p9, 1991
Environmental Data Services, London

ENDS (1992, 1993, 1994) 73
Report No 214, p42, 1992
Report No 215, p41, 1992
Report No 217, p43, 1993
Report No 219, p44, 1993
Report No 228, p45, 1994
Environmental Data Services, London

ENDS (1993a) 74
Report No 227, p43, 1993
Environmental Data Services, London

ENDS (1993b) 75
Report No 219, p4, 1993
Environmental Data Services, London

ENDS (1994) 76
Report No 232, p3, 1994
Report No 230, p44, 1994
Environmental Data Services, London

ENVIRONMENT AGENCY (1995) 77
Water pollution incidents in England and Wales - 1995
NRA Report: Water Quality series No 21, HMSO

EUROPEAN CENTRE FOR TOXICOLOGY AND ECOTOXICOLOGY OF CHEMICALS (ECTEC) (1991) 78
Environmental hazard assessment of substances
Report No. 51, ECTEC

EUROPEAN INSURANCE COMMISSION (CEA) (1993) 79
Draft recommendations for the fire protection of stores containing hazardous substances
CEA Sub-Committee 42P

FIRE PROTECTION ASSOCIATION (FPA) (1989) 80
Information sheets on hazardous materials
In: Compendium of Fire Safety, Vol. 4
The Fire Protection Association, London

FIRE PROTECTION ASSOCIATION (FPA) (1995) 81
Bund walls for flammable liquid storage tanks
In: Compendium of Fire Safety, Vol. 2, FS6027
The Fire Protection Association, London

HARRISON, T E (1992) 82
Early age thermal crack control in concrete
CIRIA Report 91, 2nd. edition

HASELTINE and TUTT (1991) 83
Brickwork retaining walls
Brick Development Association

HEALTH AND SAFETY EXECUTIVE (1977) 84
The storage of flammable liquids in fixed tanks (exceeding 10,000 m^3 total capacity)
Health and Safety series booklet HS(G)52, HMSO, London

HEALTH AND SAFETY EXECUTIVE (1984) 85
A guide to the Control of Industrial Major Accident Hazards Regulations 1984
HSE, London, rev. 1991

HEALTH AND SAFETY EXECUTIVE (1987) 86
Dangerous substances in harbour areas - approved Code of Practice
HMSO, London

HEALTH AND SAFETY EXECUTIVE (1990a) 87
The storage of flammable liquids in fixed tanks (up to 10,000 m^3 total capacity)
Health and Safety series booklet HS(G)50, HMSO, London

HEALTH AND SAFETY EXECUTIVE (1990b) 88
The storage of inflammable liquids in containers
Health and Safety series booklet HS(G)51: HMSO, London

HEALTH AND SAFETY EXECUTIVE (1992) 89
The Control of Industrial Major Accident Hazards Regulations 1984 (CIMAH): further guidance on emergency plans
Health and Safety series booklet HS(G)25, HMSO, London

HEALTH AND SAFETY EXECUTIVE (1993a) 90
Guidance on the bunding of bulk chemical storage vessels
Specialist Inspectors Report No. 39, HSE

HEALTH AND SAFETY EXECUTIVE (1993b) 91
The control of fire-water runoff from CIMAH sites to prevent environmental damage
Guidance Note EG 70, HSE

HEALTH AND SAFETY EXECUTIVE (1994) 92
Risk audits
Contractors Research Report CRR 66
HSE, London

HEALTH AND SAFETY EXECUTIVE (1995) 93
The control of fire-water run-off from CIMAH sites to prevent environmental damage
Draft guidance issued by the Hazardous Installations Policy Unit (HIPU) of the HSE, London

HMIP (1991) 94
Chief Inspector's guidance to inspectors
HMSO, London

HMSO (1975) 95
Reservoirs Act
Chapter 23, HMSO

HMSO (1980) 96
The Safety Signs Regulations 1980
HMSO, London

HMSO (1988) 97
The Control of Substances Hazardous to Health Regulations 1988
HMSO, London

HMSO (1991) 98
The Control of Pollution (Silage, Slurry and Agricultural Fuel Oil) Regulations 1991
SI No 324
HMSO, London

HOME OFFICE (1991) 99
Manual of firemanship
Home Office Fire Department, Home Office, London

HOME OFFICE/ SCOTTISH OFFICE (1994) 100
Guide to fire safety for used tyre storage in the open air
Draft guidance
HMSO, August, London

HR WALLINGFORD (1986) 101
Design and analysis of urban storm drainage - The Wallingford Procedure
National Water Council, London

ICI (1986) 102
Water supply and systems for fire fighting and fire protection
Fire Protection Guide No. 5, ICI

INGOLD and MILLER (1990) 103
Geotextiles handbook
Thomas Telford Ltd.

INSTITUTE OF CIVIL ENGINEERS (ICE) (1993) 104
Site investigation for construction
ICE Report, Thomas Telford Ltd.

INSTITUTE OF PETROLEUM (IP) (1981) 105
Marketing safety code
Model code of safe practice in the petroleum industry: Part 3
Wiley & Sons

INSTITUTE OF PETROLEUM (IP) (1993) 106
Fire precautions at petroleum refineries and bulk storage installations
Model code of safe practice in the petroleum industry: Part 19
Wiley & Sons

INSTITUTE OF PETROLEUM (IP) (1996a) 107
Environmental guidelines for petroleum distribution installations
The Institute of Petroleum, London

INSTITUTE OF PETROLEUM (IP) (1996b) 108
Bund wall construction standards
The Institute of Petroleum

INSTITUTION OF STRUCTURAL ENGINEERS (1985) 109
Manual of the design of reinforced concrete building structures
I. Struct. Eng., London

INTERNATIONAL STANDARDS ORGANISATION (1991) 110
Ductile iron pipes, fittings and accessories for pressure pipelines
ISO2531: ISO Switzerland, UK Distribution: BSI, London

JEWELL, R A (1996) 111
Soil reinforcement with geotextiles
CIRIA Special Publication 123, London

JOHNSON et al (1991) 112
Waterproofing and repairing reservoir roofs
CIRIA Technical Note 145, London

LINSLEY and FRANZINI (1992) 113
Water resources engineering
McGraw Hill

LOSS PREVENTION COUNCIL (LPC) (1992) 114
Pollutant industries
Report SHE 8, Loss Prevention Council, London

MASON, P A (1992) 115
Farm waste storage - guidelines for construction
CIRIA Report 126, London

MASON, P A *et al* (1996) 116
Construction of bunds for oil storage tanks
CIRIA Report FR/IP/20, London

MINISTRY OF AGRICULTURE, FISHERIES AND FOOD (1990) 117
Code of Practice for suppliers of pesticides to agriculture, horticulture and forestry
MAFF Publication PB0091, MAFF, London

NATIONAL RIVERS AUTHORITY (1989) 118
Earthworks on landfill sites
NRA, North West Region, Revised 1989

NATIONAL RIVERS AUTHORITY (1991) 119
Policy and practice for the protection of groundwater
NRA, London

NATIONAL RIVERS AUTHORITY (1993) 120
The quality of rivers and canals in England and Wales (1990 to 1992)
HMSO, London

NATIONAL RIVERS AUTHORITY (1994a) 121
General guide to the prevention of pollution of controlled waters
PPG1, NRA

NATIONAL RIVERS AUTHORITY (1994b) 122
Above ground oil storage tanks
PPG2, NRA

NATIONAL RIVERS AUTHORITY (1994c) 123
The use and design of oil separators in surface water drainage systems
PPG3, NRA

NATIONAL RIVERS AUTHORITY (1994d) 124
Safe storage and disposal of used oils
PPG8, NRA

NATIONAL RIVERS AUTHORITY (1994e) 125
The prevention of pollution of controlled waters by pesticides
PPG9, NRA

NATIONAL RIVERS AUTHORITY (1994f) 126
Pollution from highway depots
PPG10, NRA

NATIONAL RIVERS AUTHORITY (1994g) 127
Preventing pollution on industrial sites
PPG11, NRA

NATIONAL RIVERS AUTHORITY (1994h) 128
Retail foodstores and similar sites
PPG15, NRA

NATIONAL RIVERS AUTHORITY (1994I) 129
Pollution prevention measures for the control of spillages and fire fighting run-off
PPG18, NRA

NEW CIVIL ENGINEER (1994) 130
New Civil Engineer - Computer Supplement
In: New Civil Engineer, Thomas Telford Ltd.

OECD (1989) 131
Environmental Data Compendium
The Organisation for Economic Cooperation and Development

PADFIELD and MAIR (1984) 132
Design of retaining walls embedded in stiff clay
CIRIA Report 104, London

PORTLAND CEMENT ASSOCIATION (1969) 133
Structural analysis of large circular and rectangular tanks
Portland Cement Association, (rev 1981) Illinois, 60077, USA

PRITCHARD, P (1994) 134
Managing environmental risks and liabilities
Technical Communications (Publishing) Ltd

PRIVETT, K D *et al* (1996) 135
Barriers, liners and cover systems for containment and control of land contamination
CIRIA Special Publication 124, London

RANKILOR (1986) 136
International directory of geotextiles
Manstock Ltd.

REYNOLDS and STEADMAN (1991) 137
Reinforced concrete designer's handbook (10th edition)
E & FN Spon, Chapman & Hill

RICHARDSON (Editor) (1988) 138
Toxic hazard assessment of chemicals and risk assessment of chemicals in the environment
The Royal Society of Chemistry, London

RICHARDSON and GANGOLLI (Editors) (1995) 139
Dictionary of substances and their effects (DOSE)
The Royal Society of Chemistry, London

SHARMA and LEWIS (1994) 140
Waste containment systems, waste stabilisation and landfills: design and evaluation
John Wiley & Sons Inc.

SHAW, J D N (1984) 141
Concrete repair - materials' selection
Civil Engineering, 53-58, August

SHAW, J D N (1981) 142
Concrete decay - causes and remedies
Proceedings, Irish Concrete Society Seminar on Concrete Repair, Nov 1981

STANSFIELD, R A (1983) 143
The diquat incident at Woodkirk, Yorkshire
J Inst Water Eng & Sci, Vol 37

STATE OF HESSE, GERMANY (1982) 144
Regulation concerning installations for storing, filling and transferring water-endangering substances and the licensing of specialist operators
No 5 Legal and Regulatory Bulletin for the State of Hesse, Part 1

STEEL CONSTRUCTION INSTITUTE (1992) 145
Steel designers manual - fifth edition
Clerkenwell Scientific Publications

THE ROYAL SOCIETY (1992) 146
Risk: analysis, perception and management
Royal Society Study Group Report

TWORT, HOATER and LAW (1993) 147
Water supply
Arnold (Whittaker)

US ENVIRONMENTAL PROTECTION AGENCY (EPA) (1978) 148
Report No 281-090/IGA
Buckley & Weiner

US ENVIRONMENTAL PROTECTION AGENCY (EPA) (1980) 149
Lining of waste impoundment and disposal facilities
Matrecon Inc. , Municipal Environmental Research Laboratory, Cincinnati,

VERBAND DER CHEMISCHEN INDUSTRIE e.V.(VCI) (1988a) 150
Fire protection concept for chemical warehouses with regard to water protection
VCI, April 1987, English Edition March 1988

VERBAND DER CHEMISCHEN INDUSTRIE e. V. (VCI) (1988b) 151
Stoffliste sortiert nach CAS number
VCI

WALLINGFORD and BARR 152
Tables for hydraulic design of pipes, sewers and channels
Thomas Telford Ltd.

WARD, R (1984) 153
Fire prevention
Fire Protection Association, London

WATER RESEARCH CENTRE (WRC) (1990) 154
Sewers for adoption
WRC

WILKINSON, A (1991) 155
Bund overtopping - the consequences following catastrophic failure of large volume liquid storage vessels
Report by UKAE Safety and Reliability Directorate
HSE, London

WATER RESEARCH CENTRE (1983) 156
Sewage rehabilitation
WRC, Swindon